Environmental Assessment of Estuarine Ecosystems

A Case Study

Environmental and Ecological Risk Assessment

Series Editor
Michael C. Newman
College of William and Mary
Virginia Institute of Marine Science
Gloucester Point, Virginia

Published Titles

Coastal and Estuarine Risk Assessment
Edited by
Michael C. Newman, Morris H. Roberts, Jr., and Robert C. Hale

Risk Assessment with Time to Event Models
Edited by
Mark Crane, Michael C. Newman, Peter F. Chapman, and John Fenlon

Species Sensitivity Distributions in Ecotoxicology
Edited by
Leo Posthuma, Glenn W. Suter II, and Theo P. Traas

Regional Scale Ecological Risk Assessment:
Using the Relative Risk Method
Edited by
Wayne G. Landis

Economics and Ecological Risk Assessment:
Applications to Watershed Management
Edited by
Randall J.F. Bruins

Environmental Assessment of Estuarine Ecosystems:
A Case Study
Edited by
Claude Amiard-Triquet and Philip S. Rainbow

Environmental Assessment of Estuarine Ecosystems

A Case Study

Edited by
Claude Amiard-Triquet
Philip S. Rainbow

CRC Press
Taylor & Francis Group
Boca Raton London New York

CRC Press is an imprint of the
Taylor & Francis Group, an **informa** business

Cover photo of the mouth of the Loire River by Claude Amiard–Triquet.

CRC Press
Taylor & Francis Group
6000 Broken Sound Parkway NW, Suite 300
Boca Raton, FL 33487-2742

© 2009 by Taylor & Francis Group, LLC
CRC Press is an imprint of Taylor & Francis Group, an Informa business

No claim to original U.S. Government works
Printed in the United States of America on acid-free paper
10 9 8 7 6 5 4 3 2 1

International Standard Book Number-13: 978-1-4200-6260-1 (Hardcover)

Library of Congress Cataloging-in-Publication Data

Environmental assessment of estuarine ecosystems : a case study / editors,
    Claude Amiard-Triquet and Philip S. Rainbow.
        p. cm. -- (Environmental and ecological risk assessment)
    Includes bibliographical references and index.
    ISBN 978-1-4200-6260-1 (alk. paper)
        1. Estuarine ecology. 2. Estuarine pollution. 3. Ecological risk assessment. I.
    Rainbow, P. S. II. Amiard-Triquet, C. III. Title. IV. Series.

    QH541.5.E8E48 2009
    577.7'86--dc22   2008040756

Visit the Taylor & Francis Web site at
http://www.taylorandfrancis.com

and the CRC Press Web site at
http://www.crcpress.com

Contents

v

Preface

Estuaries are areas of high productivity, crucial in the life histories of many fish, invertebrates, and birds, for example, and the sustainability of estuarine biodiversity is vital to the ecological and economic health of coastal regions. On the other hand, estuarine ecosystems are exposed to toxic anthropogenic effluents transported by rivers from remote and nearby conurbations and industrial and agricultural concerns. It is important, therefore, to have techniques that enable society to assess the degrees of exposure of estuaries to anthropogenic toxic contamination and the significance of this exposure to the ecology of the biota living there, especially the effects on biota of commercial significance. This book describes a comparative multidisciplinary ecotoxicological study of two contrasting estuaries in France, using the results of this study to make generalisations on how different techniques might be used and interpreted in future studies assessing the ecotoxicological status of vital coastal ecosystems.

Multidisciplinary research has been carried out for years on the environmental status of the Seine estuary, France, which is one of the most important and most polluted estuaries in Northwest Europe. The comparatively clean Authie estuary nearby is not impacted by any significant human activity and can be considered a suitable reference site. Many of the contaminants accessible to chemical analysis to date have been determined in water, sediments, and biota at different levels of the food chain. The use of biochemical and physiological biomarkers, testifying to the local exposure of biota to toxins and their ecotoxicological effects, has been tested in species representative of the water column (e.g., the planktonic copepod *Eurytemora affinis*) and the sediment (the burrowing polychaete worm *Nereis diversicolor*). Further effects of contamination have been examined in different constituents of the biota: the abundance of cadmium and mercury-resistant bacteria in mudflats of the Seine; the community structures and photosynthetic capacities of microphytobenthos diatom communities; the abundance, diversity, and genus assemblage structures of foraminiferans and nematodes; and the physiological status and reproduction of copepods, worms, and estuarine fish.

Chemical stress is probably not the only reason for the observed changes, at least directly. In the Seine, land reclamation and harbour extension leading to the reductions of the surface areas of mudflats in the northern part of the estuary along with chemical stress may indeed have exerted negative effects on food availability for invertebrates and fish, impacting energy metabolism and inducing cascading effects on reproduction, populations, and communities of biota.

From a reverse view, the influence of biota on the fate of contaminants has also been investigated, for example, metals and their interactions with the sulfur cycle. The molecular quantification of the dsrAB gene that codes for an enzyme responsible for the production of hydrogen sulfide has been used to determine the degree of local microbial production of sulfides. Biogeochemical transformations in the upper layers of sediments have also been examined, taking into account both inorganic forms of sulfur such as sulfides and sulfates and fatty acids used as qualitative markers of microbial activity. Modelling has shown the influence of hydrodynamism on the profiles of dissolved compounds (oxygen, sulfates, sulfides) and of biological processes in the sediments, assessing the apparently less significant effects of bioturbation due to worm burrowing in a high energy estuary such as the Seine.

The main benefits of this study for coastal zone management and society include (i) the development of analytical tools for the determination of bioavailable forms of metals in interstitial waters; (ii) the validation of biochemical and physiological biomarkers in representative estuarine species; and (iii) recommendations for a comprehensive methodology to assess the health status of estuarine ecosystems. The outcome represents important new developments, particularly related to the application of the European Water Framework Directive. This work has been funded by the European Community's INTERREG, the French Ministry of Environment, and local partners, as well as by research institutions (CNRS, IFREMER, the Center for Estuarine and Marine Ecology of The Netherlands, The Natural History Museum of London, and several universities). This combination of funding sources underlines the double relevance of this book to both academic researchers and applied end users. It is our hope that this book will also serve as an important source of concrete examples for use in environmental science courses.

Claude Amiard-Triquet
Philip S. Rainbow

Contributors

Alain Abarnou
IFREMER
Centre de Brest
Plouzané, France
alain.abarnou@ifremer.fr

Rachid Amara
Université du Littoral Côte d'Opale
Wimereux, France
rachid.amara@univ-littoral.fr

Jean-Claude Amiard
CNRS
Université de Nantes
Nantes, France
jean-claude.amiard@univ-nantes.fr

Claude Amiard-Triquet
CNRS
Université de Nantes
Nantes, France
claude.amiard-triquet@univ-nantes.fr

Thierry Berthe
Université de Rouen
Mont Saint Aignan, France
thierry.berthe@univ-rouen.fr

Brigitte Berthet
ICES
Université de Nantes
Nantes, France
brigitte.berthet@univ-nantes.fr

Christophe Bessineton
Maison de l'Estuaire
Le Havre, France
christophe.bessineton@
    maisondelestuaire.org

Gabriel Billon
Université des Sciences et Technologies
    de Lille
Villeneuve d'Ascq, France
gabriel.billon@univ-lille1.fr

Laurent Bodineau
Université des Sciences et Technologies
    de Lille
Villeneuve d'Ascq, France
laurent.bodineau@univ-lille1.fr

Dominique Boust
Laboratoire de Radioécologie de
    Cherbourg-Octeville
Institut de Radioprotection et de Sureté
    Nucléaire
Cherbourg-Octeville, France
dominique.boust@irsn.fr

Virginie Bragigand
Laboratoire Départemental
    d'Hydrologie et d'Hygiène
Angers, France
bragigand@aol.com

Hélène Budzinski
CNRS
Université de Bordeaux
Talence, France
h.budzinski@ism.u-bordeaux1.fr

Kevin Cailleaud
LEMA
Université du Havre
Le Havre, France
kevin.cailleaud@total.com

Olivier Clarisse
Université des Sciences et Technologies
  de Lille
Villeneuve d'Ascq, France
gabriel.billon@univ-lille1.fr

Anne Créach
Université des Sciences et Technologies
  de Lille
Villeneuve d'Ascq, France
anne.creach@univ-lille1.fr

Jean-Claude Dauvin
Université des Sciences et Technologies
  de Lille
Wimereux, France
jean-claude.dauvin@univ-lille1.fr

Jean-Pierre Debenay
Paléotropique IRD, Centre de Nouméa
Nouméa, Nouvelle Calédonie
jean-pierre.debenay@noumea.ird.nc

Julien Deloffre
Université de Rouen
Mont Saint Aignan, France
julien.deloffre@univ-rouen.fr

Françoise Denis
Université du Maine
Muséum National d'Histoire Naturelle
Concarneau, France
fdenis@mnhn.fr

Lionel Denis
Université des Sciences et Technologies
  de Lille
Wimereux, France
lionel.denis@univ-lille1.fr

David Devreker
Université des Sciences et Technologies
  de Lille
Wimereux, France
david_devreker@yahoo.fr

Cyril Durou
CEREA
Université Catholique de l'Ouest
Nantes, France
cdurou@yahoo.fr

Timothy J. Ferrero
The Natural History Museum
London, United Kingdom
t.ferrero@nhm.ac.uk

Jean-Claude Fischer
Université des Sciences et Technologies
  de Lille
Villeneuve d'Ascq, France
jean-claude.fischer@univ-lille1.fr

Joëlle Forget-Leray
Université du Havre
Le Havre, France
joelle.forget@univ-lehavre.fr

François Gévaert
Université des Sciences et Technologies
  de Lille
Wimereux, France
francois.gevaert@univ-lille1.fr

Patrick Gillet
CEREA
Université Catholique de l'Ouest
Angers, France
pgillet@uco.fr

Jean-Louis Gonzalez
IFREMER
Centre de Toulon
La Seyne, France
jean.louis.gonzalez@ifremer.fr

Herman Hummel
Netherlands Institute of Ecology
Centre for Estuarine and Marine
  Ecology
Yerseke, The Netherlands
h.hummel@nioo.knaw.nl

Robert Lafite
Université de Rouen
Mont Saint Aignan, France
robert.lafite@univ-rouen.fr

Pierre Le Hir
IFREMER
Centre de Brest
Plouzané, France
Pierre.Le.Hir@ifremer.fr

Christophe Minier
Université du Havre
Le Havre, France
minier@univ-lehavre.fr

Catherine Mouneyrac
CEREA
Université Catholique de l'Ouest
Angers, France
catherine.mouneyrac@uco.fr

Baghdad Ouddane
Université des Sciences et Technologies
   de Lille
Villeneuve d'Ascq, France
baghdad.ouddane@univ-lille1.fr

Fabienne Petit
Université de Rouen
Mont Saint Aignan, France
fabienne.petit@univ-rouen.fr

Laurence Poirier
Université de Nantes
Nantes, France
laurence.poirier@univ-nantes.fr

Laurent Quillet
Université de Rouen
Mont Saint Aignan, France
laurent.quillet@univ-rouen.fr

Philip S. Rainbow
The Natural History Museum
London, United Kingdom
p.rainbow@nhm.ac.uk

Jean-Baptiste Ramond
Université de Rouen
Mont Saint Aignan, France
jean-baptiste.ramond@univ-rouen.fr

Michèle Roméo
INSERM
Université de Nice Sophia Antipolis
Nice, France
romeo@unice.fr

Sami Souissi
Université des Sciences et Technologies
   de Lille
Wimereux, France
sami.souissi@univ-lille1.fr

Florence Sylvestre
IRD-CEREGE
Université Aix-Marseille
Aix-en-Provence, France
sylvestre@cerege.fr

Bénédicte Thouvenin
IFREMER
Centre de Brest
Plouzané, France
benedicte.thouvenin@ifremer.fr

1 Introduction

Claude Amiard-Triquet and Jean-Claude Dauvin

CONTENTS

1.1 ESTUARIES: CONFLICT OF BIOLOGICAL WEALTH AND ANTHROPOGENIC PRESSURE

Historically, estuaries have been areas of settlement for many human populations, resulting in a number of negative effects on the natural environment. For example, land reclamation, harbour extension, and dredging lead to decreased areas of wetlands that are very important for the protection of water quality as well as their floristic and faunistic interest. Water quality in estuaries and particularly in urbanized regions is decreasing as a consequence of anthropogenic activities, namely inputs of chemicals associated with industrial and domestic activities and pesticides and fertilizers originating from agriculture. In addition to such local contamination, estuarine ecosystems are exposed to toxic anthropogenic effluents transported by rivers constituting the whole river basin. Concomitantly, river transport is responsible for influxes of nutrients that underlie the biological wealth of estuarine areas, ensuring their role as nurseries for many commercial species. River transport of nutrients is also responsible for the high productivity of nearby coastal areas, allowing the establishment of mariculture enterprises. However the concomitant influx of nutrients and contaminants is a source of concern related to the growth and reproduction of cultivated species and represents a health risk related to the quality of seafood products. Thus estuaries are crucial in the life histories of many invertebrates and vertebrates and the sustainability of estuarine biodiversity is vital to the ecological and economic health of coastal regions.

Environmental monitoring of coastal and estuarine areas is based mainly on the measurement of chemicals that are perceived to be relatively easy to analyse (trace metals, DDT and its metabolites, γHCH, αHCH, some congeners of PCBs, and individual PAHs). These data may be useful in predicting potential biological effects, but only if contaminant levels are related to responses in biological systems. Threshold effect levels such as PNECs (predicted no-effect concentrations) may be derived from toxicological data, but a major limiting factor is that toxicological parameters are practically always determined for individual substances, without regard to potential interactions of different chemicals and classes of chemicals in the environment. In the cases of estuaries containing complex mixtures including many compounds (persistent organic pollutants) that are not yet accessible to analysis or are extremely expensive to analyse, we must develop strategies that allow us to assess whether ecosystems are under stress (Allan et al. 2006). On the other hand, ecological quality status may be determined using different biotic indices that have been recently reviewed (Dauvin et al. 2007) with a view to their use under the European Water Framework Directive (Water Framework Directive 2000).

Nevertheless, comprehensive methodologies have been proposed to determine pollution-induced degradations. The sediment quality triad approach has been proposed to assess the effects of chemical mixtures found in natural sediments (Chapman 1990). The triad includes chemistry to measure contamination, bioassays to measure toxicity, and *in situ* biological assessment to measure effects such as changes in benthic communities. A particular effort has been devoted to determining the ecotoxicities of sediments because, in aquatic environments, sediments are the main reservoirs for most organic and inorganic chemicals entering water bodies. This is also the reason this book treats sediments as key components for assessing interactions of chemicals and biota in estuarine ecosystems (Amiard-Triquet et al. 2007).

Not only do chemical analyses not provide access to all the toxic molecules of interest, but physico-chemical environmental conditions interfere with xenobiotics, modifying their chemical characteristics and thus their bioavailability. Bioassays have been widely used in recent decades, but their value in risk assessment is still a matter of concern because it is extremely complicated to extrapolate the biological responses of small numbers of standard species observed under simplified experimental conditions to many other species submitted to innumerable interactions in the field. Chapman (2002) proposed the inclusion of more "eco" in ecotoxicology and recommended a number of criteria to reach this aim including (1) the choice of the test species, ideally dominant or keystone species from the area being assessed as identified by community-based studies, for testing in laboratory or field; and (2) the selection of endpoints that are ecologically and toxicologically relevant. Few toxicological data have been obtained from estuarine species. Most bioassays were carried out with freshwater species, and some with marine species (EC 2003).

1.2 CHEMICAL CONTAMINATION AND BIOAVAILABILITY OF CONTAMINANTS

The chemical contamination of a given environment may theoretically be determined by measuring concentrations of molecules of interest in water, sediments, and organisms (Chapter 3). In the case of water samples, because of very low concentrations, extreme precautions are necessary to avoid secondary contamination and the time scales of change may be as short as diurnal. On the other hand, sediments, as the main reservoirs for many contaminants, exhibit high concentrations, are easily analysed, and represent records of past contamination. However, if surface sediments are collected, they respond to changes on time scales dictated by deposition (Chapter 2) and bioturbation rates (Chapter 15) (O'Connor et al. 1994). The use of organisms for monitoring chemical contamination is a worldwide and well established practice. In the water column, the species of interest are mainly filter-feeding bivalves among which mussels have given their name to Mussel Watch programmes (NAS 1980) that have been developed successfully in many countries (Beliaeff et al. 1998). However, the need to use biomonitors more representative of sedimentary compartments has been recognized (Bryan and Langston 1992; Diez et al. 2000; Poirier et al. 2006). Compared to water samples, the concentrations of contaminants in biomonitors are high enough to facilitate quantification. Compared to sediments, they can also play the role of integrative recorders and also reveal directly which fractions of environmental contaminants are readily available for bioaccumulation and subsequent effects.

Bioavailability is defined as the fraction of a chemical present in the environment that is available for accumulation in organisms. The environment includes water, sediments, suspended matter, and food. The questions of metal speciation and bioavailability in aquatic systems were reviewed in Tessier and Turner (1995). The distribution of metal species in different phases (sediment in suspension or deposited, interstitial water and water column), their transport, accumulation, and fate are governed by different physico-chemical and microbiological processes mainly related to carbon and sulfur cycles. Recent improvements of analytical tools (DET/DGT) now allow direct access to metal speciation, even in areas with very low levels of contamination (Chapter 5). Because sensitive analytical methods for organic contaminants were developed later, the state of their development is more restricted. Many hydrophobic organic xenobiotics (pesticides, PAHs, PCBs, etc.) have great propensities for binding to organic materials (humic acids, natural DOM) which modifies their bioavailability in water columns (see review by Haitzer et al. 1998). Bioavailability may be determined through three complementary approaches:

1. Chemical assessment of the distribution of the contaminant in different environmental compartments from which its fate would be forecast (see, for instance, Ng et al. 2005)
2. Measurements of bioaccumulated contaminants in biota exposed in the field (Chapter 3) or in the laboratory that reflect the bioavailable concentrations in the environment (a procedure that forms the bases of biomonitoring programmes such as Mussel Watch)
3. Measurements of biological responses (biochemical, physiological; see below) associated with accumulated doses in biota exposed to contaminants in the laboratory or in the field (Chapter 4)

When biological approaches are chosen, it is necessary to take into account the adaptive strategies of organisms (Chapter 7) that metabolize and/or eliminate different organic xenobiotics (Chapter 3) at different rates or store high concentrations of metals in detoxified forms (Chapter 4).

1.3 BIOACCUMULATION AND EFFECTS OF CONTAMINANTS AT DIFFERENT LEVELS OF BIOLOGICAL ORGANISATION

In addition to being affected by the physico-chemical characteristics of contaminants and their associated bioavailability, bioaccumulation depends upon a number of natural factors such as size, age, sexual maturity, and season. The influences of these factors have given rise to a number of studies based on their importance in the design of biomonitoring programmes and interpretation of biomonitoring data (NAS 1980). It is also well established that different species accumulate different contaminants to different degrees, and again the analytical techniques available to determine metals allowed earlier development of metal ecophysiology and ecotoxicology assays compared to assays for organic chemicals.

Briefly, living organisms are able to cope with the presence of metals by controlling metal uptake, increasing metal excretion, and/or detoxifying internalized metals (Mason and Jenkins 1995). Depending on the metal handling strategy, global concentrations in tissues may vary considerably, with lower concentrations generally observed in vertebrates compared to invertebrates. However, even in limited taxonomic groups (bivalves studied by Berthet et al. 1992; crustaceans studied by Rainbow 1998), strong interspecific differences have been shown. Adaptive strategies were reinforced in a number of species chronically exposed in their environment that become tolerant (Chapter 7) through physiological acclimation or genetic adaptation (Marchand et al. 2004; Xie and Klerks 2004). In vertebrates, tolerance to metals mainly results from metal binding to a detoxificatory protein such as metallothionein (MT). In invertebrates, biomineralization into insoluble form often co-exists with MT induction (see reviews by Marigomez et al. 2002; Amiard et al. 2006). It seems obvious that organisms have developed handling strategies for metals that are normally present at low doses in natural environments (several such metals are essential). However, many reports also exist of acquired tolerance in microalgae, crustaceans, and fish exposed to herbicides, organophosphorus insecticides, PCBs, PAHs, and other compounds (Amiard-Triquet et al. 2008). Numerous processes described may explain this tolerance, e.g., multi-xenobiotic resistance (Bard 2000) and induction of biotransformation enzymes (Newman and Unger 2003b). Because they are involved in increased elimination, these latter govern at least partly the concentrations of xenobiotics in biota. Both phylogeny and the chemical characteristics of contaminants influence accumulated chemical concentrations in organisms. It is generally accepted that vertebrates are more efficient than invertebrates in the biotransformation of organic xenobiotics. On the other hand, even in invertebrates, PAHs are relatively degradable, whereas the stability of PCBs and brominated flame retardants and their lipophilic characters are responsible for their bioaccumulation (Chapter 3), particularly in fatty tissues (Bernes 1998; Burreau et al. 1999; De Boer et al. 2000).

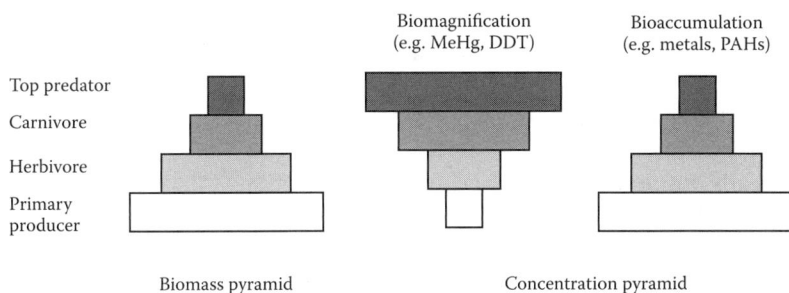

FIGURE 1.1 Biomagnification versus bioaccumulation in aquatic food chains.

One peculiar aspect of bioaccumulation is biomagnification in the food web. This has been a matter of concern since the demonstration in the 1960s that organochlorine insecticides and mercury concentrations were greatly enhanced in consumers belonging to higher trophic levels, including humans in the case of mercury (Newman and Unger 2003a; Drasch et al. 2004). In fact, the situation is variable and depends on the classes of contaminants considered (Figure 1.1). In most cases, the concentration pyramid is orientated like the biomass pyramid. This is the case for most metals along most aquatic food chains, except for elements like mercury that are, at least partly, in organometallic form in the environment and in the prey organisms. Metals in the diet contribute significantly to metal uptake in aquatic organisms (Wang 2002), but metals that are detoxified in insoluble granules are often released undigested in the faeces of predators, thus limiting transfer along food chains (Nott and Nicolaidou 1990).

Biomagnification is a situation in which the orientations of biomass and concentration pyramids are completely opposite. Due to their lipophilic characters, organic contaminants have high potentials for biomagnification but, the pattern is highly contrasted between those that are easily biodegraded (such as PAHs) and those which are very stable (such as PCBs) (Chapter 6). Among emerging contaminants, PBDEs share a number of chemical features characteristic of PCBs, and field studies have shown a clear tendency for PBDE biomagnification in food chains when the top predators are marine mammals or raptors. The pattern is not so clear when top predators are flatfish (Voorspoels et al. 2003), so it is important to increase our knowledge of the behaviour of these types of chemicals (Chapter 6).

Once incorporated into biota, chemicals can exert many different lethal or sublethal, acute or chronic responses, at different levels of biological organisation, from macromolecules to populations or communities. Recently, such a comprehensive approach has been applied to the assessment of the relative toxicity of estuarine sediments (Caeiro et al. 2005; Cunha et al. 2007). A battery of biomarkers (activities of liver ethoxyresorufin-O-deethylase, liver and gill glutathione S-transferases, muscle lactate dehydrogenase, and brain acetylcholinesterase) was examined in the fish *Sparus aurata* exposed for 10 days to sediments collected from different sites in the Sado estuary (Portugal). For all the enzymes assayed, significant differences were found among sites, allowing discrimination of different types or levels of contamination or both. The sediment ranking based upon these biomarkers agreed well with the ranking from a parallel study including chemical analysis of sediments, macrobenthic community analysis, amphipod mortality toxicity tests, and sea urchin abnormality embryo assays.

Similarly, the assessment of the chronic toxicity of estuarine sediments at different levels of biological organisation in the amphipod *Gammarus locusta* revealed a high consistency among chemical (bioaccumulation) and biochemical (metallothionein induction, DNA strand breakage, and lipid peroxidation) responses and effects on survival, growth, and reproduction (Costa et al. 2005; Neuparth et al. 2005). A similar design was used in a freshwater system to investigate effluent impacts using standard (*Daphnia magna*) and indigenous (*Gammarus pulex*) test species (Maltby et al. 2000). *In situ* bioassays carried out downstream of the discharge showed a reduction in *D. magna* survival, in *G. pulex* survival and feeding rate, and in detritus processing, consistent with biotic indices based upon macroinvertebrate community structure.

The present work involves a triad approach (Figure 1.2) based on several species: the copepod *Eurytemora affinis*, Chapter 10; the endobenthic worm *Nereis diversicolor*, Chapter 8; the European flounder *Platichthys flesus*, Chapter 11, along with higher taxa or functional groups (bacteria, Chapter 7; microphytobenthos, Chapter 7, Chapter 12; foraminiferans, Chapter 13; meiofauna, Chapter 14; macrofauna, Chapter 9) representative of the water column or the sedimentary compartment. In comparing multi-polluted and reference estuaries (see below), the objectives were (1) to establish causal relationships between bioaccumulated fractions of environmental pollutants; (2) to link biological effects at sub- and supra-individual levels; and (3) to provide tools to evaluate the health status of species important for the structure and functioning of the estuarine ecosystem.

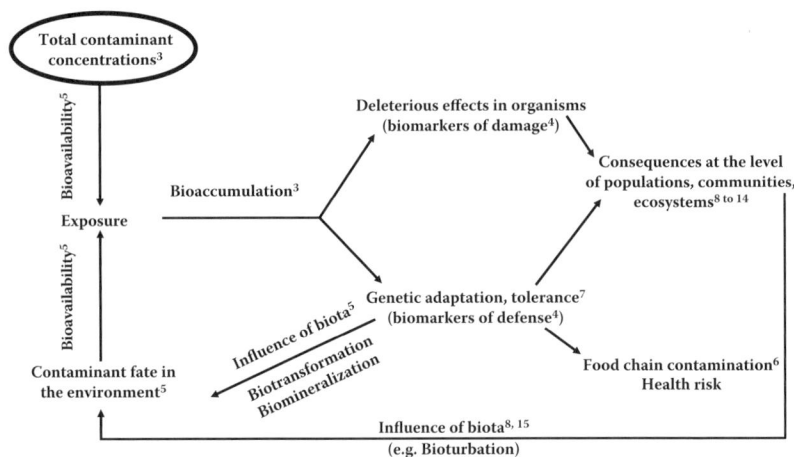

FIGURE 1.2   Links between exposure of marine organisms to contaminants, bioaccumulation, and subsequent effects at different levels of biological organization. Numbers in superscript refer to chapters.

1.4 PUTTING MORE "BIO" IN BIOGEOCHEMICAL CYCLES

The concept of the biogeochemical cycle recognizes the dynamism of multiple, complex processes that move, transform, and store chemicals in the geosphere, atmosphere, hydrosphere, and biosphere. This concept usually conjures up images of carbon, nitrogen, and phosphorus but it can be expanded to include most elements in the periodic table and even organic xenobiotics. Separate biochemical cycles can be identified for each chemical element but elements combine through chemical transformations to form compounds. Thus the biogeochemical cycle of each element or compound must also be considered in relation to the biogeochemical cycles of other elements or compounds. The biotic community may serve as an exchange pool (although it may seem more like a reservoir for some chemicals like calcium, bound in invertebrate shells over geological time scales) and serve to move chemicals from one stage of the cycle to another.

The role of bacteria in the major biogeochemical cycles was established many years ago (SCOPE 1983). The distribution of trace metal species in different phases, their transport, accumulation, and fate are controlled by different physico-chemical and microbiological processes that are mainly linked to carbon and sulfur cycles. While the importance of the carbon cycle is clearly recognized, the role of sulfur needs to be developed further. Sulfides likely play a crucial role in governing the bioavailability and toxicity of trace metals in sediments (Ankley et al. 1996; Lee et al. 2000). Sulfides, produced by the reduction of sulfates after oxidation of organic matter incorporated into sediments, react with many divalent transition metals to form insoluble precipitates (Allen et al. 1993). However, the acid-volatile sulfide (AVS) fraction may be rapidly released following changes in oxido-reduction conditions—such as oxidation due to microbial activity—that induce increased solubility and dissolved bioavailability of metals previously bound as insoluble sulfides (Svenson et al. 1998). Therefore, a multidisciplinary study was carried out to associate geochemical and microbiological expertise along with the use of fatty acids as markers of bacterial activities and of different sources of organic matter (Chapter 5).

In intertidal zones, microphytobenthos represent the major sources of primary production because turbidity restricts the development of phytoplankton in the water column. Microalgae are metal bioaccumulators and thus play a significant role in the biochemical cycle of microphytobenthos. In benthic communities of the coastal ecosystem of the Bay of Biscay, France, microphytobenthos was shown to be the main store for lead (75%) and significant for cadmium (30%) and copper (11%) (Pigeot 2001). Due to the fast succession of generations, microphytobenthos was much more important related to metal fluxes, representing 99% for lead, 98% for cadmium, 95% for copper, and 81% for zinc. Thus it was particularly important to investigate the responses of microphytobenthos to anthropogenic impacts in the Seine estuary and compare the impacts to the Authie reference site in terms of tolerance (Chapter 7) or in terms of community changes of a major microphytobenthic taxon, the diatoms (Chapter 12).

It is now accepted that bioturbation plays an important role in exchanges at the water–sediment interface. The presence of biogenic structures and the activities of endobenthic species deeply affect physical and geochemical properties of the substratum, thus influencing microbial communities and biogeochemical processes (Mermillod-Blondin et al. 2004). The impacts of benthic macrofauna on sediment mineralization rates and nutrient regeneration have served as foci of many studies (Heilskov and Holmer 2003). Numerous works have demonstrated the influence of bioturbation on denitrification through enhanced NO_3^- and O_2 supplies and coupled nitrification–denitrification (Gilbert et al. 1997, and references cited therein).

Bioturbation by infauna also affects different classes of pollutants (Ciarelli et al. 2000; Bradshaw et al. 2006; Ciutat et al. 2007). Among endobenthic species whose bioturbation activities influence the fates of contaminants, the common ragworm *Nereis diversicolor* plays an important role (Gilbert et al. 1996; Gunnarsson et al. 1999; Banta and Andersen 2003; Cuny et al. 2007; Granberg and Selck 2007). Particle mixing and burrow irrigation contribute to the transport and redistribution of pollutants. Enhancing the availability of molecular O_2 in bioturbated sediments stimulates microbial degradation of organic contaminants and, as mentioned above, through changes in oxido-reduction conditions, can influence metal speciation in relation to the sulfur cycle. Small species would have little impact on bioturbation and could not offset functions performed by larger species (Solan et al. 2004; Gilbert et al. 2007). However, meiofaunal bioturbation can affect cadmium partitioning in muddy sediments (Green and Chandler 1994).

1.5   SITES OF INTEREST

A comprehensive method for assessing the health status of estuarine ecosystems was developed on the basis of a case study carried out in the multi-polluted Seine estuary and the comparatively clean Authie estuary, both situated on the French coast of the English Channel.

1.5.1   Main Characteristics of the Seine Estuary

The Seine estuary, situated on the English Channel, is one of the most important estuaries along the French Atlantic coast, along with the Loire and Gironde estuaries in the Bay of Biscay. The geographical zone of influence of the Seine estuary runs from just upstream of the Poses dam—some 160 km upstream of Le Havre, at the limit of the tidal penetration into the estuary—to the eastern part of the Bay of Seine. This zone can be divided into three sections (Figure 1.3): the fluvial, or upstream, estuary; the middle estuary; and the marine, or downstream, estuary. The first is a freshwater zone, extending from the Poses dam to Vieux Port; the second, situated between the fluvial and marine estuaries, is a mixing zone characterized by varying

FIGURE 1.3 (*see facing page*) Sites selected for studying interactions of sediment-bound contaminants and biota. AS = Authie South. AN = Authie North. AP = Authie port, Authie estuary. HON = Honfleur. PN = Pont de Normandie, Seine estuary. Sampling sites for studies in water column (Chapter 10): ✿ Pont de Normandie; ⊕ Pont de Tancarville. White rectangle = *Nereis diversicolor* populations, 1987–2006 (Chapter 9). Position of salinity front ⥮ and estuary turbidity maximum after Dauvin (2002).

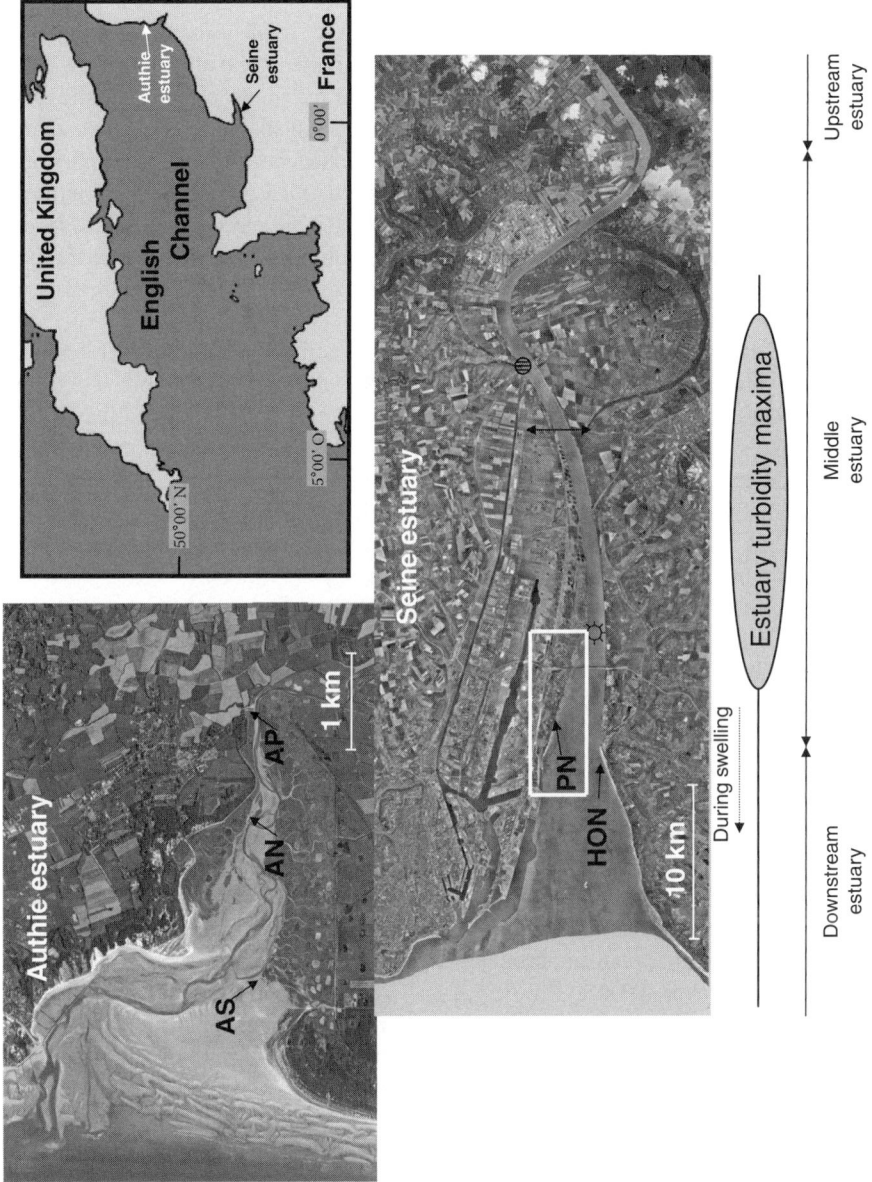

salinity levels; the third is saltwater and runs from Honfleur to the eastern part of the Bay of Seine.

The freshwater flow of the river Seine at Poses is relatively small (480 m^3.s^{-1} on average over the past 30 years), with high water volumes over 2220 m^3.s^{-1} (autumn/winter) and low water flow under 100 m^3.s^{-1} (at the end of the summer in September). The Seine estuary and its hydrodynamics are heavily influenced by tides that can reach nearly 8 m in magnitude downstream of Honfleur during the spring tides (Chapter 2). This megatidal regime causes a zone of maximum turbidity in the mixing zone (middle estuary) between the marine and fluvial sections of the estuary. This maximum turbidity zone traps suspended matter and, through the phenomena of desorption–adsorption, acts as a physico-chemical regulator for a number of elements and/or pollutants, particularly metals. It also leads to a decrease in the amount of oxygen entering the system from the oxidizing of organic matter trapped in the zone. Still, because the estuary waters are renewed by the tide, there is no anoxic zone in the downstream estuary.

The volume of seawater oscillating in this downstream section is always higher than the volume of freshwater, even during extreme floods. During the first extreme flood in autumn, most of the fine sediment and the associated contaminants that accumulate while the water flows are low are expelled into the Bay of Seine; these non-periodic paroxysmal events are, in fact, essential to the natural functioning of the estuary. Moreover, in addition to sediment arriving from upstream, sediment—mainly sand—may also be transported into the estuary from the Bay of Seine, resulting in a natural build-up of sediment in the estuary.

The estuary marks the administrative boundary between two regions (Haute-Normandie and Basse-Normandie) and three departments (Eure, Seine-Maritime, and Calvados). The Seine valley and its estuary are of major economic importance for France, notably because of the presence of two maritime ports. The Seine estuary lies at the discharge point of a watershed area covering 79,000 km^2. This area is home to 16 million people, and accounts for 50% of the river traffic in France, 40% of the country's economic activity, and 30% of its agricultural activities. In addition to the more than 10 million inhabitants of the Greater Paris area who contribute heavily to the Seine estuary's upstream inputs (mainly contaminants and purified waters), the area is also home to two other major river settlements—Rouen with 400,000 inhabitants and Le Havre with 200,000 inhabitants—and two maritime ports of international importance—Port Autonome de Rouen (PAR) and Port Autonome du Havre (PAH).

Despite the major national importance of the estuary and its highly degraded condition, it was not until the 1990s and the creation of the Seine-Aval (SA) multidisciplinary scientific programme that the knowledge base related to the Seine estuary began to grow significantly. The SA programme was designed to accomplish two important objectives: (1) to provide the knowledge needed to reveal how the estuarine ecosystem functions and (2) to develop the tools that local stakeholders require to make decisions about restoring the water quality in the Seine and preserving the natural habitats of the Seine valley. The programme was organised in three phases: SA 1 (1995–1999), SA 2 (2000–2003), and SA 3 (2004–2006) (Dauvin, 2006a and b).

Anthropogenic influences in the Seine estuary began in the mid-19th century and continue to this day. The estuary's ecosystems have become more fragile as a result of human activities and this led to the extreme compartmentalization of the biological units and a drastic reduction of the intertidal zones downstream (a loss of more than 100 km^2 between 1850 and the present). At the same time, the physico-chemical conditions of the estuarine milieu have inexorably declined for more than a century. By the 1980s, the estuary was highly contaminated; levels of metal contamination (e.g., Cd, Hg), hydrocarbons (PAHs), and polychlorinated biphenyls (PCBs) were among the highest in the world and inadequate water treatment facilities created oxygen deficits downstream of Paris and Rouen.

In the marine estuary, abnormal biological functioning led to the collapse of the fisheries sector, particularly brown shrimp (*Crangon crangon*) fisheries, while in the fluvial estuary, professional fishing stopped entirely in the 1970s due to the near-total disappearance of migratory fish species. Today, the most significant danger to the long-term functioning of the estuary comes from chemical and microbiological sources such as endocrine-disrupting chemicals (EDCs), pharmaceutical products, and antibiotic-resistant bacteria in the water.

The recent Port 2000 extension in Le Havre (2000–2005) also seriously affected the morphological and sedimentary evolution of the downstream section of the estuary. The construction project and its compensatory actions contributed to both morpho-sedimentary changes and changes in habitats and biota. It will be several years before the estuarine system establishes some kind of equilibrium in its new dimensions. The geomorphological evolution of the downstream section of the estuary, including the silting of the waterways and the advance of the banks toward the Bay, remains one of the major preoccupations for the future. However, since the Port 2000 project ended, new estuary development projects intended to enhance the economic development of this highly prized zone have begun to emerge.

Nonetheless, despite the diverse environmental assaults, the Seine estuary is still a highly favourable milieu for juveniles of commercial fish species such as sole and European sea bass, and its ornithological richness is one of the major positive aspects of its natural heritage. The richness of this natural heritage can be judged by the overabundance of regulatory measures and inventories that have sprouted over the years. The resulting growth needs to be pruned and coordinated and the "over-protection" is more apparent than real. In fact, with the exceptions of the natural reserves and the *Boucles de la Seine Normande* regional nature park, only a very small number of zones, limited in area and totally separate from one another, are adequately protected. These include the territorial acquisitions of the Coastline and Lakeshore Conservancy (Conservatoire des Espaces Littoraux et des Rivages Lacustres or CELRL), the regional nature reserves, and a few prefecture-designated biotopes. Because this fragmentation of zones rich in natural heritage is incompatible with a concept of integrated management, finding a way to restore the estuary in its totality has become an urgent matter. It is a challenge for the future that SA hopes will be accomplished by 2025, give or take a couple of years. Nowadays, one of the objectives of the Groupement d'Intérêt Public Seine-Aval project is to participate in the Global Management Plan for the period 2007–2016, focusing on estuarine habitat restoration, and tackling the perceptions of the populations involved with regard to the health of the estuary.

1.5.2 AUTHIE ESTUARY

The Authie estuary is located in the eastern part of the English Channel. The length of the River Authie is 103 km; its watershed covers 1305 km² and consists mainly of agricultural fields for breeding cattle; the area houses few industries apart from dairy operations and tourism. Agriculture began here in the 13th century, and plays an important role in the drying of the wetlands; a very dense network of small channels covers all parts of the river. About 75,200 inhabitants live in this territory, mainly in three towns including Berck sur Mer at the mouth of the estuary.

The Authie estuary covers an area of about 3000 ha, namely Authie Bay; it is a very small interface area characterized by a low freshwater input (annual mean at the mouth of estuary = 10.3 m³.s⁻¹) and freshwater volume in comparison with the volume of seawater at each spring tide. The freshwater input varies weekly through the year from a minimum of 6 m³.s⁻¹ at the end of the summer in September–October, to >30 m³.s⁻¹ in the winter. The tide runs for a length of 14 km, and the tidal range is about 9 m at the mouth of the estuary on a spring tide.

Nevertheless, the freshwater input of the Authie contributes to the formation of a low salinity zone located along the French coast (about 3 nautical miles wide) from the Bay of Seine to the Belgium coast under the influence of freshwater input from the Scheldt. This low salinity water mass body called the Fleuve Côtier is more or less distinct, depending of the quantity of freshwater input of the two main rivers, the Seine and the Scheldt, but also of a lot of smaller rivers including the Authie.

The Authie is affected by only very weak anthropogenic activity and can be considered an estuarine reference zone of near-pristine state with very low contamination. The sediments are not polluted and are nontoxic, especially by metal contaminants (Billon, 2001). Some herbicide and pesticide sources have been identified (Billon, 2001), but they are in very low concentrations. The main source of pollution is the diffuse input of nitrates coming from agriculture practices and on some occasions high levels of suspended organic matter (SOM) are present. The total input of SOM is about 12,000 t.y⁻¹. Salmon and sea trout are common in the Authie, but dams in the upper part of the river stop their upward migration.

Two main characteristics illustrate the functioning of the Authie estuary: silting and high hydrodynamism (Chapter 2). As with other bays and estuaries along the French side of the Channel, the Authie estuary is affected by strong silting up resulting from transport of sediment of marine origin, mainly sand (dominating the flood tide) that accumulates in zones with low hydrodynamics (tide and swell protection due to a south–north natural dune and erosion in the north of the estuary mouth). The important polderisation of the estuary accelerated the process of silting associated with the progression of salt marshes during the 18th and 19th centuries. The hydrodynamism arising from the megatidal regime is reinforced by swells and winds.

The natural heritage has been mainly preserved by purchases by CELRL, which holds a total of 685 ha in five sites in the north and the south of Authie Bay. Natural habitats of Annexe II of the European Habitats Directive concern a large zone from the Somme to the Authie estuaries which is included in the same Natura 2000 area.

1.5.3 COMPARISON OF THE SEINE AND AUTHIE ESTUARIES

Both sites are participants in the French Mussel Watch Programme (RNO 2006), and their levels of chemical contamination have been well documented for more than 30 years based on analyses of quarterly samples. Results of chemical analyses of mussels including metals (Cd, Cu, Hg, Pb, Zn, and more recently Ag, Cr, Ni, and V), ΣDDT (DDT + DDD + DDE), γHCH, αHCH, PCBs, and PAHs clearly contrast the pollution states of the two estuaries of interest. Contamination monitoring based on metal analysis in sediments revealed an important contribution of inputs from the Seine river into the coastal area (RNO 1995). Endocrine disruption based on the quantification of imposex in gastropod *Nucella lapillus* (Gibbs et al. 1987) has also been shown along the whole coastal area influenced by the Fleuve Côtier (see Section 1.5.2) and originating from the Seine (Huet in RNO 2004).

In the framework of the National Program of Ecotoxicology (PNETOX) under the jurisdiction of the French Ministry of Ecology and Sustainable Development, sediments as key compartments for the assessment of interactions of chemicals and biota in estuaries were studied comparatively in the multipolluted Seine estuary and the relatively clean Authie estuary (Amiard-Triquet et al. 2007).

In developed countries, pristine sites no longer exist and relatively clean sites are scarce—generally restricted to small estuaries and more likely to be spared from urbanization and industrialization. As a consequence, sites that are potentially available for use as controls have a number of natural characteristics that may differ from

TABLE 1.1

Characterization of Superficial Sediment (<1 cm deep) at Sampling Stations in the Seine and Authie Estuaries

Station	AN	AP	PN	HON
		Parameter		
Salinity:				
Mean (SD)	27 (2)	16 (5)	19 (4)	24 (3)
Fraction >250 μm				
Mean (SD)	0.9 (0.9)	0.6 (0.5)	0.3 (0.2)	1.5 (1.2)
Fraction <63 μm				
Mean (SD)	71 (14)	81 (4)	76 (13)	63 (17)
Organic matter (450°C):				
Mean (SD)	9 (2)	9 (2)	8 (4)	8 (6)

Notes: Quarterly samplings, 2002–2004. Means expressed as psu (salinity) or percentages (grain size and organic matter). SD = standard deviation. Acronyms of stations as in Figure 1.3.

those of bigger estuaries. Thus for the studies of interactions of sediment-bound chemicals and biota, particular attention has been paid to the selection of collection sites (Figure 1.3) as similar as possible in terms of sediment characteristics. Regarding superficial sediments (<1 cm deep), the relative importance of fine (<63 μm) and coarse (>250 μm) particles did not differ significantly nor did organic matter content (Table 1.1). In sediment cores collected from both sites, grain size was also very similar; silty particles were predominant (18 to 98 μm in the Seine estuary and 15 to 90 μm in the Authie estuary), while sand was present only occasionally. The total carbonates (20 to 50% in the Seine estuary and 28 to 40% in the Authie estuary) and the organic matter contents (13 to 15.5% in the Seine estuary and 12.5 to 19% in the Authie estuary) were also within the same range.

In both estuaries, salinity (measured in the field on each sampling occasion in water remaining at the mudflat surface at low tide) was of course higher at downstream sites than at upstream sites (Table 1.1). Stations AP in the Authie estuary and PN in the Seine estuary were the most directly comparable. However, the size of station AP was too restricted to endure frequent and significant sampling. Thus, for the studies dealing with the infaunal worm *Nereis diversicolor*—the biological model favoured for the sedimentary compartment—the main sampling stations were PN in the Seine and AN in the Authie. To assess the influence of salinity on the biological parameters of interest, a specific study was designed in September 2002. A complete set of samples was collected at three stations in the reference estuary along a salinity gradient (AP: 18.9 ± 2.0; AN: 29.5 ± 2.1; AS: 33.0 ± 1.0; Figure 1.3).

From an operational point of view, the purpose of this programme was to apply the findings of our research to (1) the development of analytical tools allowing the determination of bioavailable metals in porewater instead of total concentrations in sediment; (2) the validation of biological tools allowing us to distinguish between anthropogenic and natural fluctuations well before effects become evident at the level of population or community; and (3) the modelling of the impacts of bioturbating macroorganisms on the spatio-temporal dynamics of contaminants. The final aim is to provide a methodology allowing improved risk assessment, favouring a forward-looking approach, instead of a statement of environmental impairment at a stage when remediation is impossible or at best extremely difficult and expensive.

REFERENCES

Allan, I.J. et al. 2006. A "toolbox" for biological and chemical monitoring requirements for the European Union's Water Framework Directive. *Talanta* 69: 302–322.

Allen, H.E., G. Fu, and B. Deng. 1993. Analysis of acid-volatile sulfide (AVS) and simultaneously extracted metals (SEM) for the estimation of potential toxicity in aquatic sediments. *Environ. Toxicol. Chem.* 12: 1441–1453.

Amiard, J.C. et al., 2006. Metallothioneins in aquatic invertebrates: their role in metal detoxification and their use as biomarkers. *Aquat. Toxicol.* 76: 160–202.

Amiard-Triquet, C., C. Cossu-Leguille, and C. Mouneyrac. 2008. Les biomarqueurs de défense, la tolérance et ses conséquences écologiques. In *Les biomarqueurs dans l'évaluation de l'état écologique des milieux aquatiques*, J.C. Amiard and C. Amiard-Triquet, Eds., Paris: Lavoisier.

Amiard-Triquet, C. et al. 2007. A comprehensive methodology for the assessment of the health status of estuarine ecosystems. *ICES CM* 2007/I:09. http://www.ices.dk/products/CMdocs/CM-2007/I/I0907.pdf.

Ankley, G.T. et al. 1996. Technical basis and proposal for deriving sediment quality criteria for metals. *Environ. Toxicol. Chem.* 15: 2056–2066.

Banta, G.T. and O. Andersen. 2003. Bioturbation and the fate of sediment pollutants: experimental case studies of selected infauna species. *Vie Milieu* 53: 233–248.

Bard, S.M. 2000. Multixenobiotic resistance as a cellular defense mechanism in aquatic organisms. *Aquat. Toxicol.* 48: 357–389.

Beliaeff, B., T.P. O'Connor, and D. Claisse. 1998. Comparison of chemical concentrations in mussels and oysters from the United States and France. *Environ. Monit. Assess.* 49: 87–95.

Bernes, C. 1998. *Persistent Organic Pollutants. A Swedish View of an International Problem.* Stockholm: Swedish Environmental Protection Agency.

Berthet, B. et al. 1992. Bioaccumulation, toxicity and physico-chemical speciation of silver in bivalve molluscs: ecotoxicological and health consequences. *Sci. Tot. Environ.* 125: 97–122.

Billon, G. 2001. *Géochimie des métaux et du soufre dans les sédiments des estuaires de la Seine et de l'Authie.* Lille: Université des Sciences et Technologies de Lille.

Bradshaw, C., L. Kumblad, and A. Fagrell. 2006. The use of tracers to evaluate the importance of bioturbation in remobilising contaminants in Baltic sediments. *Estuar. Coast. Shelf Sci.* 66: 123–134.

Bryan, G.W. and W.J. Langston. 1992. Bioavailability, accumulation and effects of heavy metals in sediments with special reference to United Kingdom estuaries: a review. *Environ. Pollut.* 76: 89–131.

Burreau, S., D. Broman, and Y. Zebühr. 1999. Biomagnification quantification of PBDEs in fish using stable nitrogen isotopes. *Organohalog. Comp.* 40: 363–366.

Caeiro, S. et al. 2005. Assessing heavy metal contamination in Sado estuary sediment: an index analysis approach. *Ecol. Indic.* 5: 151–169.

Chapman, P.M. 1990. The sediment quality triad approach to determining pollution-induced degradation. *Sci. Tot. Environ.* 97–98: 815–825.

Chapman, P.M. 2002. Integrating toxicology and ecology: putting the "eco" into ecotoxicology. *Mar. Pollut. Bull.* 44: 7–15.

Ciarelli, S., B.J. Kater, and N.M. Van Straalen. 2000. Influence of bioturbation by the amphipod *Corophium volutator* on fluoranthene uptake in the marine polychaete *Nereis virens*. *Environ. Toxicol. Chem.* 19: 1575-1581.

Ciutat, A., M. Gerino, and A. Boudou. 2007. Remobilization and bioavailability of cadmium from historically contaminated sediments: influence of bioturbation by tubificids. *Ecotoxicol. Environ. Saf.* 68: 108–117.

Costa, F.O. et al. 2005. Multi-level assessment of chronic toxicity of estuarine sediments with the amphipod *Gammarus locusta*: II. Organism and population-level endpoints. *Mar. Environ. Res.* 60: 93–110.

Cunha, I. et al. 2007. Toxicity ranking of estuarine sediments on the basis of *Sparus aurata* biomarkers. *Environ. Toxicol. Chem.* 26: 444–453.

Cuny, P. et al. 2007. Influence of bioturbation by the polychaete *Nereis diversicolor* on the structure of bacterial communities in oil contaminated coastal sediments. *Mar. Pollut. Bull.* 54: 452–459.

Dauvin, J.C. 2002. *Patrimoine biologique et chaînes alimentaires.* Programme Scientifique Seine-Aval, fascicule 7, 46 p. Plouzané, France: Editions IFREMER.

Dauvin, J.C. 2006a. *Biological Heritage.* Seine-Aval Scientific Programme Booklet 7, 48 p. Plouzané, France: Editions IFREMER.

Dauvin, J.C. 2006b. The Seine estuary, a highly developed area. In *North Atlantic Estuaries: Problems and Perspectives*, Dauvin, J.C., Ed., Seine-Aval Special Issue. Plouzané, France: Editions IFREMER, pp. 27–32.

Dauvin, J.C. et al. 2007. The ecological quality status of the Bay of Seine and the Seine estuary: use of biotic indices. *Mar. Pollut. Bull.* 55: 241–257.

De Boer, J., K. de Boer, and J. P. Boon. 2000. Polybrominated biphenyls and diphenylethers. In *The Handbook of Environmental Chemistry*, Paasivirta, J., Ed. Berlin: Springer-Verlag, pp. 62–95.

Diez, G. et al. 2000. *Hediste (Nereis) diversicolor* as bioindicator of metal and organic chemical bioavailability: a field study. *Ecotoxicol. Environ. Restor.* 3: 7–15.

Drasch, G., M. Horvat, and M. Stoeppler. 2004. Mercury. In *Elements and their Compounds in the Environment: Occurrence, Analysis and Biological Relevance*, Merian, E. et al., Eds., Weinheim: Wiley-VCH, pp. 931–1005.

EC. 2003. *Technical Guidance Document on Risk Assessment*. EUR 20418 EN/2, European Commission Joint Research Center.

Gibbs, P.E. et al. 1987. The use of the dog whelk, *Nucella lapillus*, as an indicator of tributyltin (TBT) contamination. *J. Mar. Biol. Ass. U.K.* 67: 507–523.

Gilbert, F. et al. 1997. Hydrocarbon influence on denitrification in bioturbed Mediterranean coastal sediments. *Hydrobiologia* 345: 67–77.

Gilbert, F. et al. 2007. Sediment reworking by marine benthic species from the Gullmar Fjord (Western Sweden): importance of faunal biovolume. *J. Exp. Mar. Biol. Ecol.* 348: 133–144.

Gilbert, F., G. Stora, and J. C. Bertrand. 1996. *In situ* bioturbation and hydrocarbon fate in an experimental contaminated Mediterranean coastal ecosystem. *Chemosphere* 33: 1449–1458.

Granberg, M.E. and H. Selck. 2007. Effects of sediment organic matter quality on bioaccumulation, degradation, and distribution of pyrene in two macrofaunal species and their surrounding sediment. *Mar. Environ. Res.* 64: 313–335.

Green, A.S. and T.G. Chandler. 1994. Meiofaunal bioturbation effects on the partitioning of sediment-associated cadmium. *J. Exp. Mar. Biol. Ecol.* 180: 59–70.

Gunnarsson, J.S., K. Hollertz, and R. Rosenberg. 1999. Effects of organic enrichment and burrowing activity of the polychaete *Nereis diversicolor* on the fate of tetrachlorobiphenyl in marine sediments. *Environ. Toxicol. Chem.* 18: 1149–1156.

Haitzer, M. et al. 1998. Effects of dissolved organic matter (DOM) on the bioconcentration of organic chemicals in aquatic organisms: a review. *Chemosphere* 37: 1335–1362.

Heilskov, A.C. and M. Holmer. 2003. Influence of benthic fauna on organic matter decomposition in organic-enriched fish farm sediments. *Vie Milieu* 53: 153–161.

Lee, B.G. et al. 2000. Influence of dietary and reactive sulfides on metal bioavailability from aquatic sediments. *Science* 287: 282–284.

Maltby, L. et al. 2000. Using single-species toxicity tests, community-level responses, and toxicity identification evaluations to investigate effluent impacts. *Environ. Toxicol. Chem.* 19: 151–157.

Marchand, J. et al. 2004. Physiological cost of tolerance to toxicants in the European flounder *Platichthys flesus*, along the French Atlantic Coast. *Aquat. Toxicol.* 70: 327–343.

Marigomez, I. et al. 2002. Cellular and subcellular distribution of metals in molluscs. *Microsc. Res. Technol.* 56: 358–392.

Mason, A.Z. and K.D. Jenkins. 1995. Metal detoxication in aquatic organisms. In *Metal Speciation and Bioavailability in Aquatic Systems*, Tessier, A. and Turner, D.R., Eds., Chichester: John Wiley & Sons, pp. 479–608.

Mermillod-Blondin, F. et al. 2004. Influence of bioturbation by three benthic infaunal species on microbial communities and biogeochemical processes in marine sediment. *Aquat. Microb. Ecol.* 36: 271–284.

NAS. 1980. *The International Mussel Watch: Report of a Workshop Sponsored by the Environment Studies Board*. Washington: Natural Resources Commission of National Academy of Sciences.

Neuparth, T. et al. 2005. Multilevel assessment of chronic toxicity of estuarine sediments with the amphipod *Gammarus locusta*: I. Biochemical endpoints. *Mar. Environ. Res.* 60: 69–91.

Newman, M.C. and M.A. Unger. 2003a. *Fundamentals of Ecotoxicology*. Boca Raton: Lewis Publishers.

Newman, M.C. and M.A. Unger. 2003b. Uptake, biotransformation, detoxification, elimination, and accumulation. In *Fundamentals of Ecotoxicology*, Newman, M.C. and Unger, M.A., Eds., Boca Raton: Lewis Publishers, pp. 53–73.

Ng, T.Y. et al. 2005. Physico-chemical form of trace metals accumulated by phytoplankton and their assimilation by filter-feeding invertebrates. *Mar. Ecol. Prog. Ser.* 299: 179–191.

Nott, J.A. and A. Nicolaidou. 1990. Transfer of metal detoxification along marine food chains. *J. Mar. Biol. Ass. U.K.* 70: 905–912.

O'Connor, T.P., A.Y. Cantillo, and G.G. Lauenstein. 1994. Monitoring of temporal trends in chemical contamination by the NOAA National Status and Trends Mussel Watch Project. In *Biomonitoring of Coastal Water and Estuaries*, Kramer, K.J., Ed., Boca Raton: CRC Press, pp. 29–50.

Pigeot, J. 2001. *Approche écosystémique de la contamination métallique du compartiment biologique benthique des littoraux charentais : exemple du bassin de Marennes-Oléron*. La Rochelle: Université de La Rochelle.

Poirier, L. et al. 2006. A suitable model for the biomonitoring of trace metal bioavailabilities in estuarine sediments: the annelid polychaete *Nereis diversicolor*. *J. Mar. Biol. Ass. U.K.* 86: 71–82.

Rainbow, P.S. 1998. Phylogeny of trace metal accumulation in crustaceans. In *Metal Metabolism in Aquatic Environments*, Langston, W.J. and Bebiano, M.J., Eds., London: Chapman & Hall, pp. 285–319.

RNO. 1995. *Surveillance de la qualité du milieu marin*. Ministère de l'environnement and Institut français de recherche pour l'exploitation de la mer (IFREMER), Paris.

RNO. 2004. *Surveillance de la qualité du milieu marin*. Ministère de l'Environnement and Institut français de recherche pour l'exploitation de la mer (IFREMER), Paris.

RNO. 2006. *Surveillance de la qualité du milieu marin*. Ministère de l'Environnement and Institut français de recherche pour l'exploitation de la mer (IFREMER), Paris.

SCOPE 21. 1983. The major biogeochemical cycles and their interactions. Bolin, B. and Cook, R.B., Eds. Paris: Scientific Committee on Problems of the Environment.

Solan, M. et al. 2004. Extinction and ecosystem function in the marine benthos. *Science* 306: 1177–1180.

Svenson, A., T. Viktor, and M. Remberger. 1998. Toxicity of elemental sulfur in sediments. *Environ. Toxicol. Water Qual.* 13: 217–224.

Tessier, A. and D.R. Turner. 1995. *Metal Speciation and Bioavailability in Aquatic Systems*. Chicester: John Wiley & Sons.

Voorspoels, S., A. Covaci, and P. Schepens. 2003. Polybrominated diphenyl ethers in marine species from the Belgian North Sea and the Western Scheldt estuary: levels, profiles, and distribution. *Environ. Sci. Technol.* 37: 4348–4357.

Wang, W.X. 2002. Interactions of trace metals and different marine food chains. *Mar. Ecol. Prog. Ser.* 243: 295–309.

Water Framework Directive. 2000. Directive 2000/60/EC of European Parliament and Council. Official Journal of European Communities, Edition L327, December 2000. Brussels: European Union.

Xie, L. and P.L. Klerks. 2004. Fitness cost of resistance to cadmium in the least killifish (*Heterandria formosa*). *Environ. Toxicol. Chem.* 23: 1499–1503.

2 Sedimentary Processes on Estuarine Mudflats
Examples of the Seine and Authie Estuaries

Julien Deloffre and Robert Lafite

CONTENTS

2.1 STUDY SITES

The macrotidal Seine estuary (maximum tidal range of 8.0 m at its mouth) is located in the northwestern part of France (Figure 2.1b). It is one of the largest estuaries on the Northwestern European continental shelf, with a catchment area of more than 79,000 km². The mean annual Seine river flow, computed for the last 50 years, is 450 $m^3.s^{-1}$. Marine sand has infilled the mouth of the estuary (Avoine et al. 1981; Lesourd et al. 2003). Over the past two centuries, the Seine estuary has been greatly altered by human activity (Avoine et al. 1981; Lafite and Romaña 2001; Lesourd et al. 2001).

Intensive engineering works have been undertaken between Rouen and Le Havre to improve navigation. As a result, the Seine estuary has changed from a primarily natural system to one that is anthropogenically controlled (Lesourd et al. 2001).

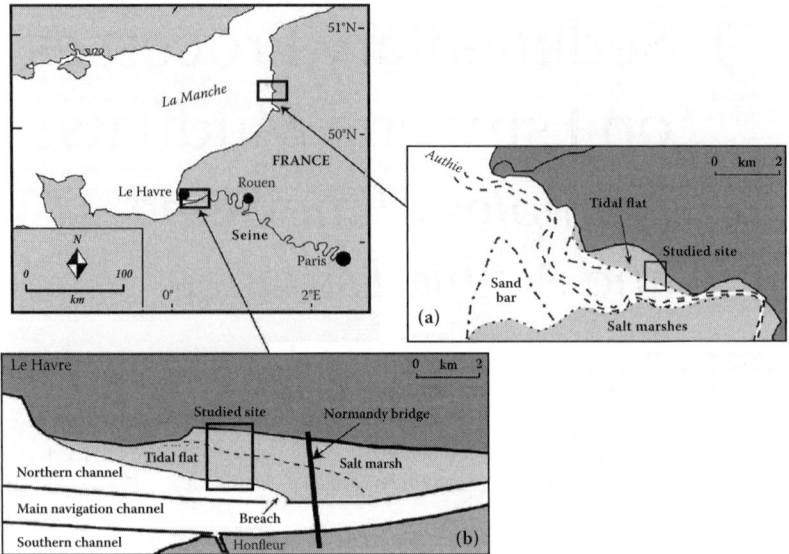

FIGURE 2.1 Locations of studied estuaries (modified from Deloffre et al. 2007). (a) Authie estuary. (b) Seine estuary.

Despite the highly dynamic nature of the system, tidal flats and salt marshes continue to develop in the lower estuary, but the intertidal surface area has drastically decreased during the past 30 years (Cuvilliez et al. 2008). The lower estuary is characterized by the presence of a distinct estuarine turbidity maximum (Avoine et al. 1981), which exerts pronounced control on the sedimentation patterns of intertidal mudflats at the estuary mouth (Deloffre et al. 2006). One of the principal hydrodynamic features in the Seine estuary is a 3-hour high-water slack period that can occur at the mouth. The funnel-shaped estuary is exposed to the prevailing SSW winds that make the intertidal regions at the mouth subject to erosion under the combined effect of waves and currents (Da Silva and Le Hir 2000; Verney et al. 2007).

The Authie estuary is also a macrotidal system (maximum tidal range of 8.5 m at the mouth) located in the northern part of France (Figure 2.1a). The mean annual discharge of the Authie River is 10 m³.s⁻¹, and it has a 985 km² catchment area. This estuarine system is rapidly filling with silt, but a major feature is the penetration of a substantial sand fraction originating from the English Channel (Anthony and Dobroniak 2000). Morphologically, the Authie consists of a bay protected by a sand bar (located in subtidal to supratidal domains) at its mouth, which shelters the estuary from storm swells (Figure 2.1a). The principal hydrodynamic feature is the rapid filling of the bay by the tide: during low tide, most of the estuary, except the main channel, is emersed, and during the flood period significant resuspension of fine sediment occurs. From a morphological view, the Authie estuary is considered a relatively natural estuary, although some polders have been constructed, inducing a seaward progression of salt marsh and increased sedimentation (Anthony and Dobroniak 2000).

2.2 SAMPLING STRATEGY

2.2.1 SEDIMENT PROPERTIES

In order to determine the relevant hydrodynamic processes and compare the evolution of the intertidal mudflats, superficial sediment properties were analysed. Surface sediments and short cores (length ~30 cm, diameter 10 cm) were sampled during each field work period (i.e., every 2 months). The physical characteristics of the sediment were determined using standard sedimentological procedures. The water content was measured using a wet–dry weight technique (water content = water weight × 100/dry weight). The grain size distribution (sand-to-clay fraction) was analysed using a Laser Beckman-Coulter LS 230. The organic matter content of the sediment was quantified by ignition loss at 525°C. Carbonate content was measured using a Bernard calcimeter.

The lithology of the cores was examined using the SCOPIX x-ray imagery method developed by the Bordeaux I University (Migeon et al. 1999). This high-resolution instrument permits the observation of millimetre-thick layers of sediment (Lofi and Weber 2001).

2.2.2 HYDRODYNAMIC MEASUREMENTS

The prevailing near-bed current velocities at the sites were measured during several spring semi-diurnal tidal cycles under low river flow conditions using a 6-MHz Nortek acoustic Doppler velocimeter (ADV) (Kim et al. 2000). The ADV measurement cell was located 15 cm from the transmitter, and was set to measure at a height of 7 cm above the sediment–water interface. This instrument measures three-dimensional current velocities near the bed at a 32-Hz frequency. These high-resolution measurements allow the calculation of bottom shear stress. The turbulent kinetic energy (TKE) method is judged to be the most suitable to estimate the turbulence generated by tidal currents and wind-induced waves on intertidal mudflats (Voulgaris and Townbridge 1998; Kim et al. 2000), but wave–current interactions are incorporated in the TKE shear stress calculations.

This study utilized the parametric Wave–Current Interaction (WCI) model proposed by Soulsby (1995). This model was applied to remove wave–current interactions in the shear stress calculations (Verney et al. 2007). The backscatter signal recorded by the ADV allowed estimation of the near-bed suspended solids concentration (SSC) (Kim et al. 2000). The relationship between ADV backscatter and SSC was derived at each site using surface sediment samples to minimize errors induced by grain-size variability (Voulgaris and Meyers 2004).

2.2.3 ALTIMETRIC MEASUREMENTS

A similar sampling strategy was used for both mudflats. A Micrel ALTUS altimeter was placed at a similar elevation in each estuary (4 to 6.5 m above the lowest sea level, i.e., on the middle slikke). This instrument measures bed elevation at high frequency (1 acoustic pulse every 10 min), with high resolution (0.2 cm) and high accuracy (0.06 cm). The altimeter includes a 2-MHz acoustic transducer that measures the time required for an acoustic pulse to travel from the mudflat surface to the transducer that was fixed at a height of ~22 cm above the sediment surface. Pairs of poles were deployed along a cross-section on each mudflat. Data collected by the altimeter deployed in the middle of the cross-section are representative of the erosion–deposition processes along the section (Bassoulet et al. 2000; Deloffre et al. 2005).

2.3 RESULTS

2.3.1 SEDIMENT PROPERTIES

The carbonate content in surface sediments ranged from 20 to 50% (Table 2.1). The organic matter content of these superficial sediments, however, was similar at each site, ranging from 12.5 to 19%. The estuaries showed little temporal variability in grain-size characteristics. The primary grain-size modes were 15, 40, and 90 µm at the Seine site and 40 and 90 µm on the Authie site (Table 2.1). The main granulo-metric difference between the sites was seen in the sand fraction: a 200-µm frac-tion over the Seine mudflat made up 5 to 15% of the sediment, while on the Authie mudflat the fine-grained sediment was usually associated with a sand fraction below 10% (modes: 200 µm and more rarely 800 µm).

The main parameter varying over an annual scale was water content. While this parameter was fairly constant over a 1-year monitoring period in the surface samples from the Authie estuary (65 to 90%), it varied widely on the Seine mudflat (75 to 250%) where fluid mud occurs during periods of sedimentation. Variations in water content in the superficial sediments of the Seine mudflat result from deposition of fluid mud (water content = 250%) on the mudflat and from dewatering processes resulting from consolidation and desiccation during neap tides. On the basis of labo-ratory experiments, Deloffre et al. (2006) estimated the impact of dewatering on the altimeter dataset; variations in bed elevation induced by dewatering have been removed from the raw altimeter dataset for the Seine estuary. The present altimeter dataset takes into account only erosion and sedimentation processes.

TABLE 2.1
Main Properties of Fine-Grained Sediment Sampled on the Studied Mudflats

	Seine Estuary	Authie Estuary
Carbonate content	20 to 50%	28 to 40%
Organic matter content	13 to 15.5%	12.5 to 19%
Grain-size modes	15, 40, 90	40, 90
Sand layer grain size	Recurrent (200 µm)	Rare (200 to 800 µm)
Water content	75 to 250%	65 to 90%

2.3.2 HYDRODYNAMICS AND SEDIMENTARY PROCESSES

An annual comparison of bed level measurements on the studied intertidal mudflats is shown in Figure 2.2. Mudflats in the Authie and Seine estuaries received net depositions of 15 to 18 cm.year^{-1} during the study. Although net sedimentation rates over an annual time scale were similar in the Authie and Seine estuaries, sedimentation rhythms were different (Figure 2.2).

On the Authie mudflat, topographical variations at a lower scale indicate that the sedimentation is controlled by the semi-lunar tidal cycle (Figure 2.3 and Figure 2.4a). Bed level increases during each spring tide and then decreases or is stabilized during neap tides when the water level is low on the mudflat or when the mudflat is emersed. The threshold between erosion and sedimentation phases corresponds to a water level of 110 cm on the mudflat, which in turn corresponds to a tidal range of 5.5 m. This pattern induces a lag of a few days between the end of deposition and

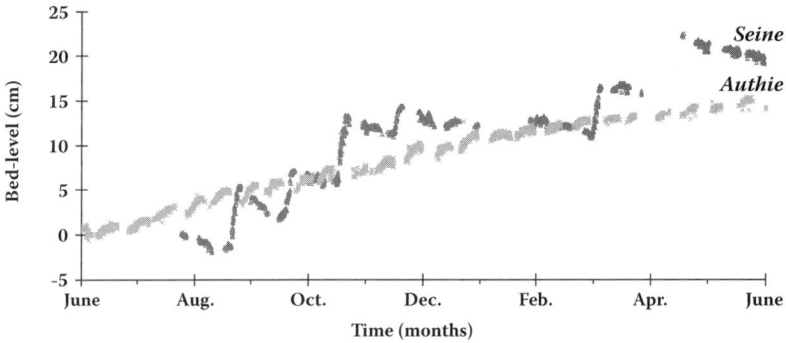

FIGURE 2.2 Topographic evolutions of Seine and Authie mudflats over one year (modified from Deloffre et al. 2007).

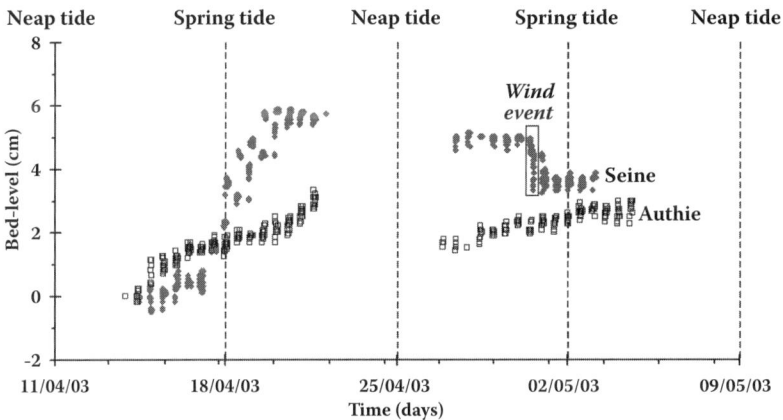

FIGURE 2.3 Topographic evolution of Seine and Authie mudflats over a lunar cycle.

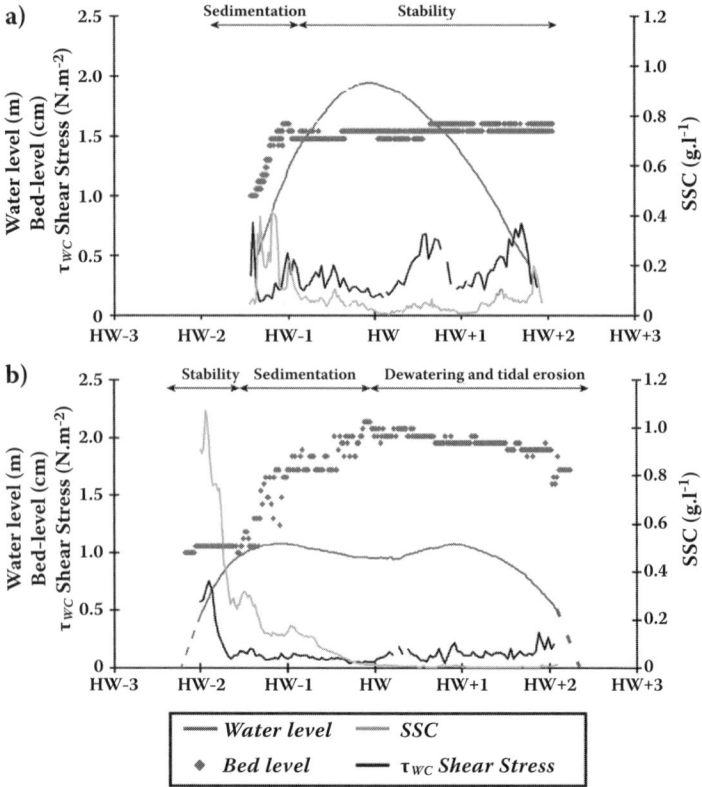

FIGURE 2.4 Sedimentary and hydrodynamic conditions on the studied mudflats during a semi-diurnal cycle (modified from Deloffre et al. 2007) (a) Authie estuary. (b) Seine estuary.

the maximum water level (Figures 2.4a and 2.4b). The sedimentation rates observed on the mudflat range from 0.1 to 0.6 cm per semi-diurnal tidal cycle, with more resuspension of fine particles in the main channel of the estuary during spring tides (when current velocities allow the reworking of fine-grained deposits), and a longer duration of immersion when a supply of fine particles was available (as opposed to during neap tides).

Processes observed at the semi-diurnal scale (Figure 2.4a) indicate that particles settle during flood periods, when the bed shear stress is low (~0.20 N.m^{-2}) and the SSC near the bed is high (~0.4 g.l^{-1}). As fine-grained sediments settle out of suspension, the SSC progressively decreases. After 1 hour of immersion, most of the sediment has settled out of suspension, resulting in a 0.6-cm-thick deposit. During the high tide slack water and ebb periods, the SSC and the bottom shear stress are low, with mean values of 0.05 g.l^{-1} and 0.25 N.m^{-2}, respectively. Twice during the survey, bottom shear stresses reached a value of 0.8 N.m^{-2} as a result of high energy events (Figure 2.4a). However, no impact on the surface of the mudflat was observed during these two events that each lasted 30 minutes. It is notable that during the second event, the water was lower on the mudflat and the SSC increased (Figure 2.4a).

This phenomenon might be linked to erosion of the upper part of the mudflat that resulted from the combined effect of waves and tidal currents. However, at the station studied, the recently settled sediment was not influenced by the waves. The Authie mudflat surface remained stable during these events. Apart from these events, all the sedimentary mechanisms recorded are related to the repetition of semi-diurnal cycles during spring tides (Figure 2.4a).

The annual sedimentation rate on the Seine estuary mudflat is 18 cm (Figure 2.3), but, in contrast to the Authie mudflat, the main deposition phase occurs during the highest spring tides, i.e., according to the lunar cycle, when the water level is >150 cm above the bed level at the station (corresponding to a tidal range of 7.1 m). During these periods, the turbidity maximum reaches high concentrations (>1.95 g.l^{-1}) and maximum volume (Le Hir et al. 2001; Lesourd et al. 2001) in both the main (navigation) channel and the northern channel, and the depositional rate on the mudflat is at a maximum (Deloffre et al. 2006). At these times, the sedimentation rate on the mudflat is high, from 0.3 to 0.8 cm per tide (Figure 2.4b).

As on the Authie mudflat, a lag between the depositional maximum and the water level maximum is observed (Figure 2.4b). During periods of lower water level (<150 cm water depth on the mudflat), the mudflat undergoes gradual erosion, with rates ranging from 0.02 to 0.085 cm during a semi-diurnal cycle. Over an annual time scale, the morphological evolution of the Seine mudflat corresponds to a few periods (6 to 10 per year) of high sedimentation, with increases in bed elevation of between 2 and 8 cm, followed by long periods of slow erosion caused by tidal currents (Figure 2.2). At the semi-diurnal scale, particle settling occurs during high water slack periods (Figure 2.4b). During flood tide when the Seine mudflat is covered, small wind waves occur, inducing a high bottom shear stress that reaches 0.8 to 1.0 N.m^{-2}. These small wind waves occur even outside storm periods. This bottom shear stress prevents deposition and the SSC in the water column remains high (up to 1 g.l^{-1}). During the early high water slack water, when the bottom shear stress decreases (~0.20 N.m^{-2}), the SSC also decreases as fine-grained material settles on the mudflat in a 1-cm thick layer (Figure 2.4b). After all the material has settled, the SSC in the water column is low. During the late slack and ebb periods, the topographic level decreases; this is interpreted to be the result of dewatering and erosion of the soft/fluid mud deposit by tidal currents. In the Seine estuary, the duration of high water slack is up to 3 hours during spring tides, with a well-developed double high tide that favours settling of fine particles and dewatering/consolidation processes just after deposition.

Altimeter measurements at a high resolution and high frequency were used to evaluate bed level changes at the tidal scale and determine the impacts of wind-generated waves on the mudflat. Compared to continuous slow tidal erosion, the wind-induced reworking of intertidal mudflats occurs rapidly.

The Authie mudflat shows no evidence of wind-generated erosion events (Figure 2.2 and Figure 2.3), consistent with the sheltered morphology of this estuary. In contrast, the Seine mudflat undergoes strong erosional phases induced by westerly to northwesterly swells and by local southerly to westerly waves in the Bay of the Seine (Lesourd et al. 2001). Such winds occur on the mudflat about 10 times per year, and are more common during the winters (Deloffre et al. 2006). At the study site, the amplitude of the wind-induced erosion was 0.2 to 2 cm, corresponding to wind speed intensities ranging from 12 to 20 m.s⁻¹ (Deloffre et al. 2006). A direct correlation between wind speed and erosion on the mudflat is difficult, however, as the consolidation state of the sediment must be taken into account. For example, for the same wind event, a fluid mud bottom (such as that found during a depositional period) will undergo more erosion than will a consolidated muddy bed (such as that found during tidal erosion periods).

2.3.3 COUPLING ALTIMETRIC MEASUREMENTS AND CORE IMAGES

The SCOPIX x-ray images of cores allow the identification of physical structures such as layers and surfaces and biological structures such as burrows, tracks, and shell remains that comprise the deposits. The images of the cores from the intertidal mudflats studied show that burrows always occur. As for physical features, the Seine and Authie mudflat deposits consist of thin layers. If only the data from the SCOPIX imagery are used, an interpretation of the sedimentary facies of the intertidal mudflats is difficult because single layers can be interpreted as the results of semi-diurnal, semi-lunar, or lunar depositional cycles. To resolve this problem for this study, we interpreted sedimentary core images in relation to the altimeter dataset. This approach allowed us to determine the duration of deposition for each layer and estimate deposition rates for each site on the basis of the number and thickness of layers.

In the Authie mudflat, a great deal of bioturbation is present, resulting primarily from the activities of polychaetes (*Nereis diversicolor*) at depth and of crustaceans (*Corophium* spp.) in the superficial subsurface. Physical facies, however, are also easily observed at this site as thin layers of fine sediment (Table 2.2). The occurrence of thin sediment layers is consistent with the observed bed level variations. At this site, where deposition is driven by the semi-lunar cycle, the depositional phases are recorded in the cores and correspond to centimetre-thick layers; however, not all the semi-lunar cycles are preserved in the cores (Table 2.2). This indicates that even in a protected setting, water current velocities are high enough to rework some deposits corresponding to fortnightly cycles and, as a result, gaps occur in the neap spring recording.

In the Seine estuary mudflat, freshly deposited sediments can clearly be identified in x-ray images of cores collected a few days after the highest spring tide period (Table 2.2). They are characterized by erosion surfaces at the bases of the elementary deposits that result from tidal- or wind-induced phases. Above these erosion surfaces, the fresh deposits are characterized by low consolidation state and low bulk density (water content of the order of 200%). The fresh deposits appear light grey in the positive x-ray images (Table 2.2). The layer thicknesses indicated by the altimeter dataset and the sedimentological variations in the cores are consistent. The lithological analysis of the uppermost part of the cores is more complicated, however, at sediment depths exceeding 10 cm, mainly because of strong mixing by bioturbation and the erosion of parts of the deposits by waves and/or tidal currents (Figures 2.2 and 2.3).

TABLE 2.2
Comparison of the Main Sedimentological Results on the Studied Mudflats

	Seine	Authie
Morphology at the mouth	Open estuary	Protected bay
Sediment supply	Turbidity maximum	Resuspended sediment inside the estuary
Forcing parameter(s)	• Strongest spring tides (TM development) • Wind (>15 m.s–1 westerlies)	Tidal cycles
Sedimentation rates at semi-diurnal scale (cm)	0.3–0.8	0.1–0.6
Main sedimentation cycles (deposit sequence)	Lunar	Fortnightly
Maximum sedimentation during one deposit episode (cm)	8	5
Number of sedimentation episodes/year	7–10	15–22
Annual sedimentation rates (cm)	18	15
Preservation rates (%)	50%	90%
Typical facies and estimated duration based on bed-level evolution		

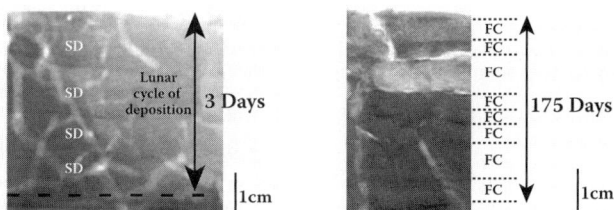

Notes: Modified from Deloffre et al. 2007. SD = semi-diurnal cycle. FC = fortnightly cycle. LC = lunar cycle. TM = turbidity maximum.

2.4 DISCUSSION AND CONCLUSIONS

The sedimentation processes on the studied tidal mudflats examined here are strongly influenced by sediment supplies and by the morphologies of the estuaries at various time scales. On the Seine and Authie mudflats, although long-term sedimentation rates are similar, the rhythms of deposition are different (Figure 2.2 and

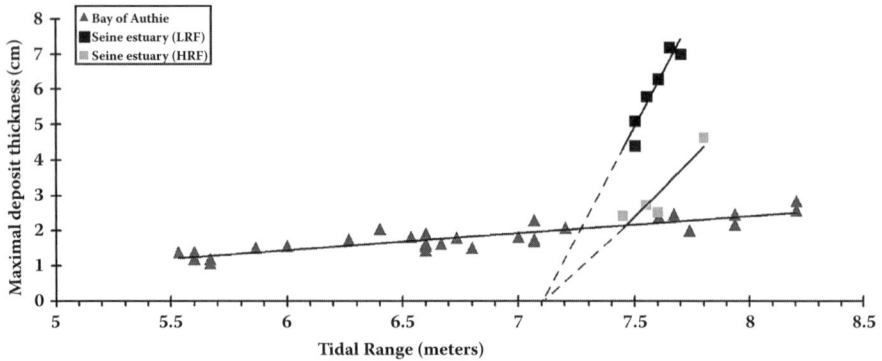

FIGURE 2.5   Relation between maximal deposit thickness and tidal range on the studied mudflats (modified from Deloffre et al. 2007). LRF = low river flow. HRF = high river flow.

Table 2.2). On the Authie mudflat, sedimentation is continuous, with rate controlled by the semi-lunar cycle. A linear relationship exists between the tidal ranges and the resulting deposit thickness, as determined from altimeter data (Figure 2.5). On the Seine estuary mudflat, no sedimentation occurs under neap to medium spring-tide conditions. Rather, sedimentation on this mudflat occurs when a tidal range threshold value of 7.1 m is reached (Deloffre et al. 2006; Figure 2.5). Sedimentation is thus discontinuous, occurring only during the higher spring tides and leading to only a few (<10) depositional episodes over the course of a year (Figure 2.2 and Table 2.2).

Sedimentation on the Authie mudflat corresponds to a semi-lunar rhythm typical of most modern sheltered mudflats. The lower Seine estuary mudflat, however, exhibits a distinct pattern of deposition–erosion. This unusual pattern is a result of the altitude of the study site; its elevation is the same as that of the Authie mudflat, and this pattern is recorded on other locations on the Seine mudflat, including at lower altitudes (Deloffre et al. 2006). The difference between the rhythms of sedimentation on the Authie and Seine mudflats is likely linked to sediment availability and sediment properties. On the Authie mudflat, the fine particles originate from the reworking of sediment from the lower parts of the slikke during the rapid filling of the estuary at flood tide. During each spring tide period, the fine material is resuspended and sedimentation occurs on the mudflat at the location studied on the middle slikke. In the Seine estuary, the delivery of sediment to the mudflat is related to the turbidity maximum (Deloffre et al. 2006). Because of the characteristics of the suspended particulates and the hydrodynamic conditions, development of the Seine estuary turbidity maximum is higher during tidal ranges that exceed 7.1 m (Avoine et al. 1981; Le Hir et al. 2001), which is the threshold value for deposition (Figure 2.5).

The control of sedimentation by the turbidity maximum also results in differences on an annual scale. Sedimentation rates are higher when river discharge is low, under which conditions the position of the turbidity maximum in the estuary is in the area of the mudflat (Le Hir et al. 2001; Lesourd et al. 2001). When river discharge is high, the turbidity maximum is expelled from the estuary into the Bay of the Seine.

During a lunar cycle (Figure 2.3), high sedimentation rates result from some specific characteristics of the Seine estuary. Because of high silt content (quartz, calcite) and low clay mineral content (Lesourd, personal communication), the settling velocities of particles in the lower Seine estuary are higher (~1 mm.s^{-1}) than in other estuarine settings (Delo and Ockenden 1992). These high settling velocities combined with the long high tide slack (up to 3 hours) in the Seine estuary lead to the high sedimentation rates observed on the mudflat during the highest spring tides. These conditions result in the settling of fluid mud, a phenomenon observed only in this estuary among the estuaries studied; thus on this mudflat, dewatering processes must be considered.

The properties of cohesive materials play an important role in controlling deposition (formation of laminae) and preservation of fine-grained sediments. Sediment properties such as grain size, water content, and settling velocities play a role in determining the thicknesses of deposits on intertidal mudflats. The depth of wind-induced erosion is related to the cohesion of surface sediments as well as to wave amplitude. As a result, erosion occurs on the open Seine mudflat where the mud is soft or even fluid (Figure 2.4b), whereas on the sheltered Authie mudflat, where the sediment is coarser-grained and less fluid, only a small amount of wave-related erosion is observed (Figure 2.4a).

Both macrotidal mudflats are highly dynamic. This implies that the mudflat surface changes during the year based on the age of the sediment (a few days to several months) and the sediment properties (especially mud state). On the other hand, the cores from both mudflats are laminated, one lamina corresponding to one deposit episode. These physical processes influence bio-geochemical processes.

ACKNOWLEDGMENTS

This work is part of the Seine Aval and the European Interreg III program RIMEW.

REFERENCES

Anthony, E.J. and C. Dobroniak. 2000. Erosion and recycling of estuary mouth dunes in a rapidly infilling macrotidal estuary, the Authie, Picardy, northern France. In *Coastal and Estuarine Environments: Sedimentology and Geoarchaelogy*, Pye, K. and Allen, J.R.L., Eds. London: Geological Society of London, Vol. 175, pp. 109–121.

Avoine, J. et al. 1981. Suspended sediment transport in the Seine estuary, France: effect of man-made modifications on estuary-shelf sedimentology. *Mar. Geol.* 40: 119–137.

Bassoullet, P. et al. 2000. Sediment transport over an intertidal mudflat: field investigations and estimation of fluxes within the Baie de Marennes-Oleron (France). *Cont. Shelf Res.* 20: 1635–1653.

Cuvilliez, A. et al. 2008. Morphological responses of an estuarine intertidal mudflat to constructions since 1978 to 2005: The Seine estuary (France). *Geomorphology*, in press.

Da Silva, J.R. and P. Le Hir. 2001. Response of stratified muddy beds to water waves. In *Coastal and Estuarine Fine Sediment Processes*, McAnally, W.H. and Mehta, A.J., Eds. Amsterdam: Elsevier Science.

Delo, E.A. and M.C. Ockenden. 1992. *Estuarine Muds Manual*, v. SR 309, Wallingford.

Deloffre, J. et al. 2007. Sedimentation on intertidal mudflats in the lower part of macrotidal estuaries: sedimentation rhythms and preservation. *Mar. Geol.* 235: 151–164.

Deloffre, J. et al. 2006. Interactions between intertidal mudflat and turbidity maximum in macrotidal estuarine context. *Mar. Geol.* 235: 151–164.

Deloffre, J. et al. 2005. Sedimentary processes on a fluvial estuarine mudflat: the macrotidal Seine example (France). *Estuar. Coast. Shelf Sci.* 64: 710–720.

Kim, S.C. et al. 2000. Estimating bottom stress in tidal boundary layer from acoustic Doppler velocimeter data. *J. Hydrol. Eng.* 126: 399–406.

Le Hir, P. et al. 2001. Fine sedimentation transport and accumulations at the mouth of the Seine estuary (France). *Estuaries* 24 (6B): 950–963.

Lafite, R. and L.A. Romaña. 2001. A man-altered macrotidal estuary: the Seine estuary (France): introduction to the special issue. *Estuaries* 24 (6B): 939.

Lesourd, S. et al. 2003. Seasonal variations in the characteristics of superficial sediments in a macrotidal estuary (the Seine inlet, France). *Estuar. Coast. Shelf Sci.* 58: 3–16.

Lesourd, S. et al. 2001. Morphosedimentary evolution of the macrotidal Seine estuary subjected to human impact. *Estuaries* 24 (6B): 940–949.

Lofi, J. and O. Weber. 2001. SCOPIX: digital processing of x-ray images for the enhancement of sedimentary structures in undisturbed core slabs. *Geo Mar. Lett.* 20: 182–186.

Migeon, S. et al. 1999. SCOPIX: a new x-ray imaging system for core analysis. *Geo Mar. Lett.* 18: 251–255.

Soulsby, R.L. 1995. Bed shear stress due to combined waves and currents. In *Advances in Coastal Morphodynamics,* Stive, M.J.F. et al., Eds. Delft: Hydraulics, Vol. 4, pp. 4–23.

Verney, R. et al. 2007. The effect of wave-induced turbulence on intertidal mudflats: impact of boat traffic and wind. *Cont. Shelf Res.* 27: 594–612.

Voulgaris, G. and S.T. Meyers. 2004. Temporal variability of hydrodynamics; sediment concentration and sediment settling velocity in a tidal creek. *Cont. Shelf Res.* 24: 1659–1683.

Voulgaris, G. and J.H. Trowbridge. 1998. Evaluation of the acoustic Doppler velocimeter (ADV) for turbulence measurements. *J. Atmos. Ocean. Technol.* 15: 272–289.

3 Quantification of Contaminants

Jean-Claude Amiard, Laurent Bodineau,
Virginie Bragigand, Christophe Minier,
and Hélène Budzinski

CONTENTS

3.1   INTRODUCTION

Both estuaries of interest, the Authie and the Seine, are participants in the French Mussel Watch Programme (RNO 1995; Claisse et al. 2006) and their levels of chemical contamination have been well documented through quarterly analysis of samples for more than 30 years. However, in order to link exposure to contaminants, incorporation into tissues of organisms, and responses of the biota, it was necessary to acquire chemical and biological data precisely at the same place and at the same moment, taking into account temporal changes within each year and between successive years. In addition, it was necessary to investigate the presence of emergent contaminants such as pharmaceuticals and brominated flame retardants as well as alkylphenol polyethoxylates for which no data were available for the Authie site.

Estuarine areas are characterized by the presence of huge quantities of sedimentary particles both deposited and in suspension (see Chapter 2). Sediments constitute the main sink for most organic and inorganic contaminants in the aquatic environment (Gagnon and Fisher 1997; Bernes 1998). Despite the fact that they live in close contact with these contaminated sediments, endobenthic species are not very widely used as filter feeders in biomonitoring programs. Previous reports provided evidence that accumulated contaminant concentrations in the endobenthic polychaete annelid *Nereis diversicolor* may be considered good biomonitoring measures of the local bioavailabilities of contaminants (Bryan and Langston 1992; Diez et al. 2000).

Thus in the present work, the quality of the environment was characterized by analyzing micropollutants in sediments and in *Nereis diversicolor* but also in water for contaminants whose persistence is limited or whose presence is mainly in the water column (dissolved phase: hydrophilic compounds).

3.2 CONTAMINANTS IN THE WATER COLUMN

For contaminants with limited persistence and/or are hydrophilic (pesticides, pharmaceuticals, alkylphenol polyethoxylates), analyses in water were conducted in the Authie estuary.

3.2.1 PESTICIDES

Pesticide concentrations determined in the Authie estuary were among the lowest that have been found in major French rivers and world systems (Le Calvez 2002) (Table 3.1).

For the Authie estuary, analyses of aqueous extracts revealed the presence of two nitrogen herbicide compounds (trifluraline and terbuthylazine), two phosphorus insecticide molecules (ethyl parathion), and a product of DDT degradation, 4,4′DDE. Cereal crops in the Authie's watershed explain the applications of trifluraline and terbuthylazine for treatment of undesirable vegetation. Due to their physical and chemical properties, the dissolved compartment served as the main reservoir of these hydrophilic compounds that were not found in particulate media such as deposited sediments and suspended particulate matter (SPM).

Lindane was quantified for each site in surface sediment and was found at very low levels (<0.1 ng.g^{-1} dry weight). This indicates that lindane in the environment is decreasing since it was banned in Europe several decades ago, now attaining an approximate background level (still detectable but at very low concentration).

3.2.2 PHARMACEUTICALS

During the 1990s, pharmaceutical compounds such as analgesics, lipid-lowering drugs, antibiotics, and hormones were detected in waste waters across Europe (Heberer 2002; Ternes 2001) and the United States (Boyd et al. 2003; Kolpin et al. 2004). Several reviews dealing with the occurrence, fate, and effects of pharmaceuticals in the environment are available (Halling-Sorensen et al. 1998; Hernando et al. 2006) and show the main influence of sewage treatment plant discharges on the contamination of aquatic systems. These compounds are not completely degraded by

TABLE 3.1

Highest Levels of Pesticides Found in Major French Rivers and World Systems

Location	Nitrogen Pesticides	Phosphorus Pesticides	Chlorinated Pesticides
Authie	150	24	0.3 to 0.6*
	Le Calvez 2002	Le Calvez 2002	Le Calvez 2002
Seine	1,000	—	>10
	Tronczyñski et al. 1999		Tronczyñski et al. 1999
Gironde	600	14	—
	Lartigues 1994	Lartigues 1994	
Marne	1,200	—	
	Garmouma et al. 1998		
Vilaine	29,000	—	
	Guiho 1998		
India		13,000	2,000
		Sujatha et al. 1999	Sujatha et al. 1999
USA	87,000		30*
	Bintein and Devillers 1996		Venkatesan et al. 1999
Korea	—	—	130*
			Lee et al. 2001
China	1,600	—	5*
	Gfrerer et al. 2002		Wu et al. 1999
Bering Sea	—	—	0.4*
			Yao et al. 2002

Note: Concentrations in the water column expressed as nanograms per liter for dissolved pesticides and as nanograms per gram for dry suspended sediment.
* Adapted from Le Calvez 2002.

the metabolism after human (and animal) consumption and are rejected into waste waters. More or less degraded in waste water treatment plants (Ternes et al. 2004), they enter the environment through the discharge of effluents into receiving waters. Thus their presence reveals inputs of waste water and anthropogenic pressures. These molecules were designed to be biologically active and thus can represent threats to aquatic organisms. For this reason, it is necessary to document their presence in the environment (Cleuvers 2003; Laville et al. 2004).

This study examines the presence in the Seine and Authie estuaries of compounds belonging to several chemical and therapeutic groups: antidepressants (diazepam), antiepileptics (carbamazepine), non-steroidal anti-inflammatory drugs (ibuprofen, naproxen, paracetamol, aspirin, ketoprofen, diclofenac), bronchodilatators (clenbuterol, salbutamol), and lipid regulators (gemfibrozil).

Several compounds such as diazepam, salbutamol, and terbutaline were never detected at either site. This is related to the fact that they are among the less consumed substances on the screened list. In the Authie estuary, four molecules were regularly detected and reported on every occasion in the Seine estuary (Table 3.2).

TABLE 3.2
Pharmaceutical Concentrations (ng.L⁻¹) Measured in Seine (Honfleur) and Authie Estuaries

Honfleur	Mar 2003	Nov 2003	Feb 2004	May 2004	Sep 2004	Nov 2004
Aspirin	11.0	nd	nd	nd	nd	10.2
Diclofenac	4.3	3.3	4.7	5.1	nd	9.2
Naproxen	10.6	nd	5.7	6.0	nd	9.5
Ibuprofen	7.4	nd	2.6	3.0	nd	5.3
Ketoprofen	4.5	6.7	nd	2.1	nd	nd
Gemfibrozil	nd	nd	nd	2.0	nd	3.0
Paracetamol	nd	—	nd	2.1	nd	3.1
Caffeine	3.3	5.1	7.5	11.6	11.4	19.4
Carbamazepine	2.1	3.8	—	6.2	7.1	7.56

Authie	Mar 2003	Nov 2003	Feb 2004	May 2004	Sep 2004	Nov 2004
Diclofenac	nd	nd	2.0	3.2	4.1	3.6
Naproxen	nd	nd	3.9	2.1	3.8	nd
Caffeine	nd	nd	3.1	2.1	4.3	3.6
Carbamazepine	nd	nd	2.0	nd	nd	nd

nd = not detected, <1 ng.L⁻¹.

They are among the most consumed compounds (such as caffeine) and also the most stable and least degraded compounds in sewage water treatment plants (diclofenac, carbamazepine). In the Seine estuary, five other compounds were occasionally reported. As a general conclusion, the Seine estuary appears more impacted than the Authie by pharmaceuticals; this may be related to anthropogenic pressures that are far less important in the Authie area. Note also that the general concentrations in both systems are very low, in the range of 2 to around 20 ng.L⁻¹, reflecting the dilution of the urban inputs upstream (the city of Rouen, for example, in the case of the Seine) in both systems. To summarize, the presence of pharmaceuticals reflects inputs of urban and domestic wastes into the studied estuaries; but the concentration levels are quite low and diluted (low chronic dose exposure). Ibuprofen concentrations in the Seine estuary (2.6 to 7.4 ng.L⁻¹) are well below the doses responsible for noxious effects in fish (Flippin et al. 2007). Similarly, gemfibrozil concentrations in the Seine (2 to 3 ng.L⁻¹) are orders of magnitude lower than the doses inducing effects in freshwater biota such as *Chlorella vulgaris* and *Daphnia magna* (Zurita et al. 2007).

3.2.3 ALKYLPHENOLS

Alkyphenol (AP) chemicals enter the environment as a consequence of their use in many herbicides, paints, and industrial and household products such as detergents, plasticizers, and UV stabilizers. Alkylphenol ethoxylates are discharged from waste

TABLE 3.3
Minimum and Maximum Concentrations
of Alkylphenols (ng.g⁻¹) Measured in the
Particulate Phase in the Authie and Seine

	Authie	Seine (near Honfleur)
NP1EO	<30 to 177	40 to 172
NP2EO	<30 to 89	20 to 227
NP1EC	<32 to 109	<32 to 266
NP	51 to 1000	68 to 3182

water treatment plants. In effluents or in the environment, they are converted into lipophilic, more persistent, and more toxic biodegradation products such as nonylphenol (NP) (see review by Ying et al. 2002).

Many reports indicate the presence of alkylphenols in natural waters (Naylor et al. 1992; Heemken et al. 2001; Inoue et al. 2002; Basheer et al. 2004; Koh et al. 2006). The estrogenic potencies of alkylphenol ethoxylates and related compounds were reviewed by Nimrod and Benson (1996). More recent studies have documented the modes of action of these xenoestrogens in fish (Meier et al. 2007; Tollefsen 2007 and references). Genotoxic effects may be suspected (Isidori et al. 2007). NP and octylphenol (OP) have also been documented to adversely affect the reproduction and embryonic developmental stages of invertebrates, and this is of great ecological importance because of the hazard at the population level (Arslan and Parlak 2007; Arslan et al. 2007).

Metabolites of APs are hydrophobic and thus tend to be accumulated by organisms and adsorb to particulates such as sediments (Table 3.3; Servos 1999). Concentrations in the particulate matter vary greatly but NP is the major compound found in both the Authie and Seine estuaries; it is the most hydrophobic metabolite. The concentrations are generally more important in the Seine estuary when expressed as nanograms per gram dry weight (dw); concentrations are more frequently below detection limits in the Authie system as opposed to the Seine system. The intersite difference is even more pronounced when concentrations are expressed as nanograms per liter, taking into account the amount of SPM, since the turbidity in the Seine estuary (113 to 533 mg.L⁻¹) is much more important than in the Authie estuary (26 to 98 mg.L⁻¹).

When considering the dissolved phase, the concentrations monitored in the Seine estuary (Figure 3.1) are much more important than those recorded for the Authie estuary and indicate a very strong input of those compounds into the Seine estuary (up to 800 ng.L⁻¹ NP1EC). The fingerprint of the Seine samples is the predominance of NP1EC showing the specific degradation in this system or specific inputs via sewage water treatment plant effluents (aerobic route; Ahel et al. 1994).

In summary we can conclude than in the case of the Seine estuary, the presence of AP is marked and may exert important chemical pressures on organisms, especially in the case of NP1EC. The two sites showed contrasting AP levels; the Authie is much less impacted.

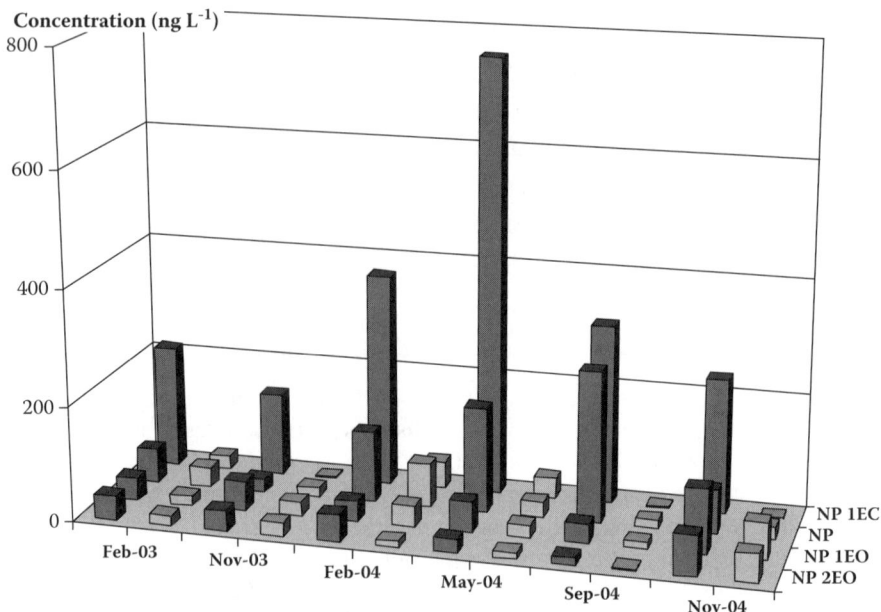

FIGURE 3.1 Concentrations of alkylphenols in the dissolved phase in the Seine estuary near Honfleur (dark grey) and the Authie (clear grey) estuary. NP2EO = nonylphenol diethoxylate. NP1EO = nonylphenol monoethoxylate. NP = nonylphenol. NP1EC = nonylphenol monoethoxycarboxylate.

3.3   ORGANIC AND INORGANIC CHEMICALS IN SEDIMENTS

Metals, PCBs, and PAHs were analysed in superficial sediments (<1 cm deep) on four occasions each year (February 2002 to November 2004). Samples were collected in the intertidal zone at low tide. The preservation and pretreatment of samples were carried out according to the recommendations to prevent losses or secondary contamination of samples (Charlou and Joanny 1983; Budzinski et al. 2000). Analyses were carried out using the best procedures currently available.

Levels of Cu and Zn were determined by flame atomic absorption spectrophotometry (FAAS). Ag, Cd, and Pb levels were analysed by electrothermal atomic absorption spectrophotometry (EAAS) with the Zeeman effect (Varian SpectrAA55 and SpectrAA800 spectrophotometers). The analytical methods used were validated via international intercalibration exercises (Coquery et al. 2001) and internal quality controls with marine sediment standard reference material IAEA-SD-M-2/TM. PCBs were analyzed by a gas chromatograph coupled to an electron capture detector (GC-ECD) according to the method described by Thompson and Budzinski (1999). Polycyclic aromatic hydrocarbons were analysed by a gas chromatograph coupled to a mass spectrometer (GC-MS) according to the method described by Baumard et al. (1999a,b). The entire analytical procedure was applied several times to certified marine sediment SRM 1944 (NIST, Gaithersburg, Maryland).

The recoveries for five replicates of this SRM ranged from 80 to 100%, with reproducibility ranging from 5 to 15% depending on the compound. For more details, we refer the reader to Poirier et al. (2006) and Durou et al. (2007). PBDEs were extracted by microwave extraction. Solid phase extraction (SPE) was used for subsequent separation. Analyses of BDE-28, -47, -99, -100, -153 and -154 were quantified by using a gas chromatograph Varian 3800 operated in MS/MS mode with an on-column injection system (GC-MS/MS). BDE-209 was determined using a Varian 3800 equipped with a 63Ni electron capture detector (GC-ECD). The column used was an RTX-500. Values obtained in recovery tests carried out in duplicate with reference materials (WMF-01, Wellington Laboratories Inc.) were in good agreement (80% at minimum) with certified values (Bragigand 2005).

3.3.1 TRACE METAL CONCENTRATIONS

The mean concentrations are shown in Table 3.4. Concerning the ecotoxicological significance of the micropollutant levels recorded in sediments from the Seine and Authie estuaries, comparisons can be made of data from this study and proposed guideline values for adverse biological effects on living resources in marine and estuarine sediments, e.g., the Effects Range Low/Median (ERL/ERM) of Long et al. (1995), the Threshold Effect Level/Probable Effect Level (TEL/PEL) of Macdonald et al. (1996), and the Ecotoxicological Assessment Criteria (EAC) of OSPAR (2000).

3.3.2 ORGANIC CHEMICAL CONCENTRATIONS

The concentrations of PCBs are more important in the Seine estuary than in the Authie (maximum: 186 versus 15 $\mu g.g^{-1}$; average: 85 versus 10 $\mu g.g^{-1}$; see Table 3.4).

TABLE 3.4

Concentrations of Metals ($\mu g.g^{-1}$ dw) and Organic Contaminants ($ng.g^{-1}$ dw) in Estuarine Sediments

	Authie			Seine		
	Mean (SD)	CV	Min–Max	Mean (SD)	CV	Min–Max
Ag	0.8 (0.2)	25	0.4–1.1	1.4 (0.4)	29	0.8–2.2
Cd	0.24 (0.07)	29	0.17–0.43	0.5 (0.3)	60	0.14–1.35
Cu	7.9 (1.7)	22	4.9–10.5	20 (12)	60	6–37
Pb	10.0 (5.4)	54	4.8–24.2	29 (18)	62	9.4–65.0
Zn	41 (8)	20	31–57	84 (39)	46	40–144
Σ PCB	10.3 (3.6)	35	3.4–15.0	85 (51)	60	27–186
Σ PAH	1046 (206)	20	492–1562	1696 (1141)	67	466–3579
Methylphenanthrene	34 (7)	21	14–44	53 (35)	66	17–121

Note: Means calculated from samples collected from February 2002 to November 2004 (n = 12). Parentheses indicate standard deviations. CV = coefficients of variation (%). Min/Max = limit values.

PCB levels are generally not so high on a global scale and are comparable with low to moderately contaminated estuaries reported in several other studies (Danis et al. 2004; Fu et al. 2003; Kowalewsha et al. 2003). Authie sediments are below the TEL/PEL value range while Seine sediments range between TEL and PEL values. They never exceeded PEL values, showing moderate risk from this class of contaminants.

The concentrations of PAHs are more important in the Seine estuary than in the Authie (maximum: 3579 versus 1562 μg.g^{-1}; average: 1696 versus 1046 μg.g^{-1}) but the difference is less pronounced for PAHs than for PCBs. PAH levels are comparable with those of moderately contaminated estuaries reported in several other studies (Woodhead et al. 1999; Fu et al. 2003; Kowalewsha et al. 2003; Grabe and Baron 2004). The sediments never reached the ERL guideline values, showing a moderate risk for this class of contaminants, especially for Authie sediments.

Different molecular ratios have been proposed to provide information on the origin of PAHs. The ratio of fluoranthene concentrations to pyrene concentrations allows discrimination between pyrolytic (Fluo/Pyr >1) and petrogenic (Fluo/Pyr <1) origins of contamination (Baumard et al. 1998; 1999a). According to Broyelle (1997), BeP/BaP <2 reveals a pyrolytic source; the ratio exceeds 5 in cases of petrogenic origin. Moreover, the ratio of phenanthrene concentrations to phenanthrene derivative concentrations (methylphenanthrenes, MPs) can also be used to indicate a petrogenic origin (Phe/MP <1) characterized by a preponderance of alkylated PAHs compared to parent compounds (Garrigues et al. 1995). From the ratios calculated for the sediments of the Seine and the Authie estuaries, it may be concluded that the PAH origins are mainly pyrolytic with weak petrogenic inputs in the Seine estuary as shown by low values of Phe/MP.

Among brominated flame retardants, the analytical effort was devoted to PBDEs (Bragigand 2005). Only four quarterly samples were analysed in each estuary (July 2002 through May 2003). The most abundant congeners were BDE-47 and -209. Mean BDE-47 and -209 concentrations were higher in sediments from the Seine estuary (0.13 and 0.33 ng.kg^{-1} dw, respectively) than in sediments from the Authie estuary (0.08 and 0.11 ng.kg^{-1} dw). As noted above for other contaminants, temporal intrasite variations were higher in the Seine (coefficients of variation of 61 and 52% for BDE-47 and -209, respectively) than in the Authie (coefficients of variation of 36 and 30% for BDE-47 and -209, respectively). In the Netherlands, BDE-209 concentrations up to 0.51 ng.kg^{-1} dw were determined (De Boer et al. 2000). Even in the Seine estuary, BDE-209 concentrations remain moderate compared to values as high as 1.8 ng.kg^{-1} detected in some European rivers (Sellström et al. 1999), 3.19 in sediments from Great Britain (Allchin et al. 1999), and 11.60 in sediments from Japan (Watanabe et al. 1986).

3.3.3 SEDIMENT CONCENTRATION RATIOS

The intersite difference was partly concealed by the intrasite variability that was higher in the Seine than in the Authie estuary, with coefficients of variation always lower in the latter (Table 3.4). Thus for purposes of comparing the two estuaries of interest, the sediment concentration ratio (SCR) for a contaminant (C) was calculated as follows:

$$SCR = \frac{[CSeine]}{[CAuthie]}$$

where [CSeine] is the concentration of the contaminant in the contaminated sediment (Seine) and [CAuthie] the corresponding concentration of the contaminant in the sediment from the Authie control site (Ruus et al. 2005).

Figure 3.2 shows results. For PCBs, SCRs were calculated for the congeners that were most abundant in sediments from both sites and for total PCBs. These SCRs always exceeded 1 and reached about 10 on average. For Cu and Pb, they were also >1 and nearing 3. Results for silver were less variable and approached 2. For Cd, MP, and HAPs, SCR levels were occasionally <1. In the case of HAPs, this limited intersite difference is a consequence of high background due to atmospheric contamination. It is well established that global concentrations of sediment-bound contaminants result from degrees of anthropogenic input but are also governed by geochemical characteristics such as grain size and organic matter content (Ujevic et al. 2000; Marengo et al. 2006). Grain sizes in the present study did not differ significantly (paired t-test, $p = 0.11$) between the studied estuaries—78% of particles <63 µm in the Seine versus 71% in the Authie on average. The organic matter content was also very similar, with values of 8.2 and 8.7% on average (paired t-test, $p = 0.76$), respectively.

The temporal fluctuations of SCRs showed similar patterns for most of the organic chemicals and metals, with the exception of Ag (Figure 3.2). Contaminants were analysed in superficial sediments that were submitted to successive events of deposition and resuspension (see Chapter 2). Depending on sampling date, sediment may have been more or less recently deposited, with subsequent changes of contaminant storage. Changes in metal equilibria between geochemical fractions during sediment aging were demonstrated by spiking experiments (Fan et al. 2002). According to Ruus et al. (2005) who exposed ragworms to spiked sediments, the lack of aging of the test sediments could have been responsible for increased bioavailability of PAHs. In the Authie estuary, sedimentation is a relatively continuous process whereas in the Seine estuary, temporal changes are more erratic (see Chapter 2). The fact that SCRs are based on comparisons of the relatively stable Authie and highly fluctuating Seine sediments may explain the similarity of temporal trends for the different contaminants.

3.4 ORGANIC AND INORGANIC CHEMICALS IN BIOTA

The total concentration of a given contaminant in a sediment has little direct ecotoxicological value because only a fraction is bioavailable (see Gourlay et al. 2005; Amiard et al. 2007; Tusseau-Vuillemin et al. 2007). For example, significant differences in absorption efficiency were observed in the endobenthic bivalve *Macoma balthica* exposed to a PAH or a PCB (benzo[a]pyrene and PCB-52) bound to different

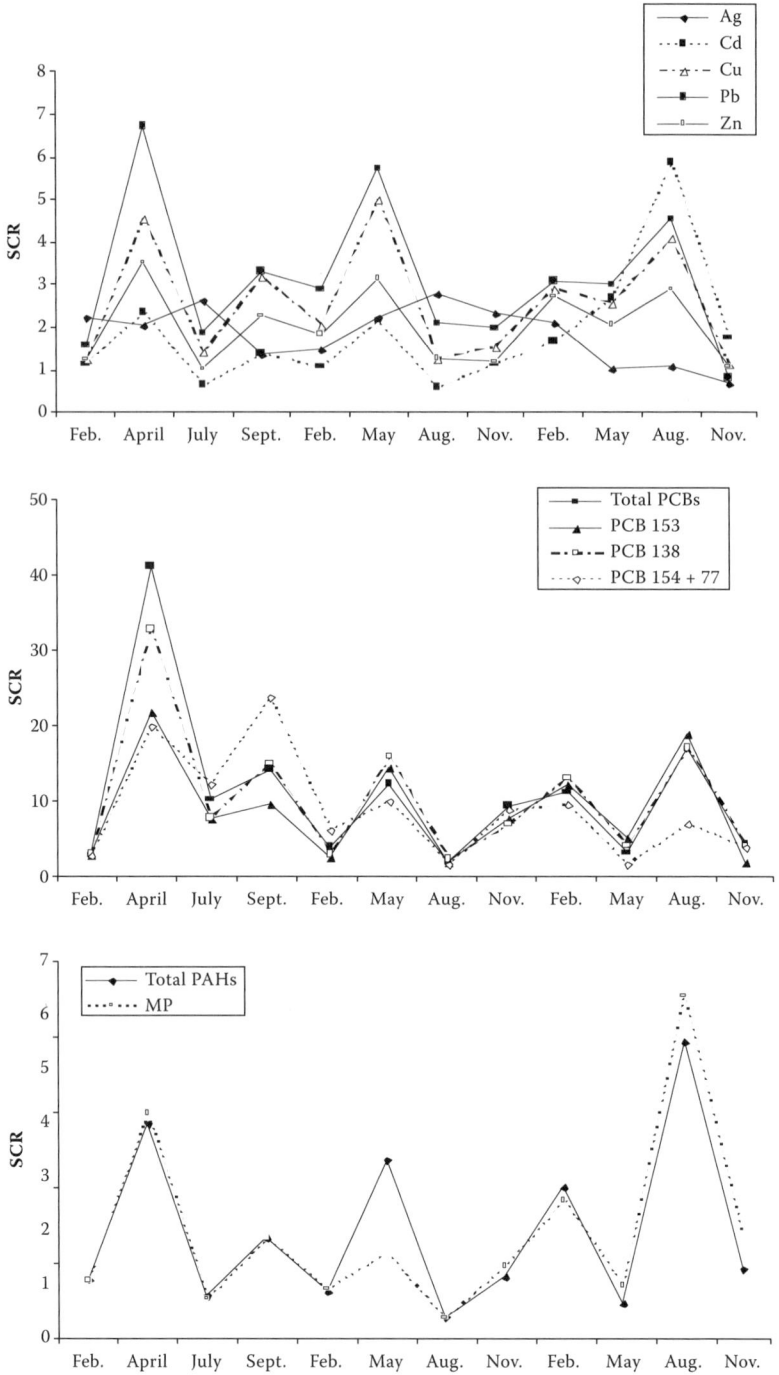

FIGURE 3.2   Ratios of mean metal and organic chemical concentrations in superficial sediments from the Seine and Authie estuaries (n = 12, February 2002 through November 2004).

carbonaceous particles collected from San Francisco Bay (McLeod et al. 2004). To link biological effects observed in worms from the polluted and the reference ecosystems, it was thus necessary to assess the quantities of contaminants that were really incorporated in these organisms depending on their sites of origin.

Worms were collected concomitantly with sediments. To allow the worms to eliminate their gut contents, they were kept for 2 days in filtered natural seawater adjusted to the salinity of their sites of origin. Chemical analyses were carried out on pooled (organic contaminants) or individual worms (20 for each site and each date for metals). Metal analyses were carried out in solutions obtained by heating the specimens with suprapure concentrated nitric acid (Berthet et al. 2003; Poirier et al. 2006). Organic contaminants were extracted from freeze-dried tissues by microwave-assisted extraction with dichloromethane (PAHs, PCBs; Budzinski et al. 1995) or by Soxhlet warm extraction (PBDEs; Bragigand et al. 2006). Subsequent separations were described for PAHs (Baumard et al. 1999a and b), PCBs (Thompson and Budzinski 1999), and PBDEs (Bragigand et al. 2006). For each chemical class, the instrumental phase was similar to the one described for sediment analyses. The analytical methods were validated through national (National Reference Laboratory; Bragigand et al. 2006) or international intercalibration exercises (Wyse et al. 2003) and use of internal quality controls (NIST-certified standard mussel tissue SRM 2976 for metals and 2974 for PAHs and PCBs; fish homogenate WMF-01, Wellington Laboratories for PBDEs).

3.4.1 Trace Metal Concentrations

Metal concentrations in ragworms are shown in Table 3.5. Ag, Cd, and Zn concentrations were significantly higher (paired t-tests) in specimens originating from the Seine estuary than in those from the Authie estuary. The same pattern was depicted for Cu but the intersite difference was not significant at the 95% level and on one occasion higher Cu concentrations were determined in ragworms from the Authie estuary (September 2002). In the case of Pb, no consistent intersite trend was shown.

No significant correlations were observed between concentrations of metals in sediments and concentrations determined in ragworms. These findings are not new and result from chemical or biological processes. As mentioned above, only a fraction of sediment-bound metals is reactive and thus potentially bioavailable. In several invertebrates including ragworms originating from different sites from France and the United Kingdom, clear relationships have been shown in many cases between labile metal concentrations and bioaccumulated concentrations (Amiard et al. 2007). On the other hand, many organisms can regulate essential metals at constant levels in their tissues while high fluctuations of metal concentrations occur in their environments (Amiard et al. 1987; Rainbow 2002).

The regulation of body Zn concentrations in the ragworm is well documented (Bryan 1976; Amiard et al. 1987) and is most probably responsible for equivalent Zn body concentrations in specimens from many sites with different levels of contamination (Berthet et al. 2003; Amiard et al. 2007). Surprisingly, in the present work, Zn concentrations were higher in worms from the Seine estuary than in those originating from the reference site.

TABLE 3.5
Concentrations of Metals (μg.g⁻¹ dw) and Organic Contaminants (ng.g⁻¹ dw) in the Ragworm *Nereis diversicolor*

	Authie			Seine		
	Mean (SD)	CV	Min–Max	Mean (SD)	CV	Min–Max
Ag	1.3 (0.8)	61	0.3–2.6	16 (14)	88	5.5–47.4
Cd	0.10 (0.06)	60	0.05–0.27	1.6 (0.4)	25	0.8–2.3
Cu	19 (17)	89	3–67	37 (16)	43	15–59
Pb	0.7 (0.7)	100	0.1–2.0	0.8 (0.7)	88	0.1–2.1
Zn	212 (97)	46	108–468	334 (118)	35	210–504
Σ PCB	91 (41)	45	44–157	605 (169)	28	432–1019
Σ PAH	66 (33)	50	26–143	274 (369)	135	72–1421
Methylphenanthrene	7.3 (5.1)	70	2–19	11.6 (6.5)	56	4–24

Note: Means calculated from samples collected from February 2002 to November 2004 (n = 12). Parentheses indicate standard deviations. CV = coefficients of variation (%). Min–Max = limit values.

In addition to exposure to higher levels of zinc in their environment, ragworms from the Seine and the Authie differed in a number of biological parameters including size and weight that were consistently higher in animals from the reference site (see Chapter 8). In the 1970s, when many studies were carried out to evaluate the influence of natural factors on metal bioaccumulation, with the objective to use bioindicators in the framework of pollution monitoring in aquatic ecosystems, the influences of size and weight were well recognized (see NAS 1980). Thus biometric data (size as total dw) and metal concentrations were determined in individual specimens collected in 2003 and 2004 (20 for each quarterly sample at each site) and the relationship between metal concentration [M] and dry weight W was established using the equation:

$$\text{Log } [M] = a * \log W + b$$

Generally weight and metal concentrations were negatively and significantly correlated with the exception of lead. Using these relationships, a correction factor was established to limit the effect of different physiological status (Poirier et al. 2006). These adjusted concentrations were calculated as follows:

$$\text{Log adjusted } [M] = \log \text{ measured } [M] + \{a * (\log \text{ mean } W - \log \text{ measured } W)\}$$

This procedure was used to confirm significant intersite differences for Ag and Cd and the absence of a significant difference for Pb. On the other hand, the difference nears the 95% level of probability for Cu ($p = 0.052$) and the intersite difference is no longer significant for Zn ($p = 0.076$).

In addition, individual adjusted concentrations were randomly extracted to generate groups of 2 to 20 individuals in worm populations collected from each estuary in

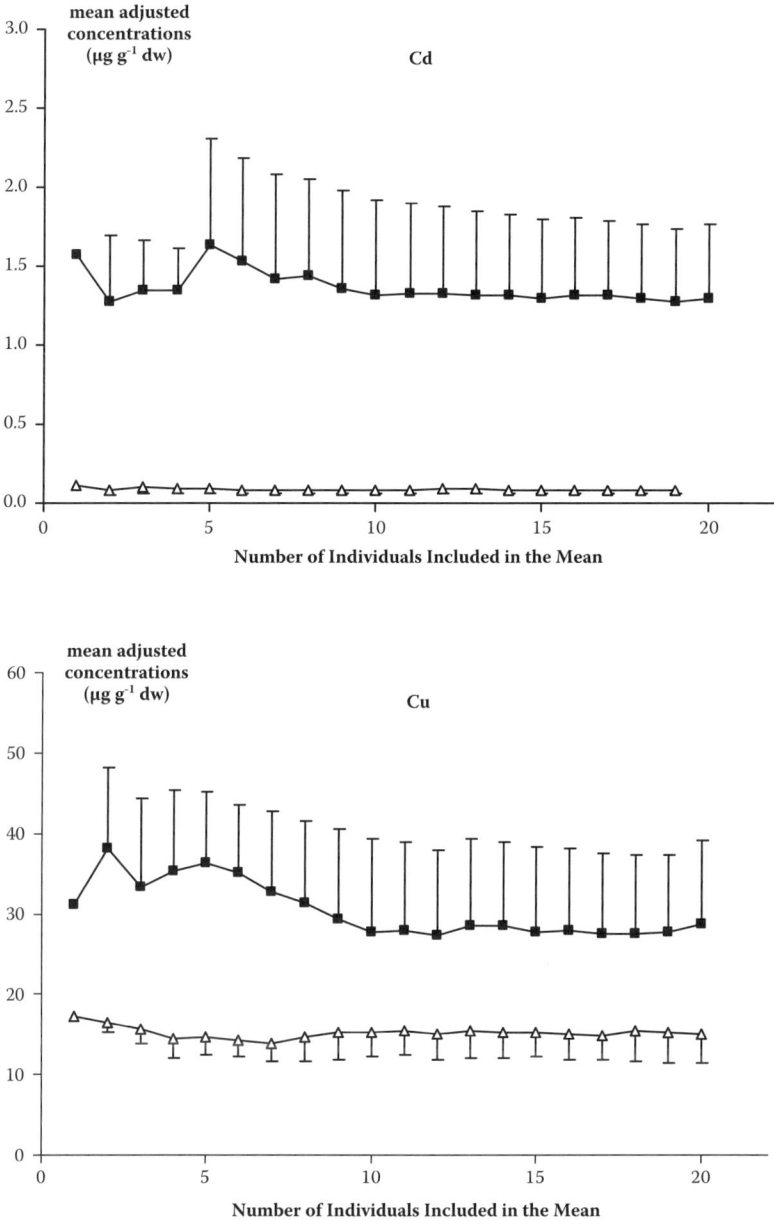

FIGURE 3.3 Means of adjusted Cd and Cu concentrations (μg.g^{-1} dw $\pm$ SD) in worms collected in February 2004. Groups of individuals on the x axis were selected at random. Seine estuary: black squares. Authie estuary: white triangles.

February 2004. Means and standard deviations were calculated for groups of increasing size as shown for Cd and Cu (Figure 3.3). For both metals, mean concentrations obtained from 5 individuals from one site or the other did not differ significantly (at the level of 95%) from the mean calculated for the whole sample (n = 20). The same was true for Ag, Pb, and Zn (for details see Poirier et al. 2006).

3.4.2 PCB CONCENTRATIONS

Because organic contaminant analyses required larger samples, it was impossible to determine individual concentrations. Thus the standard deviations depicted on Figure 3.4 were calculated for one determination in a pool of different weights at each sampling date. As shown above for sediments, PCB concentrations were higher in worms from the Seine estuary: 132, 97, and 70 ng.g^{-1} dw, respectively, for the three predominant congeners: CB-153, CB-138, and CB-154 plus -77, and corresponding values of 21, 14, and 10 ng.g^{-1} dw in specimens from the Authie (Figure 3.4a).

The pattern of differential accumulation of PCB congeners can be related to their individual characteristics. CB-138 and CB-153 are indeed the most stable PCBs whereas CB-50, -66, -77, and -101 are much more reactive in biotransformation processes, thus explaining their lower concentrations in worms (Safe 1993; Kannan et al. 1995). PCB concentrations in worms correlated significantly with those in sediments (for details see Durou et al. 2007). Again in agreement with their characteristics, CB-138 and CB-153 were proportionally more accumulated in the organisms

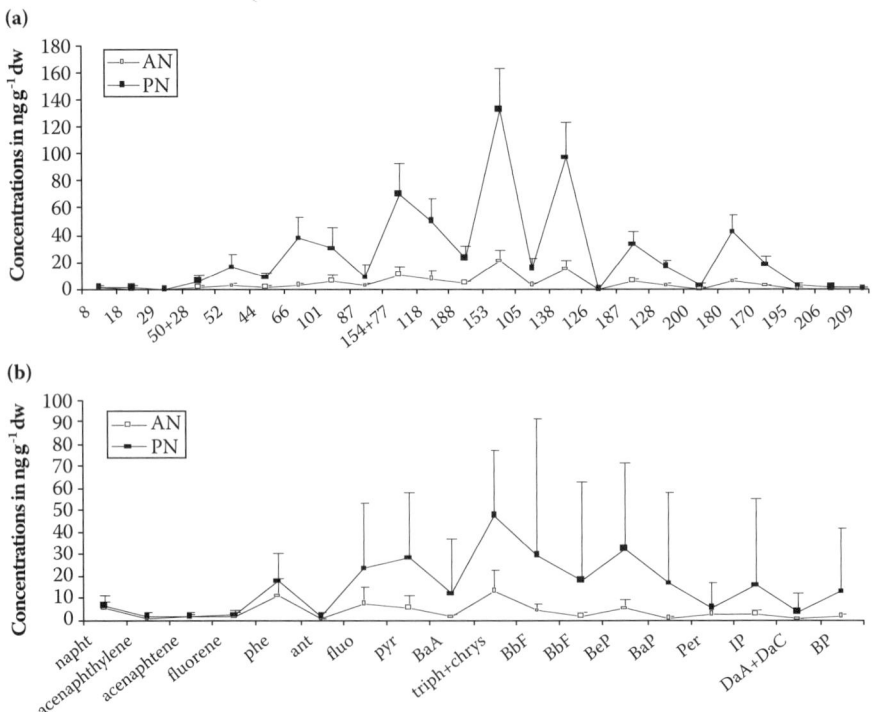

FIGURE 3.4 Concentrations of individual PCB congeners (a) and PAHs (b) in the ragworm *Nereis diversicolor*. Means and standard deviations calculated from samples collected February 2002 through November 2004 (n = 12).

than in the sediment whereas the opposite was noted for CB-66, CB-101, and CB-50 plus -28 (in specimens from the Seine only). These findings are also consistent with higher elimination half-lives for CB-153 than for CB-101 determined in *N. diversicolor* taken from the Weser estuary and exposed to PCBs in the laboratory (Goerke and Weber 1990).

3.4.3 PAH CONCENTRATIONS

Intersite differences were confirmed for concentrations of PAHs accumulated in the worms (Figure 3.4b). Triphenylene plus chrysene molecules were the most abundant in worms from both sites (47 and 14 ng.g^{-1} dw, respectively, in the Seine and the Authie), followed by benzo[e]pyrene (32 and 5 ng.g^{-1} dw), pyrene (28 and 6 ng.g^{-1} dw), fluoranthene (23 and 7 ng.g^{-1} dw), and phenanthrene (17 and 11 ng.g^{-1} dw). For ragworms exposed to control sediments from the outer Oslofjord, Ruus et al. (2005) reported concentrations of 1.2 ng.g^{-1} wet weight (ww) for triphenylene plus chrysene, 1.1 ng.g^{-1} ww for pyrene, 1.5 ng.g^{-1} ww for fluoranthene, and 0.8 ng.g^{-1} ww for phenanthrene. Considering that the ratio between wet and dry weight is about 5, the concentrations determined in specimens from the Authie estuary are very similar to these controls.

Ruus et al. (2005) also exposed ragworms to contaminated sediments collected from different areas in Kristiansand harbor (Norway) and calculated the ratio between the concentrations reached in exposed specimens and the control values mentioned above (bioaccumulation ratios = BARs). The highest BARs determined for the most polluted area reached 34.1 for triphenylene plus chrysene, 213.5 for pyrene, 64 for fluoranthene, and 9.3 for phenanthrene. BARs were calculated using the mean PAH concentrations for the same molecules in specimens from the Seine and the Authie; the maximum value was determined for pyrene (BAR = 4.6)—about two orders of magnitude lower than in the site studied by Ruus et al. (2005).

PAH concentrations in worms significantly correlated with concentrations in sediments (for details see Durou et al. 2007) but the triphenylene plus chrysene compounds showed proportionally higher accumulations in organisms than in sediments at both sites. These authors found similar patterns for phenanthrene and naphthalene in the Authie and BeP in the Seine. Opposite trends were observed for BaA, BkF, BaP, IP, and BP at both sites and in the Seine only for fluoranthene.

The concentration of a given molecule in animal tissue results from an equilibrium between uptake and elimination. In marine environments (especially for molluscs), biota sediment accumulation factors (BSAFs) generally decrease with increasing PAH log Kow* (Maruya et al. 1997; Ferguson and Chandler 1998). A similar trend was noted for terrestrial ecosystems (Brandt et al. 2002). These findings seem in contradiction with the generally accepted statement that bioaccumulation is favored by the lipophilic characters of chemicals. However, this physico-chemical property also affects solubility (Gómez-Gutiérrez et al. 2007) and ability to bind to sediment particles—and thus bioavailability. Uptake is the result of complex equilibria between decreased bioavailability and increased potential for bioaccumulation.

---

* Kow is the octanol–water partition coefficient that represents the lipophilicity of a molecule.

On the other hand, many organisms possess metabolic depuration pathways for eliminating excess body burdens of organic xenobiotics, particularly PAHs that are known to be metabolized by many species (Varanasi et al. 1989 and 1990).

Jørgensen et al. (2008) reviewed the mechanisms of biotransformation and elimination of PAHs in marine polychaetes. Briefly, phase I reactions involve hydrolysis, reduction, or oxidation of organic xenobiotics, rendering them slightly more water soluble. Phase II involves conjugation, i.e., the addition of endogenous groups (generally polar and readily available *in vivo*) to xenobiotics. The resulting increase in water solubility and decreased lipid solubility make the organic contaminants more prone to elimination (Newman and Unger 2003). Phase I enzymes belonging to the cytochrome P450 (CYP) family along with a few phase II enzymes have been identified in marine polychaetes (Jørgensen et al. 2008). *Nereis diversicolor* has been found to metabolize benzo[a]pyrene extensively (Driscoll and Mc Elroy 1996). The mechanisms of elimination of pyrene by *N. diversicolor* have been described (Christensen et al. 2002; Giessing et al. 2003); 1-hydroxypyrene and 1-hydroxypyrene glucuronide are the primary phase I and phase II metabolites of pyrene. The closely related species *N. virens* exhibited a high biotransformation capacity for PAHs with inducible pyrene hydroxylase activity, glucuronidation serving as the prominent phase II pathway (Jørgensen et al. 2005).

3.4.4 PBDE Concentrations

Among PBDE congeners, BDE-209, which was abundant in sediments, was generally not detectable in biota in which the most common congener was BDE-47. In the worm *N. diversicolor*, concentrations of BDE-47 varied from 6.2 to 8.0 ng.g^{-1} lipid in specimens from the Authie estuary and 8.8 to 10.5 in those originating from the Seine estuary. In another endobenthic species, the bivalve *Scrobicularia plana*, lower concentrations were generally determined in specimens from the Authie (2.6 to 3.2 ng.g^{-1} lipid) than in bivalves from the Seine (Honfleur = 2.9 to 8.6; Pont de Normandie = 3.9 to 5.6 ng.g^{-1} lipid).

3.5 TOXICITY IDENTIFICATION EVALUATION (TIE) OF XENO-ESTROGENS

The main reason to carry out environmental monitoring of chemical compounds is their potential toxicity to living organisms including humans. Indeed, many toxic compounds are released into the environment and the subsequent toxicities of waters and sediments are associated with adverse ecological effects such as population declines and changes in community structure (Swartz et al. 1982 and 1994; Anderson et al. 1987; Bailey et al. 1995; Hartwell et al. 1997; Hatakeyama and Yokoyama 1997; Toppari et al. 1998; WHO 2002).

However, chemical measurements alone may not be sufficient for some environmental management decisions. More than 80,000 chemical compounds in daily use may be released into waterways. Although common chemical screens may detect dozens or even hundreds of chemicals, they will leave thousands of potential toxicants unmeasured. Choosing *a priori* the compounds to be investigated is analogous to betting on only one horse in a race run by thousands of horses. Furthermore, even if the adequate techniques existed for measuring each compound with a high sensitivity, it would not be possible to monitor all of them at reasonable cost. Sound environmental monitoring would require regular measurements that society cannot possibly afford.

Arguably, the main reason for not relying on chemical measurements alone is that they do not provide any meaning regarding the toxicity in the field. While contaminants in water and sediment can be strong indicators of potential ecological risk when concentrations are elevated, the presence of a given chemical does not imply that toxicity may ensue as a result of its presence. Some contaminants may be present without being bioavailable and thus exert no effects on living organisms. The factors that control the bioavailability of a chemical to an organism are not known for every chemical and all environmental conditions, creating difficulties in defining the concentration of a chemical that would be expected to cause adverse effects. Furthermore, because waters that ultimately end up in estuaries receive contaminants from diverse sources, both waters and sediments from these zones are contaminated by mixtures of compounds whose concentrations are highly correlated with one other. Accordingly, it is nearly impossible to prove a causal link of chemical concentration and effect and as a result environmental management decisions may be made on weak bases.

Combining chemical measurements with biological tests can greatly improve the relevance of environmental monitoring programmes. This combination corresponds to the basic concept in toxicity identification evaluation (TIE) and other biologically directed fractionation approaches. Rather than using a physical detector to identify the presence of a chemical, a biological test is used as an indicator to determine whether a contaminant is biologically active within a sample and may lead to measurable effects. Chemical techniques can then help to characterise the studied sample. Physico-chemical properties of the active compounds can be used to prepare subsamples from the active sample and biological tests will indicate which subsample contains the chemical of interest. Ultimately, fractionation could produce samples that will allow identification of the chemical structure of an active compound.

TIE methodology has been used to search for xeno-estrogens in the Seine estuary. Previous studies found that male fish living in the Seine estuary and Seine bay were "feminised." Roach (*Rutilus rutilus*), gudgeon (*Gobio gobio*), and chub (*Leuciscus cephalus*) captured in the upper estuary, European smelt (*Osmerus eperlanus)* from the lower estuary, and flounder (*Platichthys flesus*) sampled in the bay exhibited various degrees of intersexuality. Histological analyses of testes indicate that some cells or parts of the organs differentiate into oocytes leading to ovotestes (Minier et al. 2000a and b).

High levels of vitellogenin have been measured in plasma samples from the same fish. Vitellogenin is a phospho-glycolipoprotein synthesized by the livers of fish under the stimulation of 17β-estradiol. This protein, after release into the blood, accumulates in growing oocytes and is used to build up energy reserves for use by the offspring. This "female" protein has been measured at high concentrations (1 $\mu g.mL^{-1}$) in 49% of the male fish analysed (Denier et al. 2008a).

Because feminisation can be driven by xeno-estrogens, a biological test specifically designed to assess compounds that may elicit estrogenic responses has been used. The yeast estrogen screen (YES) is a gene reporter assay. The yeast strain has been genetically engineered to express the human estrogen receptor and the estrogen response element. Incubating growing yeasts in a medium containing estrogenic compounds leads to synthesis of a galactosidase that hydrolyses a chromogenic substrate and allows a simple spectrophotometric measurement of the resulting enzymatic activity that is proportional to the amount of estrogenic compound present. Conducting the YES test using 17β-estradiol as positive control allows the expression of the potencies of tested samples as estradiol equivalents (EEq) per unit.

Water and sediments from the Seine estuary have been used to assess xeno-estrogens. Organic compounds were extracted in methanol (with previous solid phase extraction for water samples) and tested for total estrogenicity via the YES assay. Water sampled in the vicinity of the sewage treatment plant (STP) of Le Havre contained up to 100 ng EEq.mL^{-1}). High levels may be measured within the waters of the port but the concentration drops to a few nanograms or even below the detection limit (<0.04 ng EEq.mL^{-1}) in sites facing the open sea. The same levels of activity were recorded in sediment extracts from the same sites in Le Havre (up to 60 ng EEq.g^{-1} dw). However, results were always negative in sediment extracts from the coast line. Analyses of sediment extracts were also performed in different zones of the Seine estuary. A number of sites in the upper part of the estuary exhibited levels below detection limits and a few were in the nanogram range (Table 3.6). The effects of effluents from the city of Rouen were clearly shown; all studied sites from the lower part were positive. Results from the site nearest to Rouen and in proximity to the STP revealed estrogenic compound concentrations up to 10 ng EEq.g^{-1} dw.

Results from the YES assay clearly indicate that compounds that are estrogenic via the estrogen receptor are contaminating the sediments of the Seine estuary (Peck et al. 2007). Positive samples were used to determine the identities of the active compounds. Samples were fractionated by RP-HPLC to produce estrogenic profiles and GC-MS was used to identify compounds in fractions showing estrogenic activities. Sixty fractions were collected after fractionation and subjected to the YES assay. Only a few selected fractions showed estrogenic activity (Figure 3.5). The two main contributors to the estrogenic response were identified as the estrone (E1) and estradiol (E2) natural hormones. A third choice was sometimes apparent in samples from the vicinity of the STP of Le Havre but the identity of the compound could not be elucidated.

TABLE 3.6
Total Estrogenic Activity (ng EEq.g^{-1} dw) of Sediment Extracts Sampled in Seine Estuary and Seine Bay

Sampling Site	Upper Estuary	Lower Estuary	Seine Bay
Total estrogenicity	0.67 (± 0.51)	4.48 (± 3.51)	<DL

Note: Upper and lower estuaries delimited by city of Rouen. Confidence intervals indicated in brackets. <DL = below detection limit.

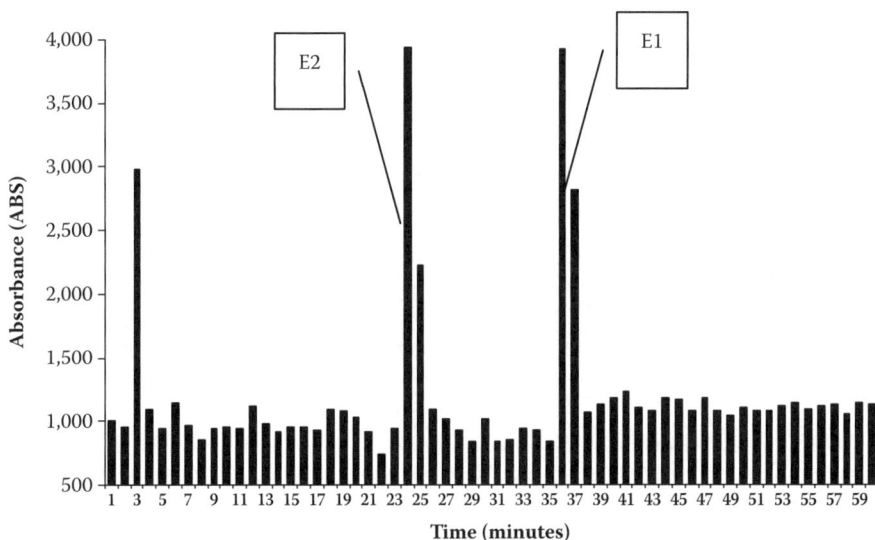

FIGURE 3.5 Representative HPLC profiles of water samples. E1 = estrone. E2 = estradiol.

Because chemicals were extracted in methanol, the results did not indicate that the identified compounds were bioavailable. To study the interactions of potential xeno-estrogens with living organisms, extractions from mussel *Mytilus edulis* tissues were performed. Total estrogenic activity was as high as 100 ng EE2.g^{-1} ww in saponified samples. Saponification was performed because mussels can readily metabolise estradiol into C16-E2 esters (Labadie et al. 2007) and the resulting esters are not estrogenic in the YES test.

The same experiment was conducted with zebra mussels (*Dresseina polymorpha*) from the upper estuary. Total estrogenic activity was 45 ng EE2.g^{-1} ww for saponified extracts. More than 98% of the estrogenic activity in the mussels was attributed to E2, although nonylphenol (NP) was also detected in concentrated fractions (Peck et al. 2007). The HPLC profiles of estrogenic substances in male and female *D. polymorpha* were similar. These results indicate that the predominant estrogenic substances in mussels collected from sites on the River Seine are E2 and NP, whereas estrogenic contaminants in the surrounding sediments included E1 as well as E2 and NP. The presence of NP in the tissues clearly demonstrates that NP in the water or sediments of the River Seine is taken up by resident mussels.

Roach and flounder bile samples were also used to study exposure of these fish species to estrogenic compounds. Unlike molluscs in which steroid synthesis and function are still questionable (Ketata et al. 2008), analyses were performed only in males to avoid detection of endogenous female steroids. Bile samples were first deconjugated and then tested for estrogenicity using YES. Results showed that more than 80% of the estrogenicity was due to natural estrogens (E2 and E1). NP and alkylpolyphenol ethoxylates (APEs) were also found (Figure 3.6). In addition to natural steroids, ethynyl estradiol (EE2) derived from contraceptive pills was present in most analysed samples.

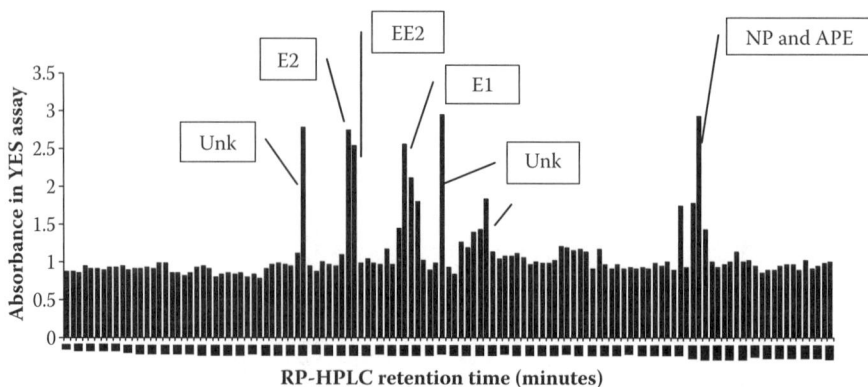

FIGURE 3.6   Representative HPLC profile of male deconjugated flounder bile. E1 = estrone. E2 = estradiol. EE2 = ethynyl estradiol. NP = nonylphenol. APE = alkylphenol polyethoxylates. Unk = unidentified compound.

In conclusion, the results showed that xeno-estrogens are contaminating the estuarine environment and that sediments may act as sinks for these compounds (Peck et al. 2004). Mussels and fish are taking up xeno-estrogens from both water and sediment and these compounds may be responsible for vitellogenin accumulation in fish blood and intersexuality. Among the identified compounds, natural hormones are mainly responsible for the estrogenicity.

This study is a good example of the use of a strategy that combines both biological and chemical tools. Nevertheless, not all questions can be answered by using one biological test or one chemical procedure.

In TIE experiments, sample preparation obviously influences the results and should be performed with caution. The nature of the biological test is also very important and the question to be answered must be well formulated before the study is undertaken. There are also some limitations to the procedure. Fractionating a sample may change a response. In the YES assay, results of the total estrogenicity of a given sample are usually higher than the sum of the responses given by isolated fractions. Furthermore, metals can potentiate the estrogenic response of E2 in the YES test (Denier et al. 2008b). Unravelling what may be the effects of all interactions among compounds in a complex mixture will require many more studies.

3.6   CONCLUSIONS

Even if the Authie estuary cannot be described as a pristine site, a clear contrast exists between levels of most chemicals determined in environmental samples in comparison with levels found in the Seine estuary. These results, in addition to information provided by the French Mussel Watch Programme (RNO 1995; Claisse et al. 2006) allow us to use the Authie estuary as a reference site.

For most of the chemicals determined in the Seine samples, concentrations are not among the highest found in estuarine areas elsewhere. However, significant effects already observed clearly indicate that the Seine estuary is an environment at risk. Embryotoxicity and *in vitro* genotoxicity (SOS Chromotest) assays revealed that organic extracts from sediments taken from the Seine estuary are potentially hazardous for species living or feeding in the area (Cachot et al. 2006). Among European flounder (*Platichthys flesus*) collected at sites near the Seine estuary in 1998, 8% of the males were intersex and another 10% exhibiting external morphological characteristics of males had gonads with only female tissues (Minier et al. 2000). Thus it seems that the Seine estuary is representative of the conditions prevailing in many estuarine areas in developed countries that (1) have made efforts to reduce chemical inputs in the environment but have not yet been totally successful with detergents, pharmaceuticals, and other molecules; (2) suffer the persistence of a number of regulated chemicals such as PCBs; and (3) must face inputs of emergent contaminants such as pharmaceuticals and brominated flame retardants.

As the ragworm *N. diversicolor* collects food, it also ingests considerable amounts of sediment and the gut content can represent up to 6% of body weight on a wet weight basis (Amiard-Triquet et al. 1984). Invertebrates including the ragworm can accumulate PAHs and metals from sediments (Clements et al. 1994; Ruus et al. 2005; Granberg and Selck 2007). For all sites, PCB concentrations in worms were nearly one order of magnitude higher than concentrations in sediment, whereas the sums of PAH concentrations in worms were dramatically lower than in sediment (Tables 3.4 and 3.5). The differences in results for these two classes of organic contaminants are usually attributed to a greater ability of organisms to transform and eliminate PAHs than PCBs. Because the ragworm is an important constituent of the estuarine food web, a subsequent transfer to higher trophic levels is expected (McElroy and Sisson 1989; Clements et al. 1994; Rainbow et al. 2004; Palmqvist et al. 2006).

Polychaetes (*Armandia brevis*) exposed for 4 weeks to field or spiked sediments (PAHs, chlorinated compounds) were fed to juvenile English sole (*Pleuronectes vetulus*). Fish growth was generally lower in groups fed contaminant-exposed worms than growth in reference groups. Exposure to contaminated food induced increased expression of CYP1A but hepatic DNA adducts were observed only in fish exposed to BaP-contaminated worms (Rice et al. 2000). Thus, polychaetes may play a key role in the transfer of contaminants and their toxicity from sediment to biota, particularly as food sources for bottom fish and wading birds.

ACKNOWLEDGMENTS

The Toxicity Identification Evaluation (TIE) of Xeno-Estrogens was part of the European Intereg IIIa titled Risk Analysis Associated with Endocrine Disruption in the Manche Regions (RAED).

REFERENCES

Ahel, M., W. Giger, and C. Schaaner. 1984. Behavior of alkylphenol polyethoxylate surfactants in the aquatic environment. I. Occurrence and transformation in sewage treatment. *Water Res.* 28: 1131–1142.

Allchin, C.R., R.J. Law, and S. Morris. 1999. Polybrominated diphenylethers in sediments and biota downstream of potential sources in the UK. *Environ. Pollut.* 105: 197–207.

Amiard, J.C. et al. 1987. Comparative study of the patterns of bioaccumulation of essential (Cu, Zn) and non-essential (Cd, Pb) trace metals in various estuarine and coastal organisms. *J. Exp. Mar. Biol. Ecol.* 106: 73–89.

Amiard, J.C. et al. 2007. Relationship between the lability of sediment-bound metals (Cd, Cu, Zn) and their bioaccumulation in benthic invertebrates. *Estuar. Coast. Shelf Sci.* 72: 511–521.

Amiard-Triquet, C., C. Métayer, and J.C. Amiard. 1984. Technical recommendations for studying the biogeochemical cycle of trace metals. *Rev. Int. Oceanogr. Med.* 73–74: 27–34.

Anderson, J. et al. 1987. *Biological Effects. Bioaccumulation and Ecotoxicology of Sediment Associated Chemicals. Fate and Effects of Sediment-Bound Chemicals in Aquatic Systems.* K. L. Dickson, A. W. Maki and W. A. Brungs (Eds.). New York: Pergamon Press, pp. 267–296.

Arslan, O. et al. 2007. Effects of nonylphenol and octylphenol on embryonic development of sea urchin (*Paracentrotus lividus*). *Arch. Environ. Contam. Toxicol.* 53: 214–219.

Arslan, O. and H. Parlak. 2007. Embryotoxic effects of nonylphenol and octylphenol in sea urchin *Arbacia lixula. Ecotoxicol.* 16: 439–444.

Basheer, C., H.K. Lee, and K.S. Tan. 2004. Endocrine disrupting alkylphenols and bisphenol-A in coastal waters and supermarket seafood from Singapore. *Mar. Pollut. Bull.* 48: 1161–1167.

Bailey, R. C. et al. 1995. Macroinvertebrate community structure and sediment bioassay results from nearshore areas of North American Great Lakes. *J. Great Lakes Res.* 21: 42–52.

Baumard, P., H. Budzinski, and P. Garrigues. 1998. Polycyclic aromatic hydrocarbons (PAHs) in sediments and mussels of the westrern mediterranean sea. *Environ. Toxicol. Chem.* 17: 765–776.

Baumard, P. et al. 1999a. Polycyclic aromatic hydrocarbons in recent sediments and mussels (*Mytilus edulis*) from the Western Baltic Sea: occurrence, bioavailability and seasonal variations. *Mar. Environ. Res.* 47: 17–47.

Baumard, P. et al. 1999b. Polycyclic aromatic hydrocarbon (PAH) burden of mussels (*Mytilus* sp.) in different marine environments in relation with sediment PAH contamination, and bioavailability. *Mar. Environ. Res.* 47: 415–439.

Bernes, C. 1998. *Persistent Organic Pollutants: A Swedish View of an International Problem.* Stockholm: Swedish Environmental Protection Agency.

Berthet, B. et al. 2003. Accumulation and soluble binding of Cd, Cu and Zn in the polychaete *Hediste diversicolor* from coastal sites with different trace metal bioavailability. *Arch. Environ. Contam. Toxicol.* 45: 468–478.

Bintein, S. and J. Devillers. 1996. Evaluating the environmental fate of atrazine in France. *Chemosphere* 32: 2441–2456.

Boyd, G.R. et al. 2003. Pharmaceuticals and personal care products (PPCPs) in surface and treated waters of Louisiana, USA and Ontario, Canada. *Sci. Total Environ.* 311: 135–149.

Bragigand, V. 2005. *Recherches écotoxicologiques sur les retardateurs de flamme bromés dans les écosystèmes estuariens (estuaires de Loire et de Seine).* Nantes: University of Nantes.

Bragigand, V. et al. 2006. Influence of biological and ecological factors on the bioaccumulation of polybrominated diphenyl ethers in aquatic food webs from French estuaries. *Sci. Total Environ.* 368: 615–626.

Brandt, C.A., J.M. Becker, and A. Porta. 2002. Distribution of polycyclic aromatic hydrocarbons in soils and terrestrial biota after a spill of crude oil in Trecate, Italy. *Environ. Toxicol. Chem.* 21: 1638–1643.

Broyelle, I. 1997. Contribution à l'étude de la matière organique en milieu estuarien : cas des deltas du Rhône et de la Léna. Thesis, Université de Paris 6, Paris, 256 p.

Bryan, G.W. 1976. Heavy metal contamination in the sea. In *Marine Pollution*, Johnston, R., Ed. London: Academic Press, pp. 185–302..

Bryan, G.W. and W.J. Langston. 1992. Bioavailability, accumulation and effects of heavy metals in sediments with special reference to United Kingdom estuaries: a review. *Environ. Pollut.* 76: 89–131.

Budzinski, H. et al. 2000. Combined protocol for the analysis of polycyclic aromatic hydrocarbons (PAHs) and polychlorobiphenyls (PCBs) from sediments using focussed microwave assisted (FMWW) extraction. *Fres. J. Anal. Chem.* 367: 165–171.

Budzinski, H. et al. 1995. Extraction assistée par chauffage micro-ondes focalisées (MOF) à pression ambiante des composés organiques dans les matrices naturelles: application à l'analyse des composés aromatiques. *C. R. Acad. Sci. Paris* 321: 69–76.

Cachot, J. et al. 2006. Evidence of genotoxicity related to high PAH content of sediments in the upper part of the Seine estuary (Normandy, France). *Aquat. Toxicol.* 79: 257–267.

Carpentier, S. et al. 2002. Quality of dredged material in the River Seine basin (France). I. Physico-chemical properties. *Sci.Total Environ.* 295: 101–113.

Charlou, J.L. and M. Joanny. 1983. Dosage du mercure et d'autres métaux (Pb, Zn, Cu, Cd, Co, Mn) dans les sédiments marins par absorption atomique. In *Manuel des analyses chimiques en milieu marin*, Aminot, A. and Chaussepied, M., Eds. Brest: CNEXO, pp. 285–295.

Christensen, M., O. Andersen, and G.T. Banta. 2002. Metabolism of pyrene by the polychaetes *Nereis diversicolor* and *Arenicola marina. Aquat. Toxicol.* 58: 15–25.

Claisse, D. et al. 2006. *Ligne de base: Les contaminants chimiques dans les huîtres et les moules du littoral français.* Nantes: IFREMER.

Clements, W.H., J.T. Oris, and T.E. Wissing. 1994. Accumulation and food chain transfer of fluoranthene and benzo[a]pyrene in *Chironomus riparus* and *Lepomis macrochirus. Arch. Environ. Contam. Toxicol.* 26: 261–266.

Cleuvers, M. 2003. Aquatic ecotoxicity of pharmaceuticals including the assessment of combination effects. *Toxicol. Lett.* 142: 185–194.

Coquery, M., S. Azemard, and S.J. de Mora. 2001. *The Analytical Performance Study for the Medpol Region: Determination of Trace Elements and Methylmercury in Estuarine Sediment Sample.* Monte Carlo: IAEA, Monaco, p. 405.

Danis, B. et al. 2004. Contaminant levels in sediments and asteroids (*Asterias rubens* L., Echinodermata) from the Belgian coast and Scheldt estuary: polychlorinated biphenyls and heavy metals. *Sci. Tot. Environ.* 333: 149–165.

De Boer, J., K. de Boer, and J.P. Boon. 2000. Polybrominated biphenyls and diphenylethers. In *The Handbook of Environmental Chemistry*, Paasivirta, J., Ed., Berlin: Springer-Verlag, pp. 62–95.

Denier X. et al. 2008a. Investigation of estrogenic contamination in the Seile estuary: Plasma vitellogénine and estrogenicity identification evaluation of flounder (*Platichthys flesus*) bile content. *Mar. Environ. Res.* Submitted.

Denier X. et al. 2008b. Estrogenic activity of cadmium, copper and zinc in the yeast-estrogen screen. *Toxicol. In Vitro,* Submitted.

Diez, G. et al. 2000. *Hediste (Nereis) diversicolor* as bioindicator of metal and organic chemical bioavailability: a field study. *Ecotoxicol. Environ. Restor.* 3: 7–15.

Driscoll, S.K. and A.E. McElroy. 1996. Bioaccumulation and metabolism of benzo[a]pyrene in three species of polychaete worms. *Environ. Toxicol. Chem.* 15: 1401–1410.

Durou, C. et al. 2007. Biomonitoring in a clean and a multi-contaminated estuary based on biomarkers and chemical analyses in the endobenthic worm *Nereis diversicolor. Environ. Pollut.* 148: 445–458.

Fan, W. et al. 2002. Cu, Ni, and Pb speciation in surface sediments from a contaminated bay of northern China. *Mar. Pollut. Bull.* 44: 820–826.

Ferguson, P.L. and G.T. Chandler. 1998. A laboratory and field comparison of sediment polycyclic aromatic hydrocarbon bioaccumulation by the cosmopolitan estuarine polychaete *Streblospio benedicti* (Webster). *Mar. Environ. Res.* 45: 387–401.

Flippin, J.L., Huggett, D., and Foran, C.M. 2007. Changes in the timing of reproduction following chronic exposure to ibuprofen in Japanese medaka, *Oryzias latipes*. *Aquat. Toxicol.* 81: 73–78.

Fu, J. et al. 2003. Persistent organic pollutants in environment of the Pearl River Delta, China: an overview. *Chemosphere* 52: 1411–1422.

Gagnon, C. and N.S. Fisher. 1997. Bioavailability of sediment-bound Cd, Co, and Ag to the mussel *Mytilus edulis*. *Can. J. Fish. Aquat. Sci.* 54: 147–156.

Garmouma, M. et al. 1998. Spatial and temporal variations of herbicide (triazines and phenylureas) concentrations in the catchment basin of the Marne river (France). *Sci. Tot. Environ.* 224: 93–107.

Garrigues, P. et al. 1995. Pyrolitic and petrogenic inputs in recent sediments: A definite signature through phenanthrene and chrysene compound distribution. *Polycyclic Aromatic Compounds* 7: 275–284.

Gfrerer, M. et al. 2002. Triazines in the aquatic systems of the Eastern Chinese Rivers Liao-He and Yangtse. *Chemosphere* 47: 455–466.

Ghosh, U., J. Zimmerman, and R. Luth. 2003. PCB and PAH Speciation among particle types in contaminated harbor sediments and effects on PAH bioavailability. *Environ. Sci. Technol.* 37: 2209–2217.

Giessing, A.M.B., L.M. Mayer, and T.L. Forbes. 2003. 1-Hydroxypyrene glucuronide as the major aqueous pyrene metabolite in tissue and gut fluid from the marine deposit-feeding polychaete *Nereis diversicolor*. *Environ. Toxicol. Chem.* 22: 1107–1114.

Goerke, H. and K. Weber. 1990. Population-dependent elimination of various polychlorinated biphenyls in *Nereis diversicolor* (Polychaeta). *Mar. Environ. Res.* 29: 205–226.

Gomez-Gutiérrez, A. et al. 2007. Influence of water filtration on the determination of a wide range of dissolved contaminants at parts-per-trillion levels. *Anal. Chim. Acta* 583: 202–209.

Gourlay, C. et al. 2005. The ability of dissolved organic matter (DOM) to influence benzo[a] pyrene bioavailability increases with DOM biodegradation. *Ecotoxicol. Environ. Saf.* 61: 74–82.

Grabe, S.A. and J. Barron. 2004. Sediment contamination, by habitat, in the Tampa Bay estuarine system (1993–1999): PAHs, pesticides and PCBs. *Environ. Monit. Assess.* 91: 105–144.

Granberg, M.E. and H. Selck. 2007. Effects of sediment organic matter quality on bioaccumulation, degradation, and distribution of pyrene in two macrofaunal species and their surrounding sediment. *Mar. Environ. Res.* 64: 313–335.

Guiho, M. 1998. Les réseaux de surveillance des produits phytosanitaires dans la ressource en eau et leurs résultats. In L'eau, les pesticides et la santé. Actes du colloque régional de Quéven.

Guosheng, C. and P.A. White. 2004. The mutagenic hazards of aquatic sediments: a review. *Mut. Res.* 567: 151–225.

Halling-Sorensen, B. et al. 1998. Occurrence, fate and effects of pharmaceutical substances in the environment: a review. *Chemosphere* 36: 357–393.

Hartwell, S. I. et al. 1997. Correlation of measures of ambient toxicity and fish community diversity in Chesapeake Bay, USA, tributaries — urbanizing watersheds. *Environ. Toxicol. Chem.* 16: 2556–2567.

Hatakeyama, S., and N. Yokoyama. 1997. Correlation between overall pesticide effects monitored by shrimp mortality test and change in macrobenthic fauna in a river. *Ecotox. Environ. Safety* 36: 148–161.

Heberer, T. 2002. Occurrence, fate and removal of pharmaceutical residues in the aquatic environment: a review of the recent research data. *Toxicol. Lett.* 131: 5–17.

Heemken, O.P. et al. 2001. The occurrence of xenoestrogens in the Elbe river and the North Sea. *Chemosphere* 45: 245–259.

Hernando, M.D. et al. 2006. Environmental risk assessment of pharmaceutical residues in wastewater effluents, surface waters and sediments. *Talanta* 69: 334–342.

Inoue, K. et al. 2002. Determination of phenolic xenoestrogens in water by liquid chromatography with coulometric array detection. *J. Chromatogr. A* 946: 291–294.

Isidori, M. et al. 2007. Influence of alkylphenols and trace elements in toxic, genotoxic, and endocrine disruption activity of wastewater treatments plants. *Environ. Toxicol. Chem.* 26: 1686–1694.

Jørgensen, A. et al. 2005. Biotransformation of the polycyclic aromatic hydrocarbon pyrene in the marine polychaete *Nereis virens*. *Environ. Toxicol. Chem.* 24: 2796–2805.

Jørgensen, A. et al. 2008. Biotransformation of polycyclic aromatic hydrocarbons in marine polycheates. *Mar. Environ. Res.* 65: 171–186.

Kannan, N. et al. 1995. Chlorobiphenyls: modal compounds for metabolism in food chain organisms and their potential use as ecotoxicological stress indicators by application of the metabolic slope concept. *Environ. Sci. Technol.* 29: 1851–1859.

Ketata I. et al. 2008. Endocrine-related reproductive effects in molluscs. *Comp. Biochem. Physiol.* 147C: 261–270.

Koh, C.H. et al. 2006. Characterization of trace organic contaminants in marine sediment from Yeongil Bay, Korea: 1. Instrumental analyses. *Environ. Pollut.* 142: 39–47.

Kolpin, D.W. et al. 2004. Urban contribution of pharmaceuticals and other organic wastewater contaminants to streams during differing flow conditions. *Sci. Total Environ.* 328: 119–130.

Kowalewska, G. et al. 2003. Transfer of organic contaminants to the Baltic in the Odra Estuary. *Mar. Pollut. Bull.* 46: 703–718.

Labadie P. et al. 2007. Identification of the steroid fatty acid ester conjugates formed in vivo in *Mytilus edulis* as a result of exposure to estrogens. *Steroids.*72: 41–49.

Lartiges, S. 1994. *Analyse et devenir de pesticides organophosphorés et organoazotés dans l'environnement aquatique.* Bordeaux: Université de Bordeaux 1.

Laville, N. et al. 2004. Effects of human pharmaceuticals on cytotoxicity, EROD activity and ROS production in fish hepatocytes. *Toxicology* 196: 41–55.

Le Calvez, N. 2002. *Mise au point d'une méthodologie analytique appliquée au devenir de contaminants organiques dans l'environnement aquatique.* Lille: Université de Lille 1.

Lee, K.T., S. Tanabe, and C.H. Koh. 2001. Distribution of organochlorine pesticides in sediments from Kyeonggi Bay and nearby areas, Korea. *Environ. Pollut.* 114: 207–213.

Long, E.R. et al. 1995. Incidence of adverse biological effects within ranges of chemical concentrations in marine and estuarine sediments. *Environ. Manage.* 19: 81–97.

Macdonald, D.D. et al. 1996. Development and evaluation of sediment quality guidelines for Florida coastal waters. *Ecotoxicology* 5: 253–278.

Marengo, E. et al. 2006. Investigation of anthropic effects connected with metal ions concentration, organic matter and grain size in Bormida river sediments. *Anal. Chim. Acta* 560: 172–183.

Maruya, K.A., R.W. Risebrough, and A.J. Horne. 1997. The bioaccumulation of polynuclear aromatic hydrocarbons by benthic invertebrates in an intertidal marsh. *Environ. Toxicol. Chem.* 16: 1087–1097.

McElroy, A.E. and J.D. Sisson. 1989. Trophic transfer of benzo[a]pyrene metabolites between benthic marine organisms. *Mar. Environ. Res.* 28: 265–269.

McLeod, P.B. et al. 2004. Effects of particulate carbonaceous matter on the bioavailability of benzo[a]pyrene and 2,2',5,5'-tetrachlorobiphenyl to the clam, *Macoma balthica*. *Environ. Sci. Technol.* 38: 4549–4556.

Meier, S. et al. 2007. Effects of alkylphenols on the reproductive system of Atlantic cod (*Gadus morhua*). *Aquat. Toxicol.* 81: 207–218.

Minier, C. et al. 2000a. Flounder health status in the Seine Bay: a multibiomarker study. *Mar. Environ. Res.* 50: 373–377.

Minier C. et al. 2000b. An investigation of the incidence of intersex fish in Seine-Maritime and Sussex regions. *Analusis.* 28: 801–806.

NAS. 1980. *The International Mussel Watch. Report of Workshop Sponsored by the Environment Studies Board.* Washington: Natural Resources Commission of National Academy of Sciences.

Naylor, C. et al. 1992. Alkylphenol ethoxylates in the environment. *J. Am. Oil Chem. Soc.* 69: 695–703.

Newman, M.C. and M.A. Unger. 2003. Uptake, biotransformation, detoxification, elimination, and accumulation. In *Fundamentals of Ecotoxicology*, Newman, M.C. and Unger, M.A., Eds. Boca Raton: Lewis Publishers, pp. 53–73.

Nimrod, A.C. and W.H. Benson. 1996. Environmental estrogenic effects of alkyphenol ethoxylates. *Crit. Rev. Toxicol.* 26: 335–364.

OSPAR, 2000. Report of an assessment of trends in the concentration of certain metals, PAHs and other organic compounds in the tissues of various fish species and blue mussels. OSPAR Commission, 1998. http://www.ospar.org/fr/html/qsr2000/qfc4.htm/.

Palmqvist, A., L.J. Rasmussen, and V.E. Forbes. 2006. Influence of biotransformation on trophic transfer of the PAH, fluoranthene. *Aquat. Toxicol.* 80: 309–319.

Peck M.R. et al. 2004. Sediments are major sinks of steroidal estrogens in two United Kingdom rivers. *Environ. Toxicol. Chem.* 23: 945–952.

Peck M.R. et al. 2007. Profiling of environmental and endogenous estrogens in the zebra mussel (*Dreissena polymorpha*). *Chemosphere* 69: 1–8.

Poirier, L. et al. 2006. A suitable model for the biomonitoring of trace metal bioavailabilities in estuarine sediments: the annelid polychaete *Nereis diversicolor. J. Mar. Biol. Ass. U.K.* 86: 71–82.

Rainbow, P.S. 2002. Trace metal concentrations in aquatic invertebrates: why and so what? *Environ. Pollut.* 120: 497–507.

Rainbow, P.S. et al. 2004. Enhanced food chain transfer of copper from a diet of copper-tolerant estuarine worms. *Mar. Ecol. Prog. Ser.* 271: 183–191.

Rice, C.A. et al. 2000. From sediment bioassay to fish biomarker: connecting the dots using simple trophic relationships. *Mar. Environ. Res.* 50: 527–533.

RNO. 1995. *Surveillance de la qualité du milieu marin.* IFREMER: Paris.

Ruus, A. et al. 2005. Experimental results on bioaccumulation of metals and organic contaminants from marine sediments. *Aquat. Toxicol.* 72: 273–292.

Safe, S. 1993. Toxicology, structure–function relationships, and human and environmental health impacts of PCBs: progress and problems. *Environ. Health Perspect.* 100: 259–268.

Sellström, U. et al. 1999. Brominated flame retardants in sediments from European estuaries, the Baltic Sea and in sewage sludge. *Organohalog. Comp.* 40: 383–386.

Servos, M.R. 1999. Review of the aquatic toxicity, estrogenic responses and bioaccumulation of alkylphenols and alkylphenols polyethoxylates. *Water Qual. Res. J. Can.* 34: 123–127.

Sujatha, C.H., S.M. Nair, and J. Chacko. 1999. Determination and distribution of endosulfan and malathion in an Indian estuary. *Water Res.* 33: 109–114.

Swartz, R. C. et al. 1982. Sediment toxicity and the distribution of amphipods in Commencement Bay, Washington, USA. *Mar. Poll. Bull.* 13: 359–364.

Swartz, R. C. et al. 1994. Sediment toxicity, contamination and amphipod abundance at a DDT and dieldrin contaminated site in San Francisco Bay. *Environ. Toxicol. Chem* 13: 949–962.

Ternes, T.A. 2001. Analytical methods for the determination of pharmaceuticals in aqueous environmental samples. *TrAC* 20: 419–434.

Ternes, T.A. et al. 2004. Detailed report of POSEIDON project. http//www.eu-poseidon.com.

Thompson, S. and H. Budzinski. 1999. Determination of polychlorinated biphenyls and chlorinated pesticides in environmental biological samples using focused microwave-assisted extraction. *Intern. J. Environ. Anal. Chem.* 76: 49–60.

Tollefsen, K.E. 2007. Binding of alkylphenols and alkylated non-phenolics to the rainbow trout (*Oncorhynchus mykiss*) plasma sex steroid-binding protein. *Ecotoxicol. Environ. Saf.* 68: 40–48.

Toppari J et al. 1995 Male reproductive health and environmental chemicals with estrogenic effects. Copenhagen: Danish Environment Protection Agency, 172 pp.

Tronczynski, J., C. Munshy, and K. Moisan. 1999. *Les contaminants organiques qui laissent des traces : sources, transport et devenir.* IFREMER: Plouzané.

Tusseau-Vuillemin, M.H. et al. 2007. Dissolved and bioavailable contaminants in the Seine river basin. *Sci. Total Environ.* 375: 244–256.

Ujevic, I., N. Odzak, and A. Baric. 2000. Trace metal accumulation in different grain size fractions of the sediments from a semi-enclosed bay heavily contaminated by urban and industrial wastewaters. *Water Res.* 34: 3055–3061.

Varanasi, U. et al. 1990. Survey of Subsistence Fish and Shellfish for exposure to oil spill from the *Exxon Valdez*, NOAA Tech. Memo. NMFS F/NWC-191. U.S. Department of Commerce, Seattle, WA.

Varanasi, U., J.E. Stein, and M. Nishimoto. 1989. Biotransformation and disposition of PAH in fish. *In*: Metabolism of Polycyclic Aromatic Hydrocarbons in the Aquatic Environment, Varanasi, U., Ed. Boca Raton: CRC Press, pp. 93–150.

Venkatesan, M.I. et al. 1999. Chlorinated hydrocarbon pesticides and polychlorinated biphenyls in sediment cores from San Francisco Bay. *Mar. Chem.* 64: 85–97.

Watanabe, I., T. Kashimoto, and R. Tatsukawa. 1986. Confirmation of the presence of flame retardant decabromobiphenyl ether in river sediment from Osaka, Japan. *Bull. Environ. Contam. Toxicol.* 36: 839–842.

WHO (World Health Organization). 2002. Global assessment of the state-of-the-science of endocrine disruptors. Damstra T, Barlow S, Bergman A, Kavlock R, Van Der Kraak G. (Eds.),WHO/IPCS/EDC/02.2. World Healh Organization, Geneva, Switzerland, 180 pp.

Woodhead, R.J., R.J. Law, and P. Matthiessen. 1999. Polycyclic aromatic hydrocarbons in surface sediments around England and Wales, and their possible biological significance. *Mar. Pollut. Bull.* 38: 773–790.

Wu, Y., J. Zhang, and Q. Zhou. 1999. Persistent organochlorine residues in sediments from Chinese river/estuary systems. *Environ. Pollut.* 105: 143–150.

Wyse, E.J., S. Azemard, and S.J. de Mora. 2003. World-wide intercomparison exercise for the determination of trace elements and methylmercury in fish homogenate, IAEA-407. Report AL/144, IAEA/MEL/72. Monte Carlo: IAEA.

Yao, Z.W., G.B. Jiang, and H.Z. Xu. 2002. Distribution of organochlorine pesticides in seawater of the Bering and Chukchi Seas. *Environ. Pollut.* 116: 49–56.

Ying, G.G., B. Williams, and R. Kookana. 2002. Environmental fate of alkylphenols and alkylphenol ethoxylates: a review. *Environ. Int.* 28: 215–226.

Zurita, J.L. et al. 2007. Toxicological effects of the lipid regulator gemfibrozil in four aquatic systems. *Aquat. Toxicol.* 81: 106–115.

4 Biomarkers Based upon Biochemical Responses

Michèle Roméo, Laurence Poirier,
and Brigitte Berthet

CONTENTS

Coastal and estuarine environments are subjected to numerous disturbances, among which chemical pollution (metals, PCBs, PAHs, HCHs, etc.) is more or less easy to quantify. Toxicological data on these substances allow the establishment of threshold effect levels such as PNECs (predicted no-effect concentrations), but these are generally obtained on pollutants studied individually, neglecting their potential interactions (antagonisms or synergies). In estuaries that contain complex mixtures of contaminants, several of which are not recognizable by known techniques or can be measured only by expensive analytical methods, biomarkers are effective as early warnings in revealing overall disturbances (Lam and Gray 2001 and 2003; Allan et al. 2006).

Although sediments are the main reservoirs for most organic and inorganic chemicals (Gagnon and Fisher 1997; Bernes 1998), infauna and flora are little used as biological monitors. Chapman and Wang (2001) noted that most methods of evaluating the toxicities of sediments concern freshwater or marine species, but few take into account the specific characteristics of estuaries. This chapter therefore focuses on a key estuarine sentinel species, the endobenthic worm *Nereis diversicolor*. In fact, this species has been recognized as a good complement to the use of filter-feeding organisms representative of the quality of the water column, thus providing a global *in situ* assessment of the bioavailabilities of sediment-bound contaminants over relatively long periods (Durou 2006; Durou et al. 2007; Moreira et al. 2006; Poirier et al. 2006).

Following Depledge (1993), a biomarker may be defined as "a biochemical, cellular, physiological or behavioural variation that can be measured in tissue or body fluid samples or at the level of whole organisms (either individuals or populations) that provides evidence of exposure to and/or effects of one or more chemical pollutants and/or radiations." The different definitions and classifications of a biomarker were reviewed by Roméo and Giamberini (2008). In this chapter discussing biomarkers based upon biochemical responses, we intend to follow the division proposed by De Lafontaine et al. (2000): biomarkers of defence and biomarkers of damage. Induction of metallothioneins (MTs) and other metal-binding ligands, and increases in the activities of the glutathione S-transferase (GST) enzyme and the anti-oxidant catalase enzyme provide defence reactions in organisms exposed to different chemicals in their environment. Inhibition of acetylcholinesterase (AChE) activity and increased thiobarbituric acid reactive substances (TBARS) reflect alterations caused to organisms by some pollutants.

We will discuss the data obtained with biomarkers measured in *Nereis diversicolor* collected in two French estuaries (the Seine and the Authie) under the framework of the PNETOX Program (Durou et al. 2007; Poirier et al. 2006).

4.1   MATERIALS AND METHODS

Four times annually from February 2002 to February 2004 worms were collected from the intertidal areas of both estuaries (stations AN and PN; see Figure 1.3 in Chapter 1) at low tide, transported, and treated as described by Poirier et al. (2006) and Durou et al. (2007). For the measurement of enzyme activities and determination of TBARS concentrations, samples were immediately frozen in the field using liquid nitrogen and then stored at −80°C until analysis. For metal-binding protein analysis and ultrastructural studies, worms were placed in filtered natural seawater adjusted to the salinity of the site of origin for 24 hours to allow them to eliminate their gut contents. They were then fixed for electron microscopy or frozen at −20°C until analysis.

Pooled worms (about 2 g total wet weight per pool) were homogenized to recover the cytosol, an aliquot of which was heat-treated to isolate MTs since these proteins are heat-stable. To characterize metal-binding ligands, their molecular masses were determined by gel permeation of total cytosol. The presence of sulfhydryl groups as tracers of MTs was determined in heat-denatured cytosol. Sulfhydryl (SH) groups were quantified by differential pulse polarography, as described by Thompson and Cosson (1984). The quantification of metallothionein-like proteins in whole worms was also based on the determination of sulfhydryl in heat-denatured cytosol (for details see Bragigand and Berthet 2003; Poirier et al. 2006).

Ultrastructural studies and microanalysis were performed on the teguments and gut walls of five specimens from the Seine estuary collected in April 2003.

We already had data on worms from another clean site (Mouneyrac et al. 2003), so no specimens from the Authie estuary were examined. The tissues were fixed as described by Poirier et al. (2006), observed on a Phillips 201 transmission electron microscope (TEM). X-ray EDS (energy-dispersive spectrometry) microanalysis was carried out on a Jeol JEM 2010 UHR TEM equipped with a Princeton Gamma Technology Model IMIX PTS at 200 KV, using 200-s counting periods and a 70-μm condenser diaphragm. These conditions allowed very good localization (50 nm at 100,000× magnification) but only qualitative *in situ* analysis.

Analyses of GST, catalase, and AChE enzymes and TBARS were carried out on eight pools of three specimens (Seine and small worms from the Authie) or ten individuals (large worms from the Authie). All procedures were carried out at –4°C. Tissues were homogenized as described by Durou et al. (2007). The homogenates were then centrifuged for 30 min at 9000 *g*. Aliquots of the supernatant (S9 fraction) were frozen at –80°C until use. All determinations were performed on S9 fractions by spectrophotometric techniques: GST (Habig et al. 1974), catalase (Clairbone 1985), and AChE (Ellman et al. 1961; Galgani and Bocquené 1991; Scaps and Borot 2000). Lipid peroxidation was estimated by the formation of TBARS. TBARS, consisting mainly of malondialdehyde (MDA), were quantified by reference to MDA absorbance. Results were not expressed as MDA concentrations because TBA can react with a range of chemicals (Csallany et al. 1984) but as TBARS levels.

4.2 BIOMARKERS OF DEFENCE

Numerous animal and plant species can survive in contaminated areas (Chapter 7). To remain alive in environments subjected to various pollutants, several of them have developed mechanisms to detoxify metals, resist oxidative stress, and biotransform xenobiotics. These mechanisms can be used as biomarkers of defence.

4.2.1 METALLOTHIONEINS AND CYTOSOLIC HEAT-STABLE THIOLIC COMPOUNDS (CHSTCs)

Metallothioneins (MTs) are proteins that have low molecular weights and high cysteine contents, are heat-stable, and lack aromatic amino acids and enzymatic functions. MTs are generally considered to play a role in essential metal (Cu, Zn, etc.) homeostasis and detoxification of non-essential metals (see review by Amiard et al. 2006). In numerous species, including annelids, induction of MTs by metals (Ag, Cd, Cu, Hg) has been demonstrated. As a result, MTs are recommended as biomarkers of defence in metal-contaminated areas. MTs also play a role of antioxidant defence as shown by the work of several authors cited by Amiard et al. (2006).

Gel permeation of extracts of *N. diversicolor* from the Seine estuary showed cytosolic heat-stable thiolic compounds, binding Ag, Cu, and Zn (Figure 4.1) with molecular masses consistent with the characteristics of monomers (6 to 7 kDa) and dimers (12 kDa) of MTs (Poirier et al. 2006). Cd (not illustrated) showed a distribution pattern very similar to the one for Cu. Evidence also indicated the presence of SH residues in chromatography fractions with molecular masses as low as 3 to 4 kDa.

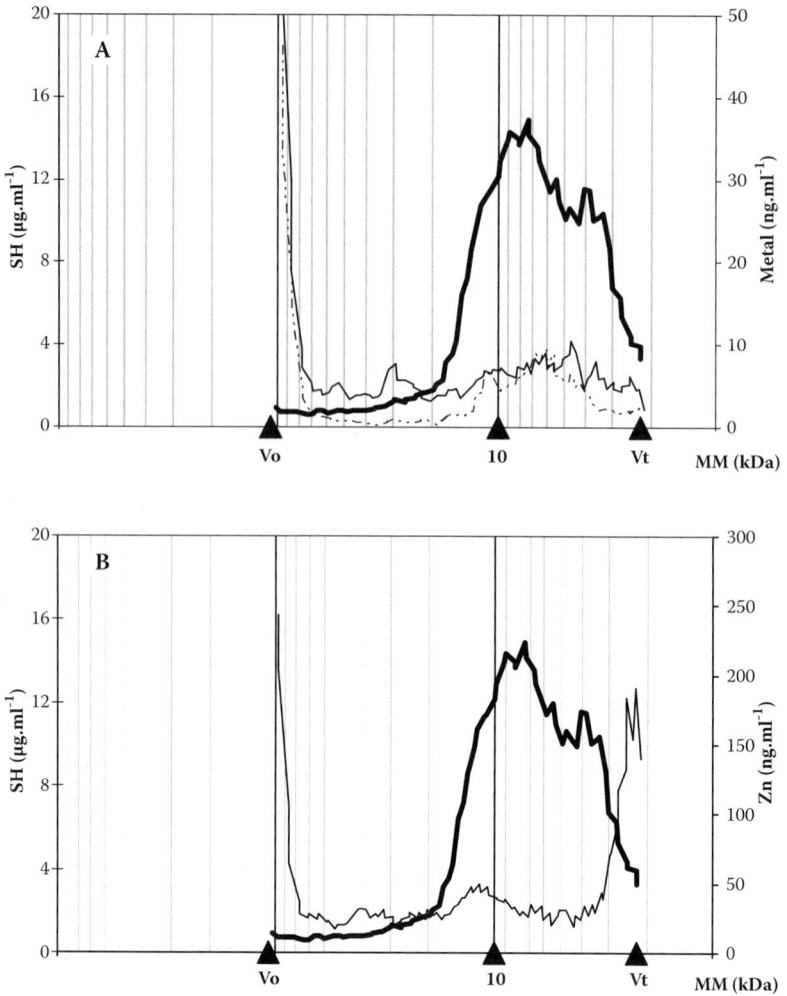

FIGURE 4.1 Distributions of metals (in gel permeation fractions of total cytosol) and sulf-hydryl compounds SH (in fractions of heat-stable cytosol) extracted from *Nereis diversicolor* as a function of molecular mass. Running means obtained from two replicates for worms collected from the Seine estuary in July 2002. (A) SH, thick and continuous line; Ag, dotted line; Cu, thin and continuous line; (B) SH, thick and continuous line; Zn, thin and continuous line. Vo = void volume. Vt = total bed volume. 10 = expected elution position of compounds of 10 kDa MM (molecular mass). (*Source:* Poirier, L. et al. 2006. *J. Mar. Biol. Ass. U.K.* 86: 71–82. With permission.)

In the remainder of this chapter will use the term *cytosolic heat-stable compounds* (CHSTC) instead of the more restrictive *MT* term.

No results of gel permeation are shown for worms from the Authie estuary because metal concentrations in fractions collected were too close to detection limits,

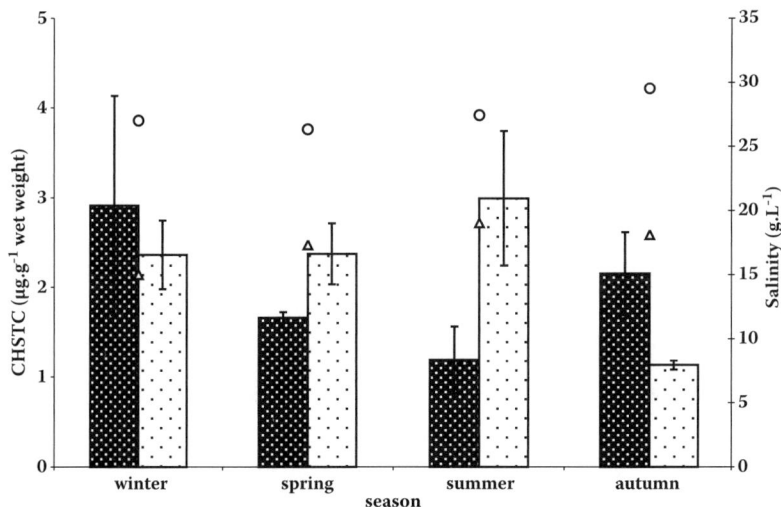

FIGURE 4.2 CHSTC concentrations in micrograms per gram (mean ± SD) in worms from Authie (dark bars) and Seine (white bars) estuaries collected in 2002 (two pools of ~2 g total weight for each site); salinity in grams per liter measured on each sampling occasion. ○ = Authie. △ = Seine.

reflecting the very low concentrations observed compared to results from worms of the Seine estuary (Chapter 3). The distribution of the SH compounds was similar.

Figure 4.2 depicts CHSTC concentrations in worms from the Seine and Authie estuaries in 2002. No consistent intersite differences were observed. CHSTC values were higher in *N. diversicolor* from the Seine estuary compared to the Authie estuary in autumn and winter, whereas the reverse was observed in spring and summer. In *N. diversicolor*, MTs represent most of the CHSTCs and bind several metals. However, even if they are often considered as core biomarkers, MTs cannot be used as biomarkers in this species because no significant relationship was established between CHSTC concentrations and accumulated metal concentrations.

4.2.2 Metal-Containing Granules

Viarengo and Nott (1993) and George and Olsson (1994) suggested that metal binding to MTs may be an intermediate step in the physiological detoxification of accumulated metals. Metals bound to MTs would be transported to lysosomes and submitted to proteolysis, and thus incorporated into the lysosomal system, leading to the build-up of insoluble lysosomal residual bodies.

This is in agreement with our ultrastructural studies of teguments (Figure 4.3) and gut walls (Poirier et al. 2006) of *N. diversicolor* from the Seine estuary. The tegument is composed of a layer of epidermal cells, surmounted by fibrillar collagen and external infolded epicuticle containing numerous dense extracellular granules. Energy-dispersive microanalysis (EDS) indicated that the granules contained silver,

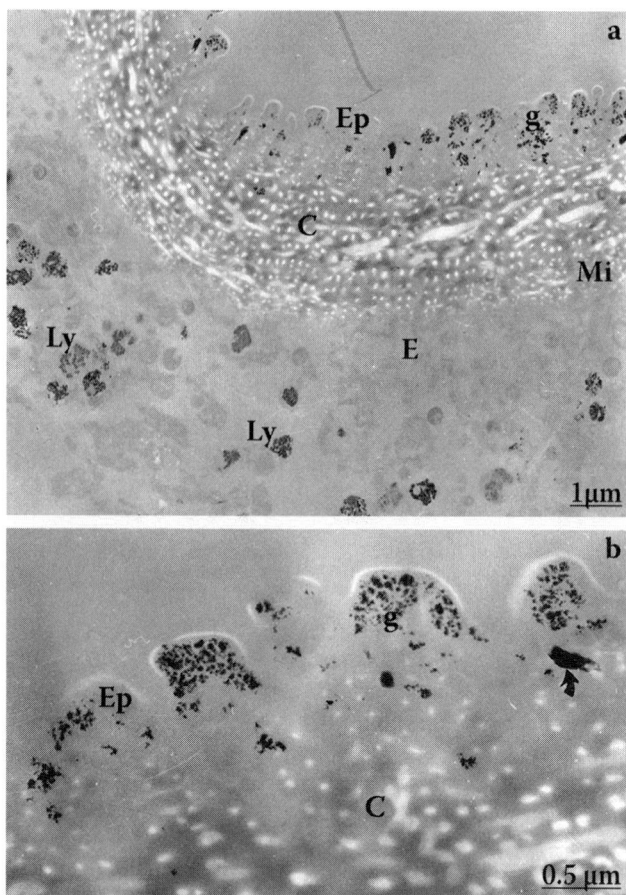

FIGURE 4.3   Ultrastructural aspect of tegument of *N. diversicolor* from the Seine estuary shown by transmission electron microscopy. Ultrathin sections, non-osmicated and non-stained. (a) The epicuticle (Ep) is loaded with numerous dense granules (g). In the epidermal cells (E), the mineralized lysosomes (Ly) are numerous (×9000). Mi = microvilli of epidermal cells. (b) Enlargement of epicuticle (×30,000). C = collagen. Arrow = epidermal extension storing mineral elements. (*Source:* Poirier, L. et al. 2006. *J. Mar. Biol. Ass. U.K.* 86: 71–82. With permission.)

copper, and sulfur (Figure 4.4). Several mineralised lysosomes in which Ag, Cu, and S were also present were observed in the apical areas of the epidermal cells (Figures 4.3 and 4.4).

In contrast, Mouneyrac et al. (2003) found only very small extracellular granules containing only S and Cu in the epicuticles of *N. diversicolor* from a reference site (Blackwater estuary, United Kingdom). In the epidermal cells, mineralized lysosomes loaded with Zn, Br, Ca, and Co were scarce. These authors also studied the same worm species taken from a far more metal-contaminated site (particularly Cu and Zn) in Restronguet Creek in the United Kingdom. In the epicuticles and apical

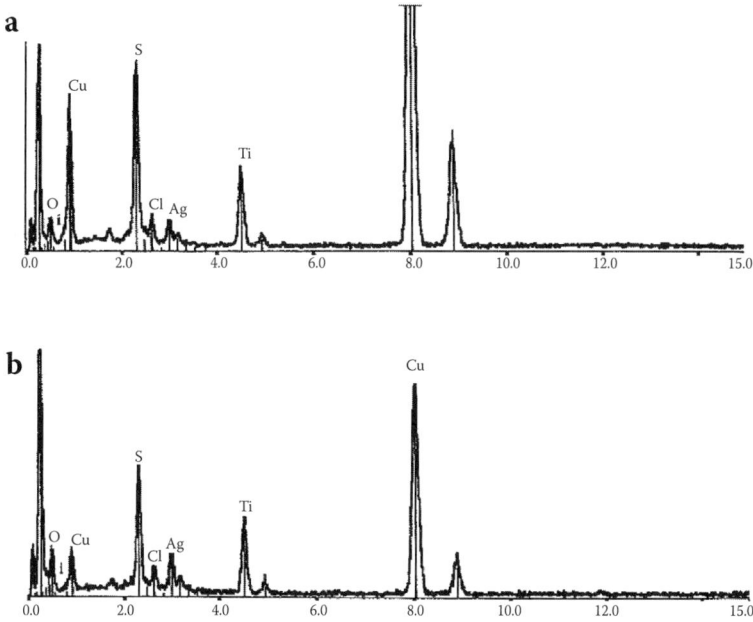

FIGURE 4.4 Energy-dispersive spectrometry microanalysis of tegument of *N. diversicolor* from the Seine estuary. Some elements such as Cu or Ti appear twice on the spectrum because of Kα and Lα x-ray emissions. The most intense peaks at the left of the spectrum are carbon and oxygen originating from the organic matrix. Ti is emitted by the grid, Cl by the resin (epon). (a) Epicuticle of the tegument. The granules contain S, Cu, and Ag. (b) The apical lysosomes of the epidermal cells contain S, Cu, and Ag. (*Source:* Poirier, L. et al. 2006. *J. Mar. Biol. Ass. U.K.* 86: 71–82. With permission.)

lysosomes of the epithelial cells, they observed numerous dense extracellular granules in which S and Cu were detected. The presence of Ag in granules and lysosomes of specimens from the Seine, and its absence in those from the Blackwater estuary reference site, is consistent with the significant Ag contamination of the Seine estuary, as shown by the French Mussel Watch Program (RNO 2001).

Numerous cytological inclusions including mineralised lysosomes and spherocrystals were present in the gut walls of worms from the Seine estuary. In both inclusions, EDS microanalysis revealed the presence of As, Fe, Zn, P, and S; Ca was detected in spherocrystals only (Poirier et al. 2006). Mouneyrac et al. (2003) noted similar compositions in spherocrystals and mineralized lysosomes from the gut walls of worms from Restronguet Creek. In the gut walls of control worms from the Blackwater estuary, they found scarce lysosomes containing the same elements, whereas spherocrystals were scanty, and contained only S and Ca.

The extracellular granules of the epicuticles, the spherocrystals of the gut walls, and the mineralized lysosomes of the epidermal cells or gut walls appear to be important detoxification stores for several metals such as Cu, Zn, or Ag, and thus may be considered as biomarkers of defence.

4.2.3 GST AND CATALASE ACTIVITIES

GSTs and catalase (CAT) enzymes intervene in the normal physiological reactions of organisms. These enzymes have been classified as biomarkers of defence because their activities increase in the presence of pollutants. GSTs constitute a multigenic family of enzymes that intervene in organic biotransformation. They have been widely used as biomarkers of defence against many chemical compounds including hydrocarbons, organochlorine insecticides, and polychlorinated biphenyls (PCBs) in molluscs (Fitzpatrick et al. 1997; Hoarau et al. 2001).

In mussels *Mytilus galloprovincialis* experimentally exposed to pp'DDE and mercury (Khessiba et al. 2001), GST activity increased compared to controls. Saint-Denis et al. (2001) exposed the terrestrial annelid *Eisenia fetida* to high lead concentrations in artificial soil and found that GST activity was affected by both dose and duration of lead acetate exposure as a function of concentration in contaminated artificial soil. Conversely, GST activities in *E. fetida* were not affected by exposure to benzo(a)pyrene (BaP) present in artificial sediments. According to the authors, the free radicals generated by BaP may be regulated by glutathione metabolism and not by enzymatic conjugation (Saint-Denis et al. 1999).

Antioxidant enzymes such as CATs belong to the cellular antioxidant system that counteracts the toxicity of reactive oxygen species (ROS). CATs are hematin-containing enzymes that facilitate the removal of the hydrogen peroxide (H_2O_2) that is metabolized to molecular oxygen (O_2) and water (Van der Oost et al. 1996). Several classes of pollutants including trace metals and organic compounds are known to enhance the formation of ROS. Variations in CAT activity along a pollutant gradient were demonstrated in several studies and CAT activity has been proposed as a biomarker of defence against oxidative stress (Porte et al. 2002; Lionetto et al. 2003; Pandey et al. 2003). Geracitano et al. (2004) and Aït Alla et al. (2006) collected two species of worms, *Laeonereis acuta* and *Nereis diversicolor*, respectively, from polluted and reference sites. They reported higher antioxidant responses (superoxide dismutase, CAT activity) in the polluted site compared with the reference site. The authors did not find any spatial difference in lipid peroxidation or TBARS levels, which means that the antioxidant defence compensates for the oxidative stress generated by the presence of diverse pollutants.

In the present work, the population structure of *N. diversicolor* is strongly contrasted in the two studied estuaries, with specimens showing greater size in the Authie estuary than in the Seine estuary (Durou et al. 2007). The influence of size was then taken into consideration for all biochemical and physiological biomarkers (Table 4.1). GST and CAT activities were not consistently influenced by the sizes of worms from the reference site. However, the smaller individuals presented lower levels of GST activities when significant differences were reported.

Figure 4.5. depicts the variations of GST activities as functions of size of individuals (large [LA] and small [SA] specimens from the Authie estuary and small [S] individuals from the Seine estuary) collected from 2002 to 2004. In February and July 2002, GST activity was lower (SA < S) in large worms collected from the Authie estuary than in the small worms from the Seine estuary. In 2003 and

TABLE 4.1
Influence of Size on Biomarker Defence (GST and Catalase) and Damage (AChE and TBARS) Levels in Different Sized Worms from the Authie Reference Site

Sampling Date	GST	Catalase	AChE	TBARS
Feb. 2003	=	=	=	=
May 2003	S < L	=	S > L	=
Aug. 2003	S < L	=	=	=
Nov. 2003	=	=	=	=
Feb. 2004	=	S > L	S > L	S > L
May 2004	=	=	S > L	S < L
Aug. 2004	=	=	S > L	S > L
Nov. 2004	S < L	=	=	=

Notes: S = small. L = large. Means were compared *post hoc* by Student's t-test after significant ANOVAs.

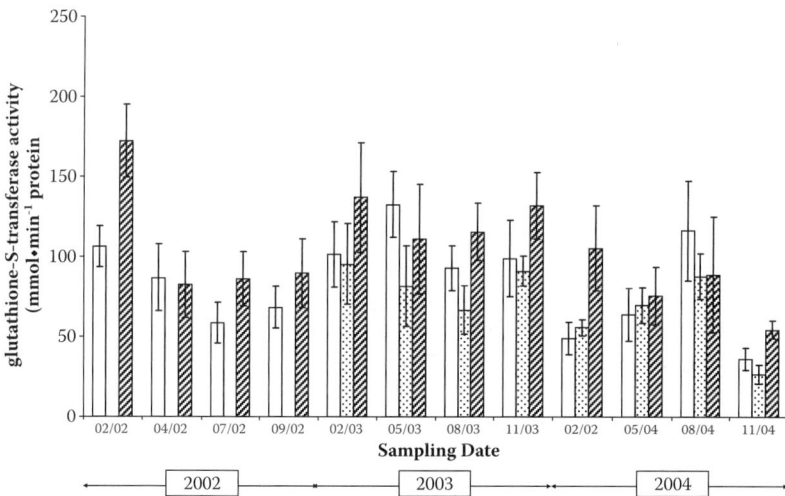

FIGURE 4.5 Influence of size and collection site on the GST (glutathione S-transferase activity) biomarker of defence measured in the worm *Nereis diversicolor*. White bars = Authie, large specimens (LA). Dotted bars = Authie, small worms (SA). Hatched bars = Seine (S).

2004, an intersite difference was generally noted when the same sized worms were considered, with GST activity higher in worms from the Seine estuary, although differences were not always significant. The results of CAT activities are not shown because they are less significant. No clear intersite differences were observed. This is not due to a confounding role of size because size generally exerted no influence on CAT activity (Table 4.1).

4.3 BIOMARKERS OF DAMAGE

The responses of organisms to pollutants may be reflected by physiological, cellular, and particularly biochemical damage.

4.3.1 TBARS CONCENTRATIONS

Anthropogenic sources of particular contaminants can increase the production of ROS in organisms by a variety of potential mechanisms including the disruption of membrane-bound electron transport and the depletion of antioxidant defences (e.g., reduced glutathione GSH). Moreira et al. (2006) reported oxidative stress in *N. diversicolor* exposed to impacted sediments, evidenced by a marked reduction in glutathione redox status (ratio of reduced GSH/oxidized GSSG) and an increase in lipid peroxide levels measured by generation of TBARS.

Elevated MDA (malondialdehyde) levels (MDA reacts with TBA) were observed in mussels exposed to paraquat, copper, and mercury (Bano and Hasan 1989) and in clams *Ruditapes decussatus* exposed to copper and mercury (Roméo and Gnassia-Barelli 1997). Regoli et al. (1998) reported that heavy metals may seriously damage the lysosomal membranes of the digestive glands in molluscs, leading to sharp decreases in MDA levels. TBARS concentrations (results not shown) in *N. diversicolor* from the Seine and Authie estuaries did not vary significantly as a function of season; no influence of size or site was noted.

4.3.2 ACHE ACTIVITIES

Esterases are known to be relevant biomarkers of damage provoked by organophosphate and carbamate pesticides in marine animals (Thompson et al. 1991; Bocquené 1996; Galgani and Bocquené 2000). Two general types of cholinesterases are recognized: those with a high affinity for acetylthiocholine and relatively non-specific esterases (Walker and Thompson 1991). In the present work, no attempt was made to distinguish the two types. Since AChE is involved in critical neural and neuromuscular functions, it may be assumed that it is more important than non-specific esterases.

Fourcy et al. 2002 reported esterase (AChE and CbE: carboxylesterase) inhibition by temephos in *N. diversicolor* exposed in areas treated for mosquito control. However, in many estuaries the natures of the AChE-inhibiting substances are unknown. Organochlorines are expected to have affinities for hydrophobic sites in enzymes including AChE, but such chlorinated compounds are not important in the inhibition of active sites (Mayer and Himel 1972). Primary candidates include carbamate and organophosphate pesticides, many of which are effective AChE inhibitors (Bocquené 1996). Payne et al. (1996) provided evidence that contaminants other than pesticides may be responsible for the depressed AChE levels. The responsiveness of AChE to heavy metals, detergents (Guilhermino et al. 1998), and algal toxins (Lehtonen et al. 2003) has been noted in marine invertebrates. AChE may thus prove

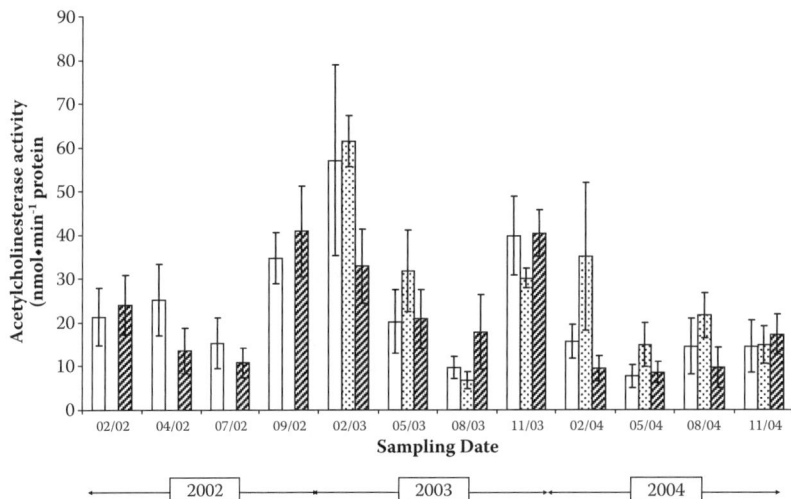

FIGURE 4.6   Influence of size and collection site on the AChE (acetylcholinesterase activity) biomarker of damage measured in the worm *Nereis diversicolor*. White bars = Authie, large specimens (LA). Dotted bars = Authie, small worms (SA). Hatched bars = Seine (S).

to be a useful biomarker of general physiological stress in aquatic organisms and in *Nereis* species in particular.

Figure 4.6 depicts the variations of AChE activities as a function of size of individual *N. diversicolor* specimens (large [LA] and small [SA] specimens from the Authie estuary and small [S] individuals from the Seine estuary : S) collected from 2002 to 2004. In 2002, lower AChE activities were determined in small worms from the Seine in spring and summer, but the pattern was not confirmed in autumn and winter (Figure 4.6). This may be due to the fact that, independently of any contamination factor, small specimens showed higher AChE activities than larger ones (Table 4.1). This seems to be a general feature in biota: even in fish, the youngest individuals often present the highest AChE activities (Bocquené 1996; Roméo et al. 2001). Subsequent analyses in 2003 and 2004 were therefore conducted with two size classes of Authie individuals, the largest (LA) allowing an interannual comparison with 2002 and the smallest (SA) resembling individuals collected in the Seine estuary. Intersite differences among individuals of comparable size were noted in most cases in which AChE activity tended to be lower in worms from the Seine estuary compared to those from the Authie.

4.4   MULTIBIOMARKER APPROACH AND POLLUTANT LEVELS COMBINED IN PRINCIPAL COMPONENT ANALYSIS

Principal component analysis (PCA) is a multivariate statistical method that allows discrimination of different parameters, aiming at a synthetic view of environment quality assessment using *N. diversicolor*. PCAs have been used in field studies involving "passive" (collection of samples) and "active" biomonitoring (transplant experiments, Roméo et al. 2001). Data used to perform the analyses were the 3-year mean results from *N. diversicolor* sampled from the Authie and Seine estuaries. The data for the different size classes were included in the PCA because size may be a factor in the responses of the whole population at each site.

Metal concentrations (Chapter 3) were totalled on a molar basis, with the exceptions of lead and zinc. The environmental source of lead was hunting at both sites. Lead elicited a low contribution in a preliminary PCA. Zinc was excluded because body concentrations of zinc are regulated in *N. diversicolor* according to Berthet et al. (2003) and the literature cited therein. PCB and PAH concentration data for worms (Chapter 3), four biochemical parameters (GST, CAT, TBARS, AChE), and three physiological markers (lipid, glycogen, and protein concentrations, Chapter 8) were also considered in the PCA (240 data items).

The first three axes represented 69% of the total variance. The coordinates of the variables for PC1/PC2 and PC1/PC3 are shown in Table 4.2. PC1 is positively correlated with GST activity and concentrations of metals and PAHs, and negatively correlated with energy reserves. Both anti- (CAT) and pro-oxidant (TBARS) forces correlate with PC2.

The plots of scores in the coordinates of principal component axes 1 (PC1) and 2 (PC2) are shown in Figure 4.7; PC1 and PC3 are shown in Figure 4.8. Figure 4.7 clearly separates the chosen study sites along the PC1 axis from negative to positive. The centroid of the cloud of Seine scores is located in the positive part of PC1, showing high GST response associated with high pollutant levels. The centroid of the Authie is located in the negative part of PC1 (correlated with higher energy reserves in Authie samples). The variables correlating with PC2 do not show significant trends since all scores are scattered more along PC1 than PC2 (Table 4.2).

TABLE 4.2
Correlations of Variables and Principal Components (PC1, PC2, and PC3)

Variable	PC1	PC2	PC3
% Total Variance	34	22	13
PAHs	**0.58**	−0.26	0.23
PCBs	0.40	−0.15	**0.80**
Metals	**0.75**	−0.26	0.08
AChE	0.38	−0.21	**−0.64**
CAT	0.49	**0.77**	−0.06
GST	**0.78**	0.08	0.22
TBARS	−0.37	**0.82**	−0.10
Glycogen	**−0.67**	0.44	0.23
Lipids	−0.50	**0.55**	0.26
Proteins	**−0.73**	−0.45	0.13

Note: Bold values indicate significant correlations.

Individuals (PC 1 & PC 2 : 56 %)

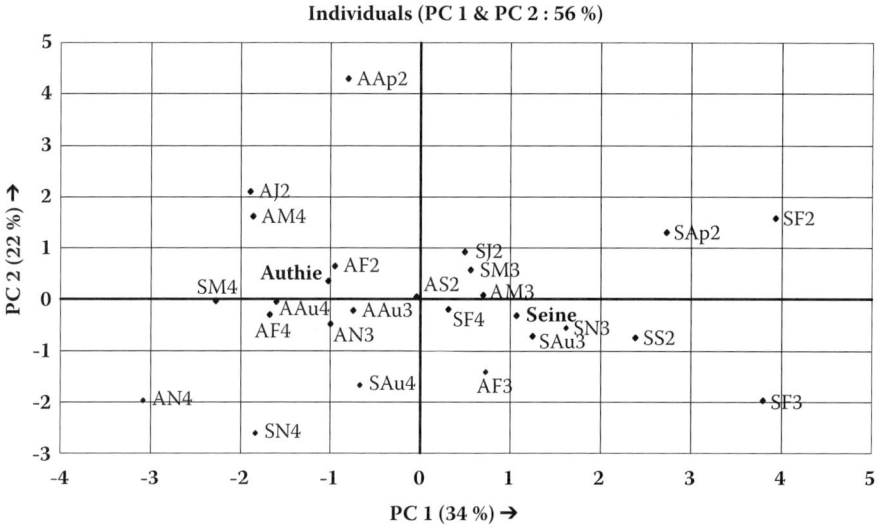

FIGURE 4.7 Representation of PC1/PC2 from principal component analysis performed with chemical and biochemical data obtained from worms collected between February 2002 and November 2004. A = Authie collection sites. S = Seine collection sites. Sampling months: F = February. Ap = April. M = May. J = July. Au = August. S = September. N = November. Years: 2 = 2002. 3 = 2003. 4 = 2004.

Individuals (PC 1 & PC 3 : 47 %)

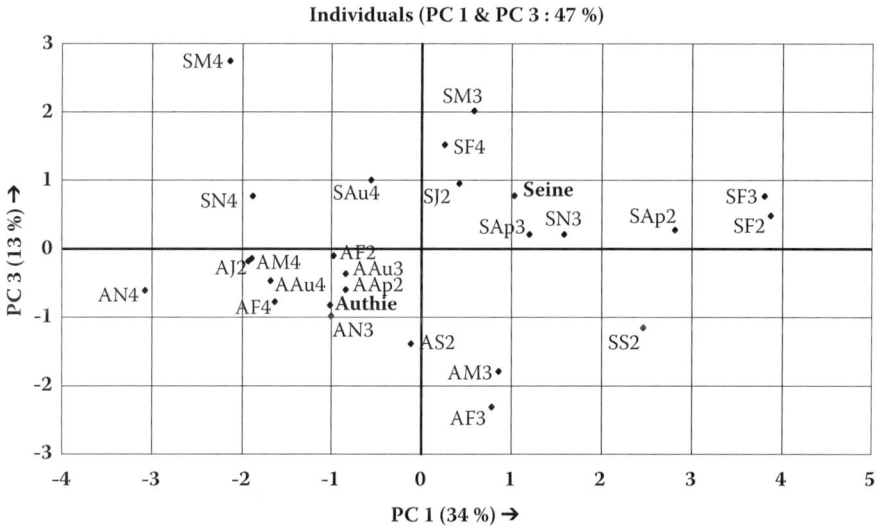

FIGURE 4.8 Representation of PC1/PC3 from principal component analysis performed with chemical and biochemical data obtained from worms collected between February 2002 and November 2004. A = Authie collection sites. S = Seine collection sites. Sampling months: F = February. Ap = April. M = May. J = July. Au = August. S = September. N = November. Years: 2 = 2002. 3 = 2003. 4 = 2004.

PC3 is positively correlated with PCB concentrations and negatively correlated with AChE activity (Table 4.2). Figure 4.8 shows discrimination along PC3 between Authie samples in the negative part with low PCB concentrations and high AChE activities, and Seine samples with high PCB and low AChE activities.

4.5 GENERAL DISCUSSION

We have studied the responses of five biomarkers measured in *Nereis diversicolor*, used here as representatives of the estuarine mudflats considered to be the main reservoirs for most of the organic and inorganic chemicals in the estuaries. A number of the studied biomarkers (MTs, catalase, and TBARS) did not yield significant responses in the Seine estuary, considered one of the most contaminated in France. On the other hand, for GST and AChE activities and metal-containing granules, significant differences were observed between the Seine estuary and the relatively unpolluted Authie estuary.

No intersite differences were shown for CHSTC (including mainly MT-like proteins) concentrations in *N. diversicolor* from the Seine and the Authie estuaries. The recent work of Ng et al. (2008) has shown that exposure of the worm *Perinereis aibuhitensis* to Cd did not increase MT concentrations but appeared to increase the rate of synthesis and breakdown (turnover) of MT. Similar phenomena can be envisaged in the case of *N. diversicolor*, the exposure to contaminants not involving an increase of the MTLP concentration, but an increase in their turnover. Another common explanation in agreement with the findings of other authors (George and Olsson 1994; Barka et al. 2001) may be proposed. When the level of stress is too high, MTLP content decreases, becoming similar to or even lower than levels in controls. This produces a bell-shaped curve to describe MTLP content in an exposed organism against exposure concentration of a potential MT-inducing metal. In the case of MTLPs, this may reflect a general alteration of protein metabolism, making the organisms unable to synthesize these proteins of defence. Similar conclusions were reached by Legras et al. (2000) who indicated that MT levels in the crab *Pachygrapsus marmoratus* were linked more to changes in general protein metabolism than to changes in exposed metal concentrations.

Dagnino et al. (2007) extended this hypothesis to other biomarkers of defence. As shown in Figure 4.9, the MTLP contents and CAT and GST activity levels of organisms follow a bell-shaped curve in a pollution gradient. The lack of clear differences in CAT activity in the Authie and Seine estuaries may correspond to an impairment of this mechanism of defence. Roméo et al. (2000) noted that CAT activity may be inhibited by very high metal exposure concentrations in fish. Pellerin-Massicotte (1997) reported that when stress overwhelms defence (antioxidant) forces such as catalase (right part of the bell-shaped curve of Figure 4.9) in mussels, a deleterious effect on membranes can be observed and an end-product of lipid peroxidation (measured as TBARS) can be measured. Increases of CAT activity and TBARS levels together may show the sensitivity of these two parameters to physiological stress (Pellerin-Massicotte 1997), which means that CAT induction is high but not sufficiently high to prevent damage to lipid membranes (TBARS increases). In worms from the Seine estuary, no consistent increases of TBARS concentrations

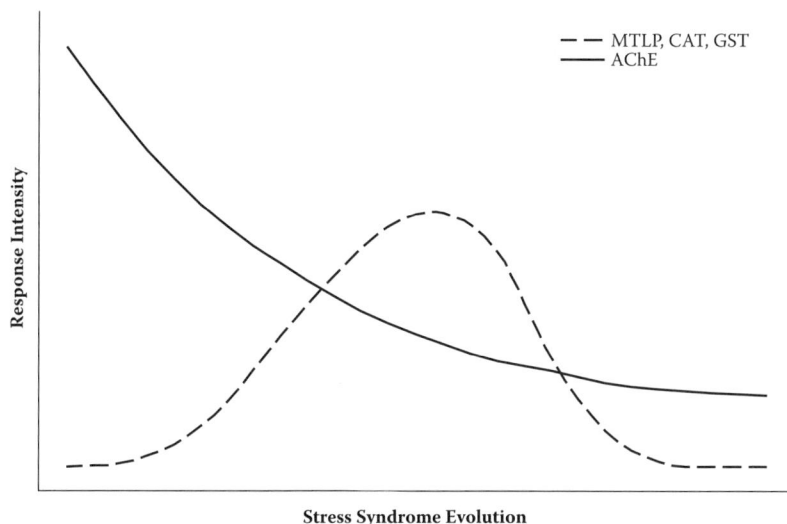

FIGURE 4.9 Responses of biomarkers along a pollution gradient. related to the development of the stress syndrome. MTLP = metallothionein-like protein content. CAT = catalase activity. GST = glutathione S-transferase enzyme activity. AChE: acetylcholinesterase activity. (*Source:* Adapted from Dagnino, A. et al. 2007. *Biomarkers* 12: 155–172.)

were established, but GST activity was increased and GST is also known to take part in anti-oxidant defences (Cossu et al. 1997). Thus, GST appears to be a biomarker of defence that may be a useful tool in assessing the environmental health status of these worms.

The metal-containing granules appear to be "good" biomarkers of defence, confirming the work of Mouneyrac et al. (2003) who showed a correlation between the tolerance of *N. diversicolor* to Cd, Cu, and Zn and the abundance of extracellular granules and mineralized lysosomes containing these metals. Similar observations were reported by Wallace et al. (1998) in Cd-resistant oligochaete worms producing metal-rich granules for Cd storage. However, in contrast to other biomarkers, the necessary analytical techniques for the ultramicroscopic detection of metal-rich granules are complicated and relatively expensive.

Despite the existence of mechanisms of defence (see Chapter 7), *N. diversicolor* does not fully cope with the stress experienced in the multipolluted Seine estuary. As theorized by Dagnino et al. (2007), AChE activity decreases along the whole range of a pollution gradient. The lower AChE activity in worms from the Seine compared to the Authie is in agreement with this pattern.

Whatever the biomarker, the influences of biological and environmental natural factors must also be considered. We have studied two: the sizes of the worms (see above) and the salinity of the sampling sites. Several authors (Leung et al. 2002 and 2005; Mouneyrac et al. 2001; Monserrat et al. 2007) have shown that MT expression can be influenced by various factors such as salinity, temperature, growth rate, etc., that make spatial comparisons difficult. We tested the influence of salinity on CHSTC concentrations in worms taken at the same date (September 2002) from sampling stations (see Figure 1.3 in Chapter 1) along a salinity gradient (AP: 18.9; AN: 29.5; AS: 33.0) within the Authie reference estuary to avoid interference by contamination. We observed increases of CHSTC as salinity increased (AP: 0.9 ± 0.1 μg.g^{-1} ww; AN: 1.17 ± 0.05 μg.g^{-1} ww; AS: 1.5 ± 0.3 μg.g^{-1} ww). However, salinity did not appear to be the major factor governing the levels of CHSTCs. Indeed, salinity was higher in the Authie than in the Seine estuary (see Fig. 4.2), whereas in autumn and winter, CHSTC concentrations were higher in the Seine worms than in the Authie estuary worms. Moreover, no within-estuary correlation was observed between salinity and CHSTC concentrations in worms at the different seasons. Thus, salinity cannot explain the lack of intersite differences in CHSTC concentrations.

GST activity has been shown to increase in bivalve larvae when salinity decreases from 30 to 20 (Damiens et al. 2004). The GST activity (Bebianno et al. 2007) measured in *Mytilus galloprovincialis* collected from the field also increased with a decrease in salinity (from 34.9 to 4). Lyons et al. (2003) demonstrated that salinity is not a key factor compared to pollution status in GST activity in field-sampled mussels. In the present work, this abiotic factor does not appear to influence GST activity measured in the worms. Salinity may exert an influence upon AChE activity because AChE was shown to increase with salinity decreases in oyster larvae and *Mytilus* sp. (Damiens et al. 2004; Pfeifer et al. 2005). If the same is true for worms, the trend of AChE activity to be lower in worms from the Seine estuary (mean salinity: 19) than in worms from the Authie estuary (mean salinity: 27) cannot be due to this abiotic factor.

4.6 CONCLUSIONS

To achieve a comparison of the polluted status of the Seine estuary and the clean status of the Authie estuary, the integration of multifactorial measurements is a basic requirement. Our results on biomarkers and chemical concentrations measured in *Nereis diversicolor* and sediments are in agreement with the recognized role of GST and metal-containing granules in the defence of organisms exposed to chemical stress and with the presence of neurotoxic compounds revealed by the inhibition of AChE activity.

By virtue of the precautionary principle, in the absence of certain contrary proof, any detection of a response of a biomarker (the activity EROD, Lopez-Barea 1994, quoted by Flammarion 2000) may be considered a sign of a not insignificant risk for the organism. Biomarkers of defence are activated in the presence of their inductors, up to a certain threshold, beyond which the toxicity is so raised that the processes of defence are inhibited. Even for chemicals which, unlike metals, are not naturally present in the environment, it seems that stress is limited in the range of concentrations over which mechanisms of defence are active. However, even if extensive data show that organisms (belonging to diverse zoological groups) chronically exposed to various contaminants are capable of physiological acclimatization or genetic adaptation to raised availability of a contaminant, it would be careless to consider that the local populations of all species are going to become tolerant.

The problem is different with the biomarkers of damage. The classic (core) biomarkers of damage are biological responses measured at the molecular, cellular, and tissue levels. This means that organism health will not be subjected systematically to danger if only a limited number of biological macromolecules or cells are affected, thus allowing the subnormal progress of physiological function. However, for certain biomarkers, the existing data allow the establishment of a threshold beyond which effects on the health of the individuals are present and populational effects are expected. In vertebrates and insects, thresholds of inhibition of AChE beyond which a supra-individual effect is likely to occur have been indicated (Bocquené et al. 1997; Fulton and Key 2001; Grue et al. 2002). Would it be possible to determine such a threshold for *N. diversicolor*?

A global approach based on multimarker measurement was used for the first time in the BIOMAR project (Narbonne et al. 1999; Narbonne et al. 2001). The procedure recently led to estimates of the relative toxicities of estuarine sediments on the basis of biomarkers (EROD in the liver, GST in the liver and gills, LDH in the muscles, cerebral AChE) measured in sea bream (Cunha et al. 2007). The resulting classification of sediments is similar to the one arising from more complex parallel studies including chemical analyses of sediments, analyses of macrobenthic communities, tests of lethal toxicity on amphipods, and tests of embryotoxicity of sea urchins. Moreover, the evaluation of the chronic toxicity of estuarine sediments at various levels of biological organisation in an amphipod (*Gammarus locusta*) showed a strong coherence of chemical (bioaccumulation) and biochemical (MT induction, DNA strand break, lipid peroxidation) data with survival, growth, and reproduction (Costa et al. 2005; Neuparth et al. 2005). In this context, the following chapters will deal in more detail with biochemical biomarkers, growth, reproduction, and the dynamics of populations of *N. diversicolor*.

In the BIOMAR project, Narbonne et al. (1999) tried to diagnose the contamination status of sites along Mediterranean coast including multibiomarker data in a scale of classification (from 1 to 5). Beliaeff and Burgeot (2002) established a simple method for summarizing biomarker responses. Their method, called the integrated biomarker response (IBR), simplifies response interpretation in biomonitoring programmes. They stress that the selection of an appropriate battery of biomarkers can avoid false negative responses obtained with a single biomarker and allow information to be summarized in the form of a multivariate data set.

According to Broeg and Lehtonen (2006), due to its mathematical basis, the IBR becomes more robust when the number of biomarkers increases. However, according to Guerlet (2007), several inconveniences limit the use of this tool: (1) a potentially significant influence of the order of the biomarkers on the value of the IBR, (2) the impossibility of its application without *a priori* information on the stress gradient because of the fluctuating characters of the types of biomarker responses (inhibition or antagonism), and (3) an overestimation of stress as a result of redundancy of the responses integrated into the IBR.

Dagnino et al. (2007) proposed an expert system that utilises a suite of biomarker tests measured in marine mussels to translate complex biological responses into a relatively simple, easy-to-understand, and objective evaluation of physiological changes induced by pollutants. Their classification was developed using a battery of nine biomarkers at cellular, tissue, and organism levels of biological organisation. Figure 4.9, adapted from Dagnino et al. (2007), describes the profile of biomarkers (MTs, CAT, GST, AChE) along a gradient of pollution. The expert system selects as a guide parameter the biomarker that shows the highest sensitivity to stress, and interprets the other data in light of the alteration level reached by the guide parameter.

More precisely, Viarengo et al. (2007), on the basis of the work of Dagnino et al. (2007), proposed a two-tier approach to assess the levels of pollutant-induced stress syndromes in sentinel organisms. Lysosomal membrane stability (LMS) assessed by either neutral red retention or a histochemical technique provides a robust Tier 1 screening biomarker for environmental impact assessment. Tier 2, consisting of bio-markers of genotoxicity and biomarkers revealing exposure (MTs, AChE, EROD, multixenobiotic resistance [MXR] transport activity, etc.), is used only for animals (mussels) sampled at sites in which LMS changes are evident and mortality is absent. In such cases, the expert system cited above is employed. However, Guerlet (2007) notes that the system provides no parallel integration of physico-chemical data and that effects are overestimated when there is redundancy of biomarkers.

In the comparison of the Seine and Authie estuaries, we chose principal com-ponent analysis (PCA) as a multivariate statistical technique to provide an over-view of the significant chemical and biological factors and determine significant trends in their variations (Figures 4.7 and 4.8). PCA presents, according to Guerlet (2007), several advantages: (1) the possibility of bringing together the biological and physico-chemical data without the latter influencing the profile of the PCA (illustra-tive variables); (2) possible application without *a priori* information on the gradient of stress; and (3) reduced effect on discriminative power arising from the addition of redundant parameters.

Figure 4.10, adapted from Guerlet (2007) provides a synthetic comparison of three tools of integration of the responses of the battery of biomarkers in aquatic organisms. With the objective of avoiding ecological disasters, as have occurred in

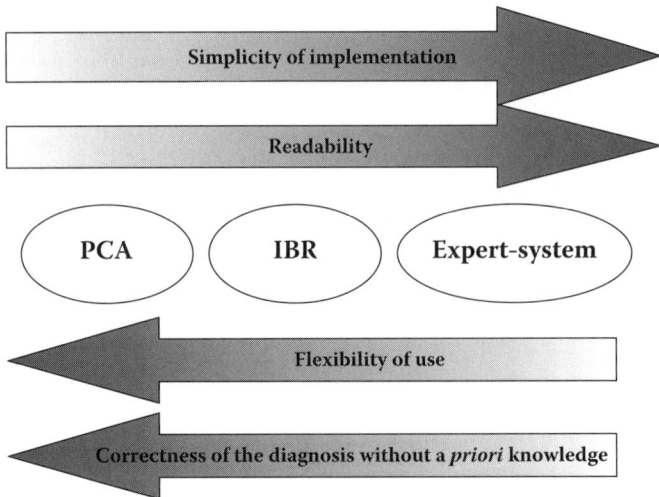

FIGURE 4.10   Synthetic comparison of three tools of integration of the responses of a bat-tery of biomarkers in aquatic organisms. PCA = principal component analysis. IRB = inte-grated biomarker response. (Source: Adapted from Guerlet, E., Doctoral Thesis, University of Metz.)

the past, it is important to balance the correctness of a diagnosis with the simplicity of the response.

The literature acknowledges that biomarkers of defence and damage can fill the gap between chemicals analyzed in the environment (sediments) and biological effects measured at population levels. Our study provided measurements of chemicals (PAHs, PCBs, metals) in the compartment where *Nereis diversicolor* lives and also diverse biomarker determinations in this sentinel species. Principal component analyses leading to a multivariate approach have allowed us to discriminate between the relatively polluted Seine estuary and the cleaner Authie estuary.

REFERENCES

Aït Alla, A. et al. 2006. Tolerance and biomarkers as useful tools for assessing environmental quality in the Oued Souss estuary (Bay of Agadir, Morocco). *Comp. Biochem. Physiol.* 143C: 23–29.

Allan, I.J. 2006. A "toolbox" for biological and chemical monitoring requirements for the European Union's Water Framework Directive. *Talanta* 69: 302–322.

Amiard, J.C. et al. 2006. Metallothioneins in aquatic invertebrates: their role in metal detoxification and their use as biomarkers. *Aquat. Toxicol.* 76: 160–202.

Bano, Y. and M. Hassan. 1989. Mercury induced time-dependent alterations in lipid profiles and lipid peroxidation in different organs of catfish *Heteropneustes fossilis. J. Environ. Sci. Health* B24: 145–166.

Barka, S. 2007. Insoluble detoxification of trace metals in a marine copepod *Tigriopus brevicornis* (Müller) exposed to copper, zinc, nickel, cadmium, silver and mercury. *Ecotoxicology* 16: 491–502.

Bebianno, M.J. et al. 2007. Glutathione S-transferase and cytochrome P450 activities in *Mytilus galloprovincialis* from the south coast of Portugal: effect of abiotic factors. *Environ. Int.* 33: 550–558.

Beliaeff, B. and T. Burgeot. 2002. Integrated biomarker response: a useful tool for ecological risk assessment. *Environ. Toxicol. Chem.* 21: 1316–1322.

Bernes, C. 1998. *Persistent Organic Pollutants. A Swedish View of an International Problem.* Stockholm: Swedish Environmental Protection Agency.

Berthet, B. et al. 2003. Accumulation and soluble binding of cadmium, copper, and zinc in the polychaete *Hediste diversicolor* from coastal sites with different trace metal bioavailabilities. *Arch. Environ. Contam. Toxicol.* 45: 468–478.

Bocquené, G. 1996. L'acétylcholinestérase, marqueur de neurotoxicité. Application à la surveillance des effets biologiques des polluants chez les organismes marins. Thesis. Ecole Pratique des Hautes Etudes, Paris.

Bocquené, G., F. Galgani, and C.H. Walker. 1997. Les cholinestérases, biomarqueurs de neurotoxicité. In *Biomarqueurs en écotoxicologie. Aspects fondamentaux*, Lagadic, L. et al., Eds. Paris: Masson, pp. 209–240.

Bragigand, V. and B. Berthet. 2003. Some methodological aspects of metallothionein evaluation. *Comp. Biochem. Physiol.* 143A: 55–61.

Broeg, K. and K.K. Lehtonen. 2006. Indices for the assessment of environmental pollution of the Baltic Sea coasts: integrated assessment of a multi-biomarker approach. *Mar. Pollut. Bull.* 53: 508–522.

Chapman, P.M. and F. Wang. 2001. Assessing sediment contamination in estuaries. *Environ. Toxicol. Chem.* 20: 3–22.

Clairbone, A. 1985. Catalase activity. In *Handbook of Methods for Oxygen Radical Research*, Greenwald, R.A., Ed. Boca Raton, FL: CRC Press, pp. 283–284.

Cossu, A. et al. 1997. Biomarqueurs du stress oxydant chez les animaux aquatiques. In *Biomarqueurs en écotoxicologie. Aspects fondamentaux*, Lagadic, L. et al., Eds. Paris: Masson, p. 149–163.

Costa, F.O. et al. 2005. Multi-level assessment of chronic toxicity of estuarine sediments with the amphipod *Gammarus locusta*: II. Organism and population-level endpoints. *Mar. Environ. Res.* 60: 93–110.

Csallany, A.S. et al. 1984. Free malondialdehyde determination in tissues by high-performance liquid chromatography. *Anal. Biochem.* 142: 277–283.

Cunha, I. et al. 2007. Toxicity ranking of estuarine sediments on the basis of *Sparus aurata* biomarkers. *Environ. Toxicol. Chem.* 26: 444–453.

Dagnino, A. et al. 2007. Development of an expert system for the integration of biomarker responses in mussels into an animal health index. *Biomarkers* 12: 155–172.

Damiens, G. et al. 2004. Evaluation of biomarkers in oyster larvae in natural and polluted conditions. *Comp. Biochem. Physiol. C.* 138: 121–128.

De Lafontaine, Y. et al. 2000. Biomarkers in zebra mussels (*Dreissena polymorpha*) for the assessment and monitoring of water quality of the St. Lawrence River (Canada). *Aquat. Toxicol.* 50: 51–71.

Depledge, M.H. 1993. The rational basis for the use of biomarkers as ecotoxicological tools. In *Nondestructive Biomarkers in Vertebrates*, Fossi, M.C. and Leonzio, C., Eds. Boca Raton, FL: Lewis Publishers, pp. 261–285.

Durou, C. 2006. *Recherche d'indicateurs de l'état physiologique de l'annélide polychète endogée* Nereis diversicolor *en relation avec la qualité du milieu*. Thèse de Doctorat, Université de Nantes.

Durou, C. et al. 2007. Biomonitoring in a clean and a multi-contaminated estuary based on bio-markers and chemical analyses in the endobenthic worm *Nereis diversicolor. Environ. Pollut.* 148: 445–458.

Ellman, G.L. et al. 1961. A new and rapid colorimetric determination of acetylcholinesterase activity. *Biochem. Pharmacol.* 7 : 88–95.

Fitzpatrick, P.J. et al. 1997. Assessment of a glutathione S-transferase and related proteins in the gill and digestive gland of *Mytilus edulis* (L), as potential organic pollution biomark-ers. *Biomarkers* 2: 51–56.

Flammarion, P. 2000. *Mesure d'un biomarqueur de pollution chez des poissons d'eau douce. Validation et optimisation.* Lyon: Cemagref Editions.

Fourcy, D. et al. 2002. Esterases as biomarkers in *Nereis* (Hediste) *diversicolor* exposed to temephos and *Bacillus thuringiensis* var. *israelensis* used for mosquito control in coastal wetlands of Morbihan (Brittany, France). *Mar. Environ. Res.* 54: 755–759.

Fulton, M.H. and P.B. Key. 2001. Acetylcholinesterase inhibition in estuarine fish and inver-tebrates as an indicator of organophosphorus insecticide exposure and effects. *Environ. Toxicol. Chem.* 20: 37–45.

Gagnon, C. and N.S. Fisher. 1997. The bioavailability of sediment-bound Cd, Co and Ag to the mussel *Mytilus edulis. Can. J. Fish. Aquat. Sci.* 54:147–156.

Galgani, F. and G. Bocquené. 1991. Semi-automated colorimetric and enzymatic assays for aquatic organisms using microplate readers. *Water Res.* 25: 147–150.

Galgani, F. and G. Bocquené. 2000. Molecular biomarkers of exposure of marine organisms to organophosphorus pesticides and carbamates. In *Use of Biomarkers in Monitoring Environmental Health,* Lagadic, L. et al., Eds. Enfield: Science Publishers, pp. 133–137.

George, S.G. and P.E. Olsson. 1994. Metallothioneins as indicators of trace metal pollution. In *Biomonitoring of Coastal Waters and Estuaries,* Kramer, K.J.M., Ed. Boca Raton, FL: CRC Press, pp. 151–171.

Geracitano, L.A. et al. 2004. Oxidative stress responses in two populations of *Laeonereis acuta* (Polychaeta, Nereididae) after acute and chronic exposure to copper. *Mar. Environ. Res.* 58: 1–17.

Grue, C.E., S.C. Gardner, and P.L. Gibert. 2002. On the significance of pollutant-induced alterations in the behaviour of fish and wildlife. In *Behavioural Ecotoxicology*, Dell'Omo, G., Ed. Chichester: John Wiley & Sons, pp. 1–90.

Guerlet, E. 2007. Utilisation de biomarqueurs cellulaires chez plusieurs espèces d'invertébrés pour l'évaluation de la contamination des milieux dulçaquicoles. Thèse de Doctorat, Université de Metz.

Guilhermino, L. et al. 1998. Should the use of inhibition of cholinesterases as a specific biomarker for organophosphate and carbamate pesticides be questioned? *Biomarkers* 3: 157–163.

Habig, W.H., M.J. Pabst, and W.B. Jakobi. 1974. Glutathione S-transferases: the first enzymatic step in mercapturic acid formation. *J. Biol. Chem.* 249: 7130–7139.

Hoarau, P. et al. 2001. Differential induction of glutathione S-transferases in the clam *Ruditapes decussatus* exposed to organic compounds. *Environ. Toxicol. Chem.* 20: 523–529.

Khessiba, A. et al. 2001. Biochemical response of the mussel *Mytilus galloprovincialis* from Bizerta (Tunisia) to chemical pollutant exposure. *Arch. Environ. Contam. Toxicol.* 40: 222–229.

Lam, P.K. and J.S. Gray. 2001. Predicting effects of toxic chemicals in the marine environment. *Mar. Pollut. Bull.* 42: 169–173.

Lam, P.K. and J.S. Gray. 2003. The use of biomarkers in environmental monitoring programmes. *Mar. Pollut. Bull.* 46: 182–186.

Legras, S.C. et al. 2000. Changes in metallothionein concentrations in response to variation in natural factors (salinity, sex, weight) and metal contamination in crabs from a metal-rich estuary. *J. Exp. Mar. Biol. Ecol.* 246: 259–279.

Lehtonen, K.K. et al. 2003. Accumulation of nodularin-like compounds from the cyanobacterium *Nodularia spumigena* and changes in acetylcholinesterase activity in the clam *Macoma balthica* during short-term laboratory exposure. *Aquat. Toxicol.* 64: 461–476.

Leung, K.M. et al. 2002. Influence of static and fluctuating salinity on cadmium uptake and metallothionein expression by the dog whelk *Nucella lapillus* (L.). *J. Exp. Mar. Biol. Ecol.* 274: 175–189.

Leung, K.M.Y. et al. 2005. Metallothioneins and trace metals in the dog whelk *Nucella lapillus* (L.) collected from Icelandic coasts. *Mar. Pollut. Bull.* 51: 729–737.

Lionetto, M.G. et al. 2003. Integrated use of biomarkers (acetylcholinesterase and antioxidant enzymes activities) in *Mytilus galloprovincialis* and *Mullus barbatus* in an Italian coastal marine area. *Mar. Pollut. Bull.* 46: 324–330.

Lyons, C. et al. 2003. Variability of heat shock proteins and glutathione S-transferase in gill and digestive gland of blue mussel, *Mytilus edulis*. *Mar. Environ. Res.* 56: 585–597.

Mayer, R.T. and C.M. Himel. 1972. Dynamics of fluorescent probe cholinesterase reactions. *Biochemistry* 11, 2082–2090.

Monserrat, J.M. et al. 2007. Pollution biomarkers in estuarine animals: critical review and new perspectives. *Comp. Biochem. Physiol.* 146C: 221–234.

Moreira, S.M. et al. 2006. Effects of estuarine sediment contamination on feeding and on key physiological functions of the polychaete *Hediste diversicolor*: laboratory and *in situ* assays. *Aquat. Toxicol.* 78: 186–201.

Mouneyrac, C. et al. 2001. Comparison of metallothionein concentrations and tissue distribution of trace metals in crabs (*Pachygrapsus marmoratus*) from a metal-rich estuary, in and out of the reproductive season. *Comp. Biochem. Physiol.* 129C: 193–209.

Mouneyrac, C. et al. 2003. Trace-metal detoxification and tolerance of the estuarine worm *Hediste diversicolor* chronically exposed in their environment. *Mar. Biol.* 143: 731–744.

Narbonne, J.F. et al. 2001. Biochemical markers in mussel, *Mytilus sp.* and pollution monitoring in European coasts: data analysis. In *Biomarkers in Marine Organisms: A Practical Approach*, Garrigues, P. et al., Eds. Amsterdam: Elsevier, pp. 216–236.

Narbonne, J.F. et al. 1999. Scale of classification based on biochemical markers in mussels: application to pollution monitoring in European coasts. *Biomarkers* 4: 415–424.

Ng, T.Y.T. et al. 2008. Decoupling of cadmium biokinetics and metallothionein turnover in a marine polychaete after metal exposure. *Aquat. Toxicol.* 89: 47–54.

Neuparth T. et al. 2005. Multi-level assessment of chronic toxicity of estuarine sediments with the amphipod *Gammarus locusta*: I. Biochemical endpoints. *Mar. Environ. Res.* 60: 69–91.

Pandey, S. et al. 2003. Biomarkers of oxidative stress: a comparative study of river Yamuna fish *Wallago attu* (Bl. and Sch.). *Sci. Total Environ.* 309: 105–115.

Payne, J.F. et al. 1996. Acetylcholinesterase, an old biomarker with a new future? Field trials in association with two urban rivers and a paper mill in Newfoundland. *Mar. Pollut. Bull.* 32: 225–231.

Pellerin-Massicotte, J. 1997. Influence of elevated temperature and air-exposure on MDA levels and catalase activities in digestive glands of the blue mussel (*Mytilus edulis*). *J. Res. Océanogr.* 22: 91–98.

Pfeifer, S., D. Schiedek, and J.W. Dippner. 2005. Effect of temperature and salinity on acetylcholinesterase activity, a common pollution biomarker, in *Mytilus* species from the southwestern Baltic Sea. *J. Exp. Mar. Biol. Ecol.* 320: 93–103.

Poirier, L. et al. 2006. A suitable model for the biomonitoring of trace metal bioavailabilities in estuarine sediments: the annelid polychaete *Nereis diversicolor*. *J. Mar. Biol. Assess. U.K.* 86: 71–82.

Porte, C. et al. 2002. Assessment of coastal pollution by combined determination of chemical and biochemical markers in *Mullus barbatus*. *Mar. Ecol. Prog. Ser.* 235: 205–216.

Regoli, F., M. Nigro, and E. Orlando. 1998. Lysosomal and antioxidant responses to metals in the Antarctic scallop *Adamussium colbecki*. *Aquat. Toxicol.* 40: 375–392.

RNO, 2001. L'argent, le cobalt, le nickel et le vanadium dans les mollusques du littoral Français. In *Surveillance du milieu marin*, Nantes: INFREMER, pp. 11–20.

Roméo, M. and M. Gnassia-Barelli. 1997. Effect of heavy metals on lipid peroxidation in the Mediterranean clam *Ruditapes decussatus*. *Comp. Biochem. Physiol.* 118C: 33–37.

Roméo, M. et al. 2000. Cadmium and copper display different responses toward oxidative stress in the kidney of the sea bass *Dicentrarchus labrax*. *Aquat. Toxicol.* 48: 185–194.

Roméo, M. et al. 2001. Evaluation of various biomarkers in the wild fish *Serranus cabrilla* collected in the NW Mediterranean Sea. In *Biomarkers in Marine Organisms: A Practical Approach*, Garrigues, P. et al., Eds. Amsterdam: Elsevier, pp. 343–356.

Roméo, M. and L. Giamberini. 2008. Historique. In *Les biomarqueurs dans l'évaluation de l'état écologique des milieux aquatiques*, Amiard, J.C. and Amiart-Triquet, C., Eds. Paris: Lovoisier, pp. 17–54.

Saint-Denis, M. et al. 1999. Biochemical responses of the earthworm *Eisenia fetida andrei* exposed to contaminated artificial soil: effects of benzo(a)pyrene. *Soil Biol. Biochem.* 31: 1837–1846.

Saint-Denis, M. et al. 2001. Biochemical responses of the earthworm *Eisenia fetida andrei* exposed to contaminated artificial soil: effects of lead acetate. *Soil Biol. Biochem.* 33: 395–404.

Scaps, P. and O. Borot. 2000. Acetylcholinesterase activity of the polychaete *Nereis diversicolor*: effects of temperature and salinity. *Comp. Biochem. Physiol.* C125: 377–383.

Thompson, H.M., C.H. Walker, and A.R. Hardy. 1991. Changes in activity of avian serum esterases following exposure to organophosphorous insecticides. *Arch. Environ. Contam. Toxicol.* 20: 514–521.

Thompson, J.A.J. and R.P. Cosson. 1984. An improved electrochemical method for the quantification of metallothionein in marine organisms. *Mar. Environ. Res.* 11: 137–152.

Van der Oost, R. et al. 1996. Biomonitoring of aquatic pollution with feral eel (*Anguilla anguilla*): II. Biomarkers: pollution-induced biochemical responses. *Aquat. Toxicol.* 36: 189–222.

Viarengo, A. and J.A. Nott, 1993. Mechanisms of heavy metal cation homeostasis in marine invertebrates. *Comp. Biochem. Physiol.* 104C: 355–372.

Viarengo, A. et al. 2007. The use of biomarkers in biomonitoring: a two-tier approach assessing the level of pollutant-induced stress syndrome in sentinel organism. *Comp. Biochem. Physiol.* 146 C: 281–300.

Walker, C.H. and H.M. Thompson. 1991. Phylogenetic distribution of cholinesterases and related esterases. In *Cholinesterase-Inhibiting Insecticides: Chemicals in Agriculture,* Vol. 2., Mineau, P., Ed. Amsterdam: Elsevier, pp. 1–17.

Wallace, W.G., G.R. Lopez, and J.S. Levinton. 1998. Cadmium resistance in an oligochaete and its effects on cadmium trophic transfer to an omnivorous shrimp. *Mar. Ecol. Progr. Ser.* 172: 225–237.

5 Biogeochemistry of Metals in Sediments

Development of Microscale Analytical Tools and Use of Indicators of Biological Activities

Baghdad Ouddane, Laurent Quillet,
Olivier Clarisse, Gabriel Billon,
Jean-Claude Fischer, and Fabienne Petit

CONTENTS

Intertidal mudflats are sites of deposition from upstream of suspended particulate materials along with an array of contaminant chemicals that reflect the industrial activities of the past 20 years. In strongly (Seine estuary) and weakly (Authie bay) anthropogenically affected sites, one way to assess the quality of the environment is by analysis of the contaminants in the sediments, although such measures do not directly predict metal toxicity effects on aquatic animals (Luoma 1983; Luoma 1989; Di Toro et al. 1990; Luoma 1995). However, studying the distribution of metals among geochemical phases is an important step in forecasting their ultimate fate, bioavailability, and toxicities (Salomons and Förstner 1980, 1984).

Metal distribution is controlled by various physico-chemical and microbiological processes mainly connected to the biogeochemical cycles of carbon and sulfur. Some metabolic capacities developed by sulfate-reducing microorganisms (SRMs) allow them to play an important role in the speciation and the mobility of metal ions in anoxic environments. In the present study, the chemical speciation and bioavailabilities of metals were studied in connection with the sulfide cycle. The chemical analysis of anoxic sediments demonstrated the trapping of trace metals by sulfides, the origins of which can be understood only by the sulfidogenic activities of certain microbial communities—the bacteria and archaea microbial communities are the most represented in these environments.

5.1   METAL MOBILITY IN THE SOLID PHASE

Sediments play an important role in influencing the metal concentrations of an overlaying water column but can also be used to record its degree of pollution over time. The remobilization of sediments can release some trace metals from the solid phase, with subsequent effects on the aquatic system, particularly during early diagenetic processes. The mobility of trace metals in sediments depends strongly on their chemical forms and interactions, e.g., the sorption and desorption processes may significantly affect the behavior, fate, and toxicity of metals in the natural environment. These metals can be present in various chemical forms, and generally reveal different physical and chemical behaviors in terms of mobility, bioavailability, and potential toxicity.

It is commonly accepted that the ecological effects of metals are related to their mobile fractions rather than the total concentrations in sediment (Adriano 2001). To achieve a more precise understanding of potential toxicity and impacts on biota, it is necessary to identify and quantify the forms in which metals are present in sediment. Total content data alone are insufficient for estimating the possible risk of mobilization and potential uptake of liberated metals by biota (Fedotov et al. 2006).

Two approaches are commonly used for leaching metals from sediment samples: single partial extractions and fractionation by sequential extractions. Fractionation is usually performed by a sequence of selective chemical extraction techniques, including the successive removal or dissolution of these phases and their associated metals (Tessier et al. 1979; Quevauviller et al. 1997; Ure and Davidson 2002). There is no general agreement on the preferred solutions for extraction of the various components in sediments, mostly because of the matrix effects involved in heterogeneous chemical processes (Martin et al. 1987; Bacon and Davidson 2008).

All factors must be carefully considered when an extractant for a specific investigation is chosen. Partial dissolution techniques should include reagents that are sensitive to only one of the various components significant in trace metal binding. Many efforts have been made by choosing relevant reference materials such as standard minerals and sediments collected from specific sites from which detailed certified analyses of metal contents are available (Canfield 1989; Quevauviller et al. 1997).

Single-step leaching extraction represents a well recognized, simple procedure that takes little time for estimating the biologically available fractions of metals to evaluate anthropogenic enrichment and discriminate pollution sources (Sutherland et al. 2004). Acid extraction is the most common procedure; it solubilises the various non-residual fractions while minimizing the dissolution of detrital minerals. Although they provide less detailed information than sequential extraction methods, these techniques are suitable for a number of environmental applications and can be particularly useful, for example, in monitoring programs and in investigations that require the analysis of a large number of samples (Bettiol et al. 2008). Roychoudhury (2006) tested single-step extractions using different extractants for targeting specific iron phases from salt marsh sediments. He concludes that an appropriate concentration of HCl is necessary to extract the metal associated with sulfides (as acid-volatile sulfides [AVS]).

Because of the large number of sediment samples collected in the Seine and Authie estuaries, we chose the single hydrochloric acid (1M) extraction procedure to estimate the labile sediment metal concentration fraction as a model of the available fraction. This leaching may be considered a proxy that simulates the digestive actions of gastric juices in the stomachs of various organisms living in association with the mudflat, and thus defines the maximum quantities of metals extractable (and presumed bioavailable) for these species (Cooper and Morse 1998). The extractabilities of the major and minor elements of the sediments, expressed in percents compared to the total concentrations (average values) measured in the Seine and Authie estuaries mudflats, are presented in Figure 5.1.

The extractability can be explained by the chemical distribution of these elements in the various fractions in the sediments (exchangeables, carbonates, hydroxides, sulfides, organics, and residual fractions). Calcium and magnesium appeared totally extractable; both elements were mainly present in the sediments in the form of carbonates (Billon et al. 2001; Clarisse 2003). In contrast, the titanium fraction extracted by hydrochloric acid was weak as a result of its presence in refractory compounds like titanium oxides. The extractabilities of Cd, Cu, Pb, and Zn appeared

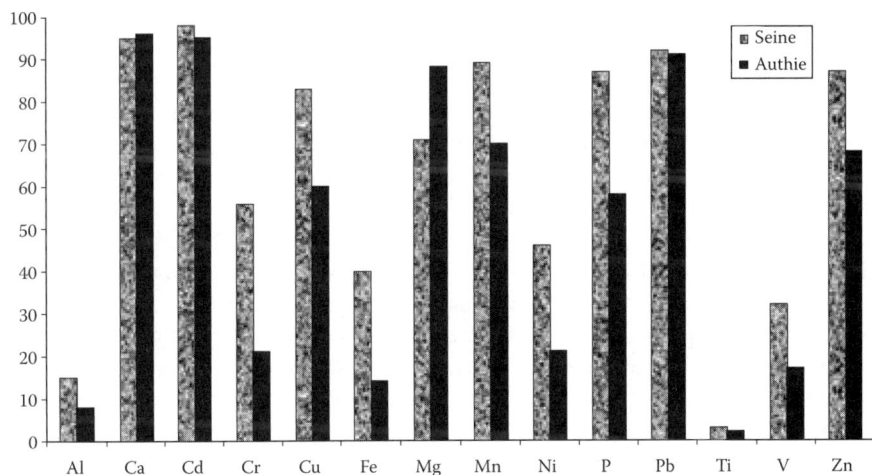

FIGURE 5.1 Extractabilities of major and minor elements in the Seine and the Authie sediments (expressed in *percent* compared to the total concentration).

relatively important; these trace metals are mainly associated with carbonate and/or sulfide fractions. It is well known that these four elements present a strong affinity for sulfides (Ankley et al. 1996; Cooper and Morse 1998; O'Day et al. 2000). The sulfur species present in anoxic sediments are available to trap these metals through the formation of insoluble metal sulfide compounds, limiting their mobility and as a result their expected bioaccumulation by living organisms.

In order to evaluate the potential sediment toxicities of trace metals (Cd, Cu, Ni, Pb, and Zn), the ratio of concentrations of simultaneously extracted metals (ΣSEM) to acid-volatile sulfides was proposed by Di Toro et al. (1990, 1992) and USEPA (2004) as an alternative to the traditional approaches of extraction and as an illustration of the capacity of the sediment to trap trace metals in the sulfide phase in any eventual potential remobilization. If ΣSEM/AVS >1, metal mobility and subsequently expected bioavailability are probable, and the sediment may be potentially toxic for benthic communities. On the other hand, if the ΣSEM/AVS ratio is <1, the metals are present as insoluble sulfides (and not bioavailable) and susceptible toxicity due to metals is unlikely. Note that many authors used this approach to estimate sediment toxicity (Li et al. 2001; Fang et al. 2005; Prica et al. 2008). However, only biological tests with living organisms can provide real estimates of sediment toxicity (Lee et al. 2000).

The procedures for AVS and ΣSEM analysis were adopted from a method described by Canfield (1989) and Allen et al. (1993) with some modification (Billon et al. 2002). For AVS, 1 g of sediment was introduced into an AVS assembly continuously flushed with nitrogen. HCl was then carefully added and the H_2S gas produced was trapped in a NaOH solution. After the extraction step, H_2S was titred by potentiometry using a solid sulfur electrode. For ΣSEM measurement, the sediment suspension was stirred overnight in 20 mL of $1M$ HCl solution at room temperature. The solution was then filtered through a 0.45-μm membrane and the concentrations of trace metals were determined with ICP-AES or ICP-MS. AVS and ΣSEM cold extractions were performed on five cores from the Seine and the Authie estuaries with contrasting sediment chemistries. The core sediments were collected from three sites (see Figure 1.3: Oissel in the upstream estuary, the northern mudflat at station PN in the middle estuary, and a subtidal mudflat in the downstream estuary) distributed along the Seine estuary to cover a large salinity range and from two sites in the Authie (river [AP] and marine [AS]). Figure 5.2 depicts the results.

Changes in ΣSEM/AVS ratios with depth were very similar in the Seine cores with the exceptions of the fresh water deeper sediment at Oissel and the surface subtidal mudflat sediment cores. The significant values of ΣSEM/AVS ratios (that are potentially toxic) arise from high contamination by metals and low values of AVS in the Oissel deeper sedimentary layers. Indeed, at these depths, the quantities of sulfides are not sufficient to trap all the trace metals present in the form of insoluble sulfides (limitation of sulfates in fresh water, the principal source of sulfide production in deposited sediment by sulfate-reducing bacteria). The potential toxicity of the sediments of the northern mudflat at PN and in the Authie estuary appeared lower. ΣSEM/AVS ratios were <1 over the entire lengths of sedimentary cores. The particulate sulfides were present in sufficient quantity to immobilize and trap metals, and the sediment did not appear potentially toxic.

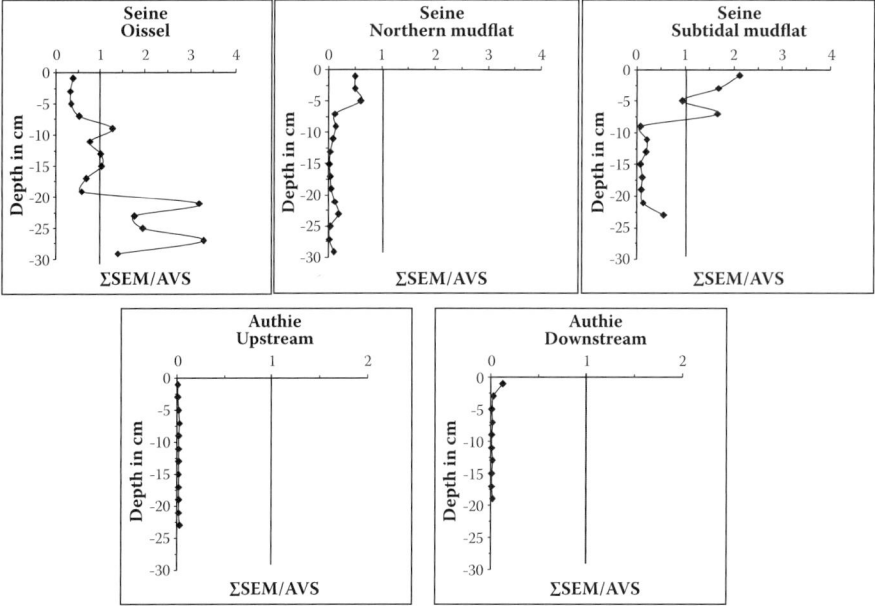

FIGURE 5.2 ΣSEM/AVS ratio in sediment cores of the Seine and Authie estuaries.

Miramand et al. (2001) studied the bioaccumulation of trace metals Cd, Cu, Pb, and Zn in the Seine sediments by larvae (*Crassostrea gigas*). Their principal conclusions reveal a relatively low bioavailability of these metals for all the sedimentary horizons tested with a maximum bioaccumulation of these contaminants by larvae from the subsurface sediments (0 to 6 cm), thus consolidating the results of ΣSEM/AVS ratios obtained for the subtidal mudflat. Only the surface sediments of this site appeared to be potentially toxic.

The ΣSEM/AVS ratio constitutes a fast tool for evaluating the potential toxicities of sediments without additional biological tests. The ratio is used by the USEPA to evaluate potential sediment toxicity as a screening-level analysis to identify potential problems that may prove not significant upon further analysis (USEPA, 1995). However, ΣSEM/AVS ratio predictions may overestimate metal bioavailability if significant quantities of metals bound to other phases can be leached out in the 1M HCl extraction. In consequence, not all sediments with ΣSEM/AVS >1 can cause increased toxicity because sediments also contain many other metal-binding phases such as organic matter and Fe and Mn oxides (Ankley et al. 1996; Leonard et al. 1996; Chapman et al. 2002).

It is also difficult to predict which of the five metals (Cd, Cu, Ni, Pb, and Zn) has the highest bioavailability for the living organisms under these conditions because bioavailability depends on a multitude of further parameters like the presence of fresh organic matter (Brown et al. 1995). Additionally, this ratio does not integrate the concept of physical remobilization of the sediment caused by tide variations, boat traffic, or dredging operations. Dredging processes will certainly increase the concentrations of bioavailable metals and disturb the sedimentation dynamics by the oxidation of the sulfur phases of anoxic sediments that affects the mobility of associated metals.

To simulate the regular resuspension of the sediments, a series of remobilization tests was performed in the laboratory. The intent was to determine the mechanism for metal release into the water column and elucidate the chemical and physical processes controlling the fates of trace metals in contaminated sediments. Clarisse (2003) carried out a series of laboratory tests by controlling several parameters according to a protocol described in his thesis. The experiments of remobilization consist of a dispersion of a few grams of sediment, preserved under anoxic conditions, in two or three liters of filtered natural water at room temperature. The monitoring of various parameters (pH, temperature, dissolved oxygen, redox potential, and concentrations of dissolved metals) as a function of time through sediment resuspension permits the prediction of the fate of a metal in an aquatic system. The investigations were carried out with fresh and sea water samples, respectively, in Oissel and Le Havre to correspond reasonably well to natural conditions of sediment resuspension in the field.

Finally, the influence of dissolved oxygen on the mechanisms of metal evolution was also studied through experiments under free and inert controlled (N_2) atmospheres to simulate oxic and anoxic conditions. At the Oissel site, the remobilization tests were carried out in fresh water. On the northern mudflat at the PN site, tests were conducted in both fresh water and sea water. Finally for the subtidal mudflat, only the influence of sea water was considered for sediment resuspension. The experiment duration selected was relatively short (one day) since desorption from the solid phase is supposed to be a fast process (Saulnier and Mucci 2000), unlike bioaccumulation tests on living organisms that require much more contact time.

Positive metal fluxes correspond to dispersion of porewaters and/or desorption and dissolution from the solid phase. Negative fluxes are indicative of opposite reactions like precipitation, co-precipitation, or adsorption of these elements on the sediment. The physico-chemical parameters measured reveal the presence of very reactive species, depending on operating conditions. For the subtidal mudflat, the physical remobilization of the sediment involves in the first minutes a consequent release of Na, Ca, and Mg into the water column, related to the partial dissolution of the reactive fraction of the sediment, followed by their disappearance from the water by adsorption, precipitation, or co-precipitation.

The changes in the concentrations of these three elements were distinct from the other sites; the calculated remobilization fluxes from the solid phase to the water column were close to zero. Calcium, however, was an exception. In the experiments carried out in fresh water in absence of oxygen, it disappeared from the water column. The disappearance over time correlated with a reduction of alkalinity, suggesting precipitation of calcium carbonates.

The behavior of manganese was relatively atypical and similar for the various sites. The dissolution of certain phases of the sediment like $MnCO_3$ or $Mn_xCa(CO_3)_{1+x}$ releases manganese into the water column and the manganese precipitates to form oxy-hydroxides (Saulnier and Mucci 2000).

The remobilization of Fe, Cu, Cd, and Zn was weak for the three sites except for the contaminated sediments of Oissel tested under free atmosphere conditions. The partial oxidation of reduced sulfur compounds seemed to involve a temporary and consequent release of Zn and to a lesser extent, Fe, Cd, and Cu. For this sediment, the ΣSEM/AVS ratio was >1, resulting in a probably high availability of some trace metals. Because Zn accounted for approximately 70% of the total amount of acid-extractable metals (ΣSEM), it appeared in greater proportion in the remobilization test. The ΣSEM/AVS ratio thus can become a predictive tool to estimate the availability of trace metals in the event of physical remobilization of the sediment (Simpson et al. 1998).

In conclusion, the availability of metal contaminants is related to their chemical environment in a sediment. The extractions with $1M$ HCl solution underlined the extractability and potential bioavailability of the elements associated with carbonates. Moreover, in anoxic sediments, sulfide compounds limited the mobility of trace metals and consequently their release from the sediment was significantly reduced. The ΣSEM/AVS ratio thus provided information on the potential trace metal toxicities of the studied sediments. The contaminated sediments of Oissel and the subtidal mudflat surface sediment cores were potentially toxic, whereas the sediments of the northern mudflat and of the Authie exhibited great capacities to trap these contaminants in sulfur phases. Finally, the mobility of trace metals was also studied in the laboratory by sediment resuspension in the water column. The results indicated low availability of the elements in the solid phase and important competition among dissolution, adsorption, and precipitation reactions.

5.2 BIOAVAILABILITY OF METALS IN POREWATERS

During the sedimentation process, water is trapped and entrained in the sediment, forming porewater (Batley and Giles 1979) where it may be considered in equilibrium with the solid phase sediment. Chemical species found in the water column and adsorbed to suspended sediment will also be trapped in the bottom sediment. Sediment–water interactions play a fundamental role in the biogeochemical cycling and transport processes of trace elements (Carignan and Lean 1991; Barbanti et al. 1992). Oxygen is exhausted near the sediment surface by microorganisms present in the deposited sediments. These microorganisms use oxidised forms such as dissolved O_2, NO_3^-, and Fe and Mn oxides present in the sediment as oxidation agents (Billon et al. 2001), thereby decreasing the redox potential of the sediment.

In fact, the major changes that occur between oxic waters and anoxic sediments can exert strong influences on the speciation and bioavailabilities of many metals. Anoxic conditions often result in the microbially mediated production of major amounts of sulfides in marine and estuarine environments (Howarth et al. 1992). Although most of the sulfides are usually reoxidised, a fraction of H_2S reacts to form metal sulfides, principally FeS, FeS_2 (pyrite), and many others (Millero 1991; Ouddane et al. 2001).

These iron sulfides co-precipitate with and/or adsorb many trace metals (Morse et al. 1987; Rickard et al. 1995). Another factor affecting trace metal speciation in sediment–water systems is pH. Indeed the pH of porewater is generally lower than the pH of surface water, due to the presence of oxidants including compounds containing Fe, N, and S in lower oxidation states that are ultimately oxidised by microbial mediation to Fe^{3+}, NO_3^-, and SO_4^{2-}, thereby releasing equivalent amounts of H^+ into the environment (Morel 1983).

In this anoxic environment, which is generally rich in organic matter, both deposition and mobilization of most trace elements are possible reactions. Due to diagenetic processes such as precipitation, adsorption, sulfide formation, remobilization, and biological degradation, the concentrations of metal chemical species in porewater and the overlaying water will not be necessarily the same (Sakata 1985; Salomons et al. 1987). Thus, knowledge of trace metal concentrations in porewater is useful for determining and evaluating sediment contamination and assessing the contribution of sediment to the pollution of the overlying water column (DiToro et al. 1990).

The cycling of trace metals in sediments is controlled by precipitation and dissolution of minerals, but the mobility and bioavailability may be also related to organic and inorganic complexation. These binding phases (sulfide, organic matter, and metal oxides) can move trace metals from and into the porewater (Morrison 1989; Förstner 1990; Fu and Allen 1992). During aerobic degradation of organic matter in the sediments, the incorporated metals are released into porewater (Fu et al. 1992) and can migrate back to the water column. In the topmost sediment layer, dissolved metals are usually enriched relative to bottom water. In deeper anoxic sediments, the dissolved metal concentrations are generally low due to the precipitation of less soluble metal sulfides (Lyons and Fitzgerald 1983; Emerson et al. 1984; Fernex et al. 1986; Shaw et al. 1990). Furthermore, in the anoxic sediment layer, trace metals bound to Fe and Mn oxides can be released into porewater following reduction of these oxides. In contrast, Fe^{2+} and Mn^{2+} diffusing into the oxic sediment layer will be oxidized to oxides.

Note that recently formed Fe and Mn oxides are very efficient scavengers for trace metals in this layer. Therefore, the cycling of Fe and Mn may play an important role in transport processes of other metals between the sediments, porewater, and overlaying water (Matsunaga et al. 1993; Morse and Arakaki 1993). Overall, the solubility and mobility of trace metals in solid–porewater systems may be affected by five major factors: pH, natural or synthetic complexing agents, salt concentrations, redox conditions, and decomposing organic matter containing trace elements (Carignan and Nriagu 1985; Boudreau 1987; Boudreau and Canfield 1988; Boudreau, 1991).

Other studies show that organic colloids or dissolved organic carbon in general increase with depth as a result of the degradation of organic matter. This may enhance the mobility and bioavailability of trace metals during early diagenesis (Douglas et al. 1986; Deflandre et al. 2002). It is well documented that a sediment–porewater system is a very reactive anoxic medium that requires careful handling and treatment due to rapid oxidation and changes in physico-chemical parameters. To assess trace metal content in porewater, it is necessary to consider some important factors. The use of proper sampling techniques on sediment from the natural environment for *ex situ* analyses is very important to avoid mixing of oxic and anoxic sediments; otherwise, trace metal speciation in the sediment and porewater will be altered due to oxidation (Bufflap and Allen 1995).

A vertical distribution (concentration as a function of depth) of some metals (Fe, Mn, Ni, Cu, Pb, V, Zn, and Cd) in the extracted porewater and their speciation from the Seine estuary were undertaken in some cores. For this purpose, the study focused on the development of a method for sample collection and handling under an inert atmosphere to prevent potential artifacts in the extracted porewater.

We adapted an analytical method for the determination of trace metal speciation (distinction between organic and inorganic forms) based on a combination of two specific resins. First, C18 bonded silica gel retains organic metal forms. In general, this resin can retain metals complexed by humic substances, amino acids, aromatic ligands, and all small organic ligands forming neutral complexes (Mills et al. 1987; Groschner and Appriou 1994; Leepipattpibtoon 1995). C18 can also isolate colloid-associated non-polar organic ligands (Landrum et al. 1987). A chelamine-chelating resin produced by the immobilisation of a pentamine ligand on an organic polymer is the second compound. Blain et al. (1993) demonstrated that this resin is capable of separating some trace metals (Cd, Cu, Mn, Ni, Pb, Zn) from sea water with high selectivity at different pH values (5.5 to 8).

Total metal concentration was determined by the fixation of metal–8-hydroxy-quinoline complexes in C18 columns at a pH range 6 to 8 (Abbasse et al. 2002)—adaptable to the pH range of porewater sampled. The trace metals fixed on the two resins (C18 and chelamine columns) for all determinations were stripped by $2M$ nitric acid. The combination of the resins and the method are intended to separate trace metals from their matrices and their preconcentrations to allow detection by inductively coupled plasma atomic emission spectroscopy (ICP-AES) or Zeeman graphite furnace atomic absorption spectroscopy (ZGF-AAS). The principal results obtained for the Seine estuary (Ouddane et al. 2004) revealed that the porewater metal profiles of Cd, Cu, Ni, Pb, V, Zn, Mn, and Fe reflect the intensive release of these metals during the decomposition of organic substances in the sedimentary column (early diagenesis), because the concentrations of the studied metals were much higher than their concentrations in the overlaying water, even in bottom sediment where the precipitation of metal sulfides was the dominant process.

As a result of the biological processes that take place in recently deposited sediments, various organic and inorganic ligands and metals released into the porewater also cause the formation of some strong metal organic complexes and other labile complexes that are retained by the C18 column and the chelating chelamine resin, respectively. The metals released, especially in the surface sediment–water systems (up to 12 cm depth), diffuse both into the anoxic zone where they precipitate as metal sulfides (leading to decreasing metal concentrations in this zone) and back into the overlaying water, where they may be precipitated or co-precipitated by Fe and Mn oxides and recycled again through other biogeochemical processes. Generally, in recently deposited sediments, there is competition among metal sulfide formation, soluble metal organic complexation, and Fe/Mn oxide adsorption. All these processes control the mobility and speciation of metals and trace metals in the sediment–water system.

5.3 USE OF GEL DIFFUSION TECHNIQUES (DGT-DET)

Most of the techniques involved in speciation measurement, e.g., electrochemistry (anodic stripping voltammetry [ASV] and cathodic stripping voltammetry [CSV]) or chromatography (extraction on C18 cartridges and ligand exchange), require analysis of samples brought to a laboratory. However, modification of metal speciation can occur during sample collection, storage, and/or treatment (Tercier and Buffle 1993; Van den Berg et al. 1994). Development of techniques to measure metal speciation *in situ* has been necessary to overcome this problem. In aqueous systems, the diffusive gradient in thin film (DGT) technique provides an *in situ* means of estimating labile trace metal species (Zhang and Davison 2000; Twiss and Moffet 2002).

In porewater, the mean labile metal concentration measured by this technique is dependent on resupply from the solid phase (Zhang et al. 1998). Moreover, the millimeter scale profiles obtained with this technique can be compared with profiles measured at the same high resolution by the diffusional equilibrium in thin films (DET) technique. Evaluation of porewater metal concentration by these two methods reveals information about metal speciation (labile and total forms measured, respectively, with DGT and DET) and consequently about toxicity and diagenetic processes in sediment. Such profiles were carried out for iron and manganese in the sediments of the Seine and the Authie.

The DGT method consists mainly of an assemblage of a well defined diffusion layer and a binding agent that accumulates solutes quantitatively (Zhang and Davison 1995). A polyacrylamide gel and membrane filter of known thickness (Δg) for gel protection that acts as an extension of the gel (Davison and Zhang 1994) commonly serve as the diffusive layer. To determine trace metal accumulation, Chelex 100 resin is incorporated into a second gel layer and serves as a binding agent. The two gels are enclosed in a small plastic device that is immersed in solution or sediment. Knowing the time of deployment (t) of the probe, the surface area (A) of the diffusive layer in contact with the solution or the sediment, and the mass of accumulated metals (M) in the associated resin layer permits the estimation of the flux (F) of metals from the sediment or solution to the DGT probe based on this equation:

$$F = \frac{M}{A.t} \tag{5.1}$$

According to Fick's first law, this flux, can also be interpreted as the mean concentration (C_i) at the surface of the probe:

$$F = D \frac{(C_i - C)}{\Delta g} \tag{5.2}$$

If the metals ions bind rapidly and efficiently to the resin, C is effectively equal to zero and the resin is not saturated. Therefore, Equation (5.2) can be simplified to:

$$F = D \frac{C_i}{\Delta g} \tag{5.3}$$

Finally, by combining Equations (5.1) and (5.3), C_i can be estimated:

$$C_i = \frac{M.\Delta g}{D.A.t} \tag{5.4}$$

It is important to note that the introduction of the DGT device in sediment perturbs the system by introducing a local sink. Metals diffuse from the porewater to the DGT device, where they are bound to the resin layer. Therefore the DGT device induces a flux from the porewater, depleting the local porewater concentration. However, this depletion is assumed to be counteracted by a resupply of the trace metal to the porewater from the sediment solid phase. The mean concentration C_i measured by DGT is therefore dependent on the resupply rate and capacity from the solid phase into the porewater. In fact, according to Zhang et al. (1998) three cases may arise with respect to DGT-measured flux and interfacial concentration:

1. Fully sustained. Metals removed from the porewater by the DGT probe are rapidly resupplied from solid phase; the flux to the device is effectively maintained at its maximal value throughout the deployment. C_i is equal to the bulk porewater concentration, C_b.

2. Diffusion only. There is no resupply from the solid phase to the porewater. The supply of metals to the DGT probe is solely by diffusion through the porewater. The DGT-measured flux decreases with deployment time due to the depletion of the porewater concentration. For a probe with a 0.8-mm thick diffusive gel, the average flux for 24-hour deployment is approximately 10% of the initially established flux and $C_i \approx 0.1 * C_b$ (Harper et al. 1998).

3. Partially sustained. There is some resupply of metals from solid phase to porewater, but it is insufficient to maintain the maximum DGT flux. Depletion in porewater concentration occurs, but to a lesser degree than for case 1 ($0.1 * C_b < C_i < C_b$).

Only in the first case can fluxes measured by DGT be interpreted as porewater concentration. In the other cases, porewater concentration is underestimated. Therefore, single DGT flux measurement in sediment appears insufficient for gaining information about porewater metal concentration or the resupply process. A comparison of porewater metal concentration measured with an alternative technique is needed to reach such conclusions (Harper et al. 1998)—a technique such as ASV that can discriminate species as DGT does.

Another method to distinguish the three cases consists of the deployment of several DGT devices. If gels of different thicknesses are deployed for the same period, the plot of DGT flux (F) versus $1/\Delta g$ will be linear in case 1 and non-linear in case 3 (Zhang et al. 1998). In the same manner, DGT fluxes measured by deployment of several same-thickness gels for different periods in the sediment will be constant only in case 1 (Hudson et al. 1990).

Three DGT probes were deployed in the Authie sediment for 24, 64, and 116 hours. After retrieval, the resin gel was cut and eluted and manganese and iron were analyzed. To determine the resupply rate from the solid phase, DGT results are first interpreted as fluxes (Figure 5.3).

Manganese profiles for the three DGT probes were comparable: maximum flux localization due to the reduction of organic matter at the first centimeter. Systematic horizontal and vertical change in porewater concentration was revealed

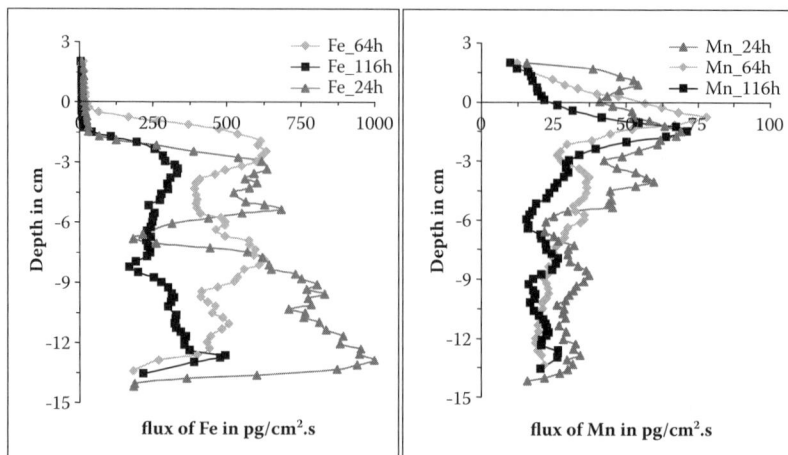

FIGURE 5.3   Flux (pg.cm^{-2}.s^{-1}) of iron and manganese induced by DGT probes for 24, 64, and 116 hours deployment in Authie sediment.

by voltammetry at high spatial resolution in two dimensions for iron and manganese (Luther et al. 1998; Taillefert et al. 2000) and by DET (Shuttleworth et al. 1999; Docekalova et al. 2002). Manganese fluxes were similar for the three DGT probes; small differences may be attributed to sampling localization. According to the theory, manganese behavior appears to belong to the fully sustained case: resupply from solid phase appears sufficient to ensure a maximal flux to the DGT device throughout the deployment. Thus, DGT measurements may be interpreted as porewater concentrations.

The behavior of iron was very different. Except for the artifact observed at 7 cm depth due perhaps to the presence of an oxic microniche related to biological activity (Zhang et al. 1998; Zhang and Davison 2000), profile shapes appeared similar and fluxes seemed to decrease with time deployment. Although porewater iron concentration depletion occurs, this profile can be divided into two depth measurements: (1) the first 7 cm where flux maxima between 24 and 64 hours of deployment are in the same range, and (2) the last 7 cm where fluxes decrease by a factor of two over time. In the first 7 cm, resupply from the solid phase occurs and seems to be sufficient to maintain the sustained case between 24 and 64 hours but not until 116 hours. For the last 7 cm depth, resupply is partial.

Iron availability from the solid phase is well known to be weaker than for manganese in sediments (Marin and Giresse 2001) and to decrease with depth when sulfide phases appear (Billon et al. 2002). Solid metal distribution may partly explain our different DGT profiles (Clarisse 2003). In the case of the flux results, iron DGT fluxes cannot be interpreted as porewater concentrations. However, we performed such calculations for a 24-hour deployment (depletion in porewater iron concentration is minimum) and compared these mean DGT porewater concentrations with those measured by DET (Figure 5.4). Manganese porewater profiles measured by these two techniques were quite different. Assuming that DGT manganese concentrations

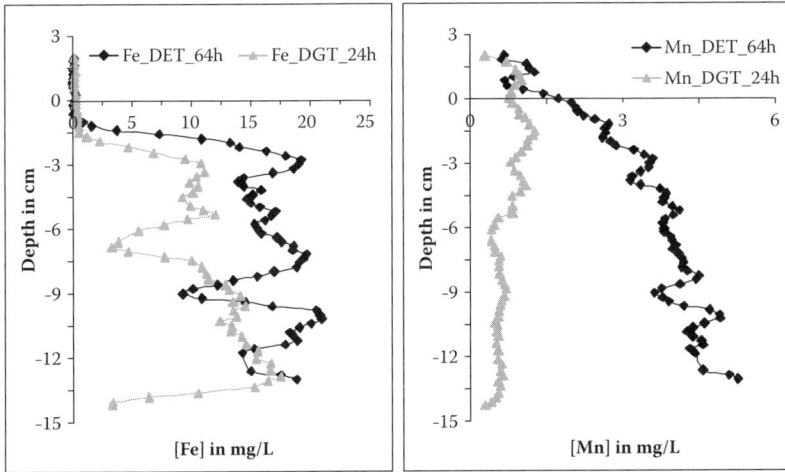

FIGURE 5.4   Porewater concentrations of iron and manganese measured by DGT and DET techniques in Authie sediments.

represent the labile form (Zhang et al. 2002), manganese appears to be mainly complexed in this sediment.

Conversely, iron profiles were very similar. The oxic microniche revealed at 7 cm depth by the DGT technique was found at 9 cm depth with the DET technique. The three-dimensional sediment structure can explain the small localization shift. At the first 7 cm depth where resupply from the solid phase appears sufficient to maintain maximal concentration through 64 hours, labile iron measured by DGT represents more than 50% of the total porewater metal concentration. At the last 7 cm depth, iron concentrations calculated from DGT flux and measured by DET were similar. Interpretations of DGT iron flux for 24-hour deployment as porewater iron concentrations appear to make sense and the main portion of iron at such depth is labile in porewater.

This study has shown some limitations of the use of DGT in sediment. Depletion in porewater concentration induced by the DGT device and three-dimensional sediment structure represent some limitations for the interpretation of DGT results. However, deployment of several DGT devices with same-thickness diffusion gel for different exposure times may at least partially solve these problems. Short deployment time (24 hours) minimizes the DGT pump effect and appears useful for achieving a good estimation of labile metal concentration. Finally, the comparison of DGT porewater concentration profiles with those measured by DET provides high resolution speciation information about porewater. To verify the efficiency of these procedures, we compared manganese speciation profiles obtained by DET and conventional sampling (classical porewater separation by centrifugation and extraction by chelamine and C18 columns) of the Authie sediment core (Figure 5.5). Comparisons of results obtained with diffusion and other speciation techniques are difficult because of differences in spatial resolution (high for DET, low for porewater extraction by centrifugation) and separation procedures (stoical exclusion for DET, chemical complexation for resin). Nevertheless, as shown in Figure 5.5, the orders of

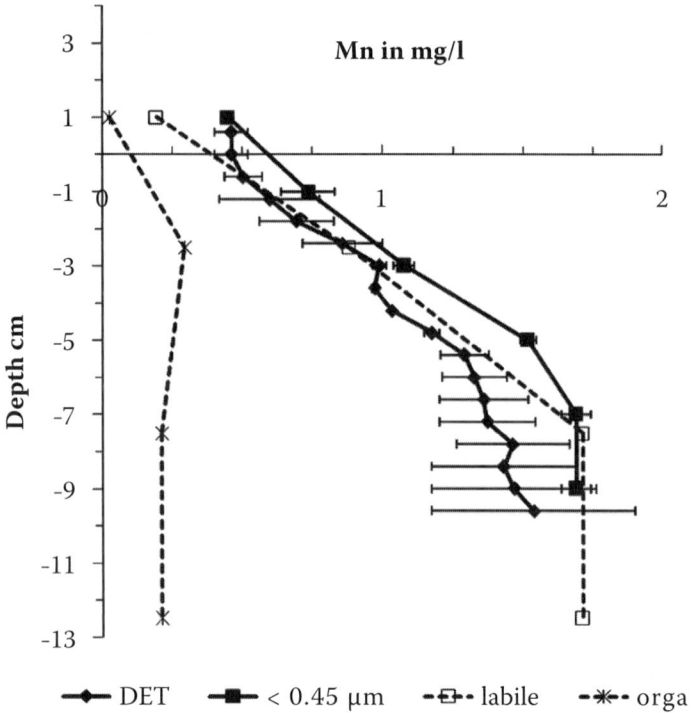

FIGURE 5.5 Comparison of Mn speciation profiles obtained from DET techniques and conventional sampling methods in the Authie sediment core.

magnitude of manganese concentration results obtained by the different techniques are very similar. DET has proved to be an efficient tool for sampling dissolved trace metals *in situ* and in speciation studies with complementary techniques (e.g., DGT) and/or conventional methods (e.g., voltammetry or chromatography extraction via chelamine and C18 columns).

5.4 COMMUNITIES OF SULFATE-REDUCING BACTERIA (*dsr*AB GENES) IN ANOXIC SEDIMENTS FROM THE SEINE AND AUTHIE ESTUARIES

Sulfate-reducing microorganisms (SRMs) play an important role in the speciation and mobility of metal ions in anoxic environments because they are capable of producing important quantities of sulfide ions during anaerobic sulfate respiration. These sulfide ions react with the metal ions present in the environment and precipitate in the form of insoluble metal sulfides. They are also capable of dissimilatory reductions of certain metal ions such as technetium (Tc VII), iron (Fe III), uranium (U IV), and chromium (Cr VI) (Lovley et al. 1993; Lloyd et al. 2001).

In anoxic sediments, trace metals become trapped by hydrogen sulfides resulting from the sulfidogenic activities of certain microbial communities, mainly bacteria and archaea in the medium of interest. These microorganisms are particularly capable of using sulfate as the final electron acceptor.

While this community constitutes a polyphyletic group, these microorganisms all possess a characteristic enzyme that catalyses the dissimilatory reduction from sulfite (HSO_3^-) to sulfide of hydrogen (H_2S), and is coded by the *dsr*AB (dissimilatory sulfite reductase) gene (Wagner et al. 1998). The molecular quantification of the *dsr*AB gene was used to determine the quantity of SRM present as an indicator of the microbial production of sulfides.

5.4.1 MOLECULAR QUANTIFICATION OF SRM

We chose to develop a tool of quantification based on the technique of quantitative competitive PCR to determine the abundance of the *dsr*AB genes of sulfate-reducing microorganisms. A comparison of the abundance of the microbial sulfidogenic *dsr*AB community was made at two sites where the main geochemical characteristics of the sediment (granulometry, salinity, organic matter, water, carbonates) were comparable: (1) the north intertidal mudflat of Authie bay (AN: reference site weakly contaminated, five campaigns, 2003–2004) and the north intertidal mudflat of the Seine estuary (PN: contaminated site, five campaigns, 2001–2002). An annual multidisciplinary study taking into account various environmental factors that may influence the development of SRMs (temperature, redox potential, presence of O_2, topography) was conducted.

The results of the quantifying SRMs in 30-cm long sediment cores showed that SRMs were present at all depths, regardless of site studied and sampling date, with preferential locations in the first 15 cm of the sediments. Generally, the results for both sites changed with the seasons. We observed a seasonal change in the abundance of the microorganisms' *dsr*AB that was far greater in the Seine estuary than in the Authie bay mudflats, with maximal values shown at the end of spring and start of summer (Figure 5.6). When we focused on the results for individual cores, we noticed in sediments of the Seine estuary a preferential location of the SRM in the first 15 cm and abundance diminished with depth. Sulfate-reducing activity was studied in the sediments and similar changes in the profiles of sulfate-reducing activity were observed (Leloup et al. 2005).

In the Authie bay sediments, the profile of distribution of these microorganisms was sometimes atypical (e.g., May 2003), with variations over some centimeters of depth that may be important, possibly because of the processes of bioturbation (e.g., the presence of the ragworm *Nereis diversicolor* that facilitates the appearance of a suboxic environment unfavourable to the development of SRMs).

SRM measurements obtained after the integration of the data for the first 10 cm of the cores were slightly higher in sediments of the Seine estuary. Leloup et al. (2005) showed that the abundance of SRM increased with the temperature and the concentration of dissolved organic carbon in the Seine sediments, but also depended on sediment erosion and deposit episodes that were responsible for the low numbers of SRMs in the August 2001 core. On the other hand in the Authie bay sediments, the seasonal change in the abundance of the sulfidogenic *dsr*AB community was less marked, probably due to bioturbation.

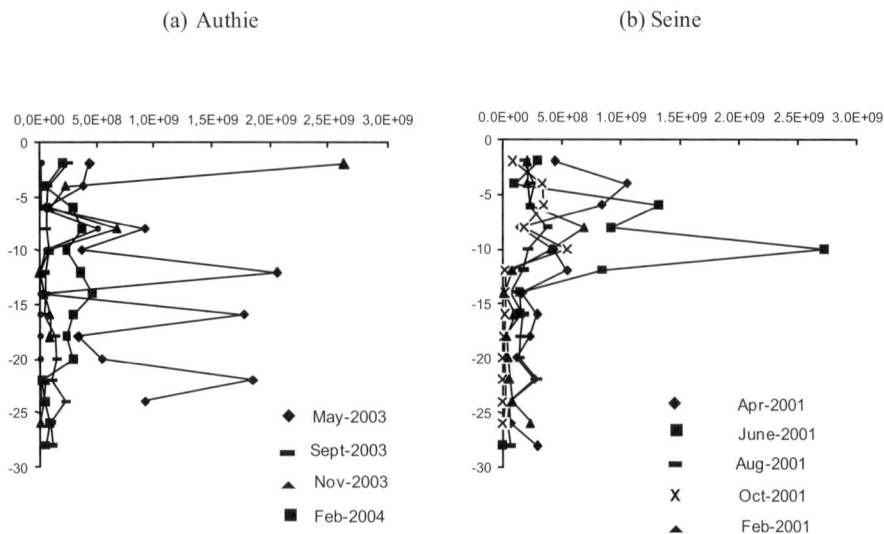

FIGURE 5.6   Vertical distribution of abundance of SRM in the Authie (a) and mixing zone Seine estuary (b) mudflats. The abundance of SRMs (number of microorganisms per gram of sediment, x-axis) was determined along a 30-cm core depth (centimeters, y-axis) collected from the sediments. Values shown as mean ± standard deviation of triplicates.

We verified that the SRMs in the cores were metabolically active by following up the expression of the dsrAB gene by RT-PCR after the extraction of total RNA was realized for both intertidal mudflats. The results show SRMs, even when weakly abundant (in particular at depths between 15 and 30 cm), are always active. We compared the results of quantification of the dsrAB gene with the quantification of freshly precipitated sulfides (AVS) produced by these microorganisms in both intertidal mudflats. The comparison did not yield significant results, particularly in the first centimeters of the sedimentary cores, probably because of the very strong chemical ability of AVS to react. Indeed AVS can be reoxidized as soon as conditions allow (e.g., presence of manganese oxide or weak concentration of O_2).

As to knowledge of the mechanisms and fine tuning of tools, this work based on microbiological and chemical studies has allowed us to better account for the speciation of metal contaminants when evaluating the risks associated with the their presence in estuarine mudflats. The sulfidogenic dsrAB community thus plays a dominant role in the metal contamination of biotopes by reducing the bioavailable metal fraction that is the potentially toxic fraction for the biocenosis.

5.4.2   SRM Phylogenetic Study in the Intertidal North Mudflat of the Seine Estuary

The diversity of the SRMs in the intertidal north mudflat of the Seine estuary (6 cm depth, June 2001) was assessed from the analysis of dsrAB sequences amplified from total DNA extracts of sediments. The abundance of SRMs was 1.3×10^9 SRM

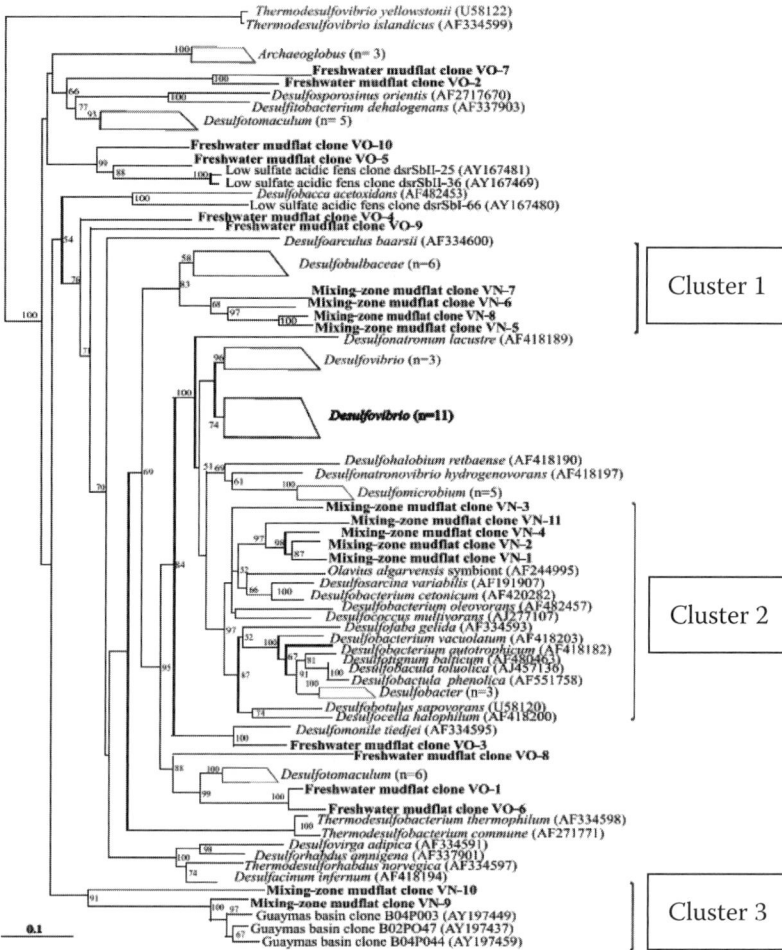

FIGURE 5.7 Phylogenetic relationships of *dsr*AB sequences from the Seine estuary mudflats. After sequencing the full length of one clone from each RFLP group, a distance matrix tree based on *dsr*AB-deduced amino acid sequences (total of 501 amino acids; gaps and ambiguous residues were excluded) was created with the PHYLIP package. Reference sequences from sulfate-reducing microorganisms are shown in italics. Notes indicate bootstrap confidence values (100 resamplings). The scale bar represents 0.1 substitutions per site.

*dsr*AB per gram. The *dsr*AB sequences were cloned on the basis of RFLP analysis. Ten RFLP groups were identified and a characteristic clone of every group was sequenced. Despite some discrepancies between the 16SrRNA and *dsr*AB phylogenies explained by possible lateral gene transfer events, topologies remained congruent for most SRM lineages (Wagner et al. 1998; Klein et al. 2001). A phylogenetic study undertaken from sequences so obtained and *dsr*AB sequences of known sulfidogenic microorganisms showed an important variety of *dsr*AB sequences with three clusters (Figure 5.7).

Most *dsr*AB sequences isolated from the mudflat of the Seine estuary (VN in Figure 5.7) are phylogenetically close to *dsr*AB sequences present in the genomes of microorganisms of the delta proteobacteria class, and more particularly the order of desulfobacteria (Gram-negative bacteria). A comparison with the diversity of *dsr*AB sequences isolated from a fresh water mudflat showed that the conditions of salinity influenced the diversity of the sequences and thus the community of the sulfidogenic microorganisms carrying this gene (Leloup et al. 2006). These results are consistent with the literature; Purdy et al. (2001) showed that *Desulfobacter* represented the dominant genus (50%) in marine sediments of the river Tama in Japan.

5.5 CONCLUSIONS

The results reported above clearly indicate a need to put more *bio* into *biogeochemical* cycles. The present study is a good example of the advancements that may be gained when geochemists and microbiologists work together in co-designed research programs. Different methodologies can involve a quantification of microorganisms whose mode of functioning is relatively well-established and/or biochemical markers such as fatty acids (FAs) that represent the bacterial imprints in the sediments (Billon et al. 2007). Different categories of FAs can be distinguished and it is possible to establish links between their nature and origin (bacteria, diatoms, higher plants) to learn more about the interactions between biota and their environments. Because oxygenation is a major factor governing the fate of microorganisms and chemical species, it may be also important to examine the potential impacts of biota on oxygen penetration into deeper levels of sediments as illustrated by the presence of oxic microniches in sediment profiles (Chapter 15).

REFERENCES

Abbasse, G., B. Ouddane, and J.C. Fischer. 2002. Determination of total and labile fraction of metals in seawater using solid phase extraction and inductively coupled plasma atomic emission spectrometry (ICP-AES). *J. Anal. Atom. Spectrom.* 17: 1354–1358.

Adriano, D.C. 2001. *Trace Elements in Terrestrial Environments: Biogeochemistry, Bioavailability, and Risks of Metals.* London: Springer.

Allen, H.E., G. Fu, and B. Deng. 1993. Analysis of acid-volatile sulfide (AVS) and simultaneously extracted metals (SEM) for the estimation of potential toxicity in aquatic sediments. *Environ. Toxicol. Chem.* 12: 1441–1453.

Ankley, G.T., D.M. Di Toro, and D.J. Hansen. 1996. Technical basis and proposal for deriving sediment quality criteria for metals. *Environ. Toxicol. Chem.* 15: 2056–2066.

Bacon, J.R. and C.M. Davidson. 2008. Is there a future for sequential chemical extraction? *Analyst* 132: 25–46.

Barbanti, A. et al. 1992. Nutrient regeneration processes in bottom sediments in a Po delta lagoon (Italy) and the role of bioturbation in determining the fluxes at the sediment–water interface. *Hydrobiologia* 228: 1–21.

Batley, G.E. and M.S. Giles. 1979. Solvent displacement of sediment interstitial waters before trace metal analysis. *Water Res.* 3: 879–886.

Bettiol, C. et al. 2008. Evaluation of microwave-assisted acid extraction procedures for the determination of metal content and potential bioavailability in sediments. *Appl. Geochem.* 23: 1140–1151.

Billon, G. et al. 2002. Depth variability and some geochemical characteristics of Fe, Mn, Ca, Mg, Sr, S, P, Cd and Zn in anoxic sediments from Authie Bay (Northern France). *Estuar. Coast. Shelf Sci.* 55: 167–181.

Billon, G., B. Ouddane, and A. Boughriet. 2001. Chemical speciation of sulfur compounds in surface sediments from three bays (Fresnaye, Seine and Authie) in northern France and identification of some factors controlling their generation. *Talanta* 53: 971–981.

Billon, G., B. Ouddane, and A. Boughriet. 2002. Artefacts in the speciation of sulfides in anoxic sediments. *Analyst* 126: 1805–1809.

Blain, S., P. Appriou, and H. Handel. 1993. Preconcentration of trace metals from seawater with chelating resin chelamine. *Anal. Chim. Acta* 272: 91–97.

Boudreau, P.B. 1987. A steady-state diagenetic model for dissolved carbonate species and pH in the porewater of oxic and suboxic sediments. *Geochim. Cosmochim. Acta* 51: 1985–1996.

Boudreau, P.B. 1991. Modelling the sulfide–oxygen reaction and associated pH gradient in porewaters. *Geochim. Cosmochim. Acta* 55: 145–159.

Boudreau, P.B. and D.E. Canfield. 1988. A provisional diagenetic model for pH in anoxic porewaters: application to the FOAM Site. *J. Mar. Res.* 46: 429–455.

Brown R.H., D.J. Baker, and W.S. Wilson. 1995. The Utility of AVS/EqP in hazardous waste site evaluations, Proceedings NOAA, Seattle.

Bufflap, S.E. and H.E. Allen. 1995. Sediment porewater collection methods for trace metal analysis: a review. *Water Res.* 29: 165–177.

Canfield, D. 1989. Reactive iron in marine sediments. *Geochim. Cosmochim. Acta* 53: 619–632.

Carignan, R. and D.R.S. Lean. 1991. Regeneration of dissolved substances in a seasonally anoxic lake: the relative importance of processes occurring in the water column and in the sediments. *Limnol. Oceanogr.* 36: 683–707.

Carignan, R. and J.O. Nriagu. 1985. Trace metals deposition and mobility in the sediments of two lakes near Sudbury, Ontario. *Geochim. Cosmochim. Acta* 49: 1753–1764.

Chapman, P.M. et al. 2002. Porewater testing and analysis: the good, the bad, and the ugly. *Mar. Pollut. Bull.* 44: 359–366.

Clarisse, O. 2003. *Approche Géochimique du Fonctionnement et de la Dynamique des Vasières de l'Estuaire de la Seine*. PhD Thesis, Université de Lille, France.

Cooper, D.C. and Morse, J.W. 1998. Extractibility of metal sulfide minerals in acidic solutions: application to environmental studies of trace metal contamination within anoxic sediments. *Environ. Sci. Technol.* 32: 1076–1078.

Davison, W. et al. 2000. Dialysis, DET and DGT: *in situ* diffusional techniques for studying water, sediments and soils. In *In Situ Monitoring of Aquatic Systems: Chemical Analysis and Speciation*, Buffle, J. and Horvai, G., Eds. Chichester: John Wiley & Sons, pp. 495–569.

Davison, W. and H. Zhang. 1994. *In situ* speciation measurements of trace components in natural waters using thin film gels. *Nature* 367: 546-548.

Deflandre, B. et al. 2002. Early diagenetic processes in coastal marine sediments disturbed by a catastrophic sedimentation event. *Geochim. Cosmochim. Acta* 66: 2547–2558.

Di Toro, D.M. et al. 1990. Toxicity of cadmium in sediments: the role of acid volatile sulfide. *Environ. Toxicol. Chem.* 9: 1497–1502.

Di Toro, D.M. et al. 1992. Acid volatile sulfide predicts the acute toxicity of cadmium and nickel in sediments. *Environ. Sci. Technol.* 26: 96–101.

Docekalova H. et al. 2002. Use of constrained DET probe for high-resolution determination of metals and anions distribution in the sediment porewater. *Talanta* 57: 145–155.

Douglas, G.S., G.L. Mills, and J.G. Quinn. 1986. Organic copper and chromium complexes in the interstitial waters of Narragansett Bay sediments. *Mar. Chem.* 19: 161–174.

Emerson, S., R. Jahnke, and D. Heggie. 1984. Sediment–water exchange in shallow water estuarine sediments. *J. Mar. Res.* 42: 709–730.

Fang, T., X. Li, and G. Zhang. 2005. Acid volatile sulfide and simultaneously extracted metals in the sediment cores of the Pearl River Estuary, South China. *Ecotoxicol. Environ. Saf.* 61: 420–431.

Fedotov, P.S. et al. 2006. A hyphenated flow-through analytical system for the study of the mobility and fractionation of trace and major elements in environmental solid samples. *Analyst* 131: 509–515.

Fernex, F.E. et al. 1986. Behavior of some metals in surficial sediments of the northwest Mediterranean continental shelf. In *Sediment and Water Interactions*, Sly, P.G., Ed. Heidelberg: Springer, pp. 353–370.

Förstner, U. 1990. Inorganic sediment chemistry and elemental speciation. In *Sediments: Chemistry and Toxicity of In-Place Pollutants*, Baudo, R. et al., Eds. Chelsea: Lewis Publishers, pp. 61–105.

Fu, G. and H.E. Allen. 1992. Cadmium adsorption by oxic sediment. *Water Res.* 26: 225–233.

Fu, G., H.E. Allen, and Y. Cao. 1992. The importance of humic acids to proton and cadmium binding in sediments. *Environ. Toxicol. Chem.* 11: 1363–1372.

Groschner, M. and P. Appriou. 1994. Three-column system for preconcentration and speciation determination of trace metals in natural waters. *Anal. Chim. Acta* 297: 369–376.

Harper, M.P., W. Davison, and W. Tych. 1997. Temporal, spatial, and resolution constraints for *in situ* sampling devices using diffusional equilibration: dialysis and DET. *Environ. Sci. Technol.* 31: 3110–3119.

Harper, M.P. et al. 1998. Kinetics of metal exchange between solids and solution in sediments and soils interpreted from DGT measured fluxes. *Geochim. Cosmochim. Acta* 62: 2757–2770.

Howarth, R.W., J.W.B. Stewart, and M.V. Ivanov. 1992. *Sulfur Cycling on the Continent: Wetlands, Terrestrial Ecosystems and Associated Water Bodies.* Chichester: John Wiley & Sons.

Hudson, R.J.M. and F.M.M. Morel. 1990. Iron transport in marine phytoplankton: kinetics of cellular and medium coordination reactions. *Limnol. Oceanogr.* 35: 1002–1020.

Klein, M. et al. 2001. Multiple lateral transfers of dissimilatory sulfite reductase genes between major lineages of sulfate-reducing prokaryotes. *J. Bacteriol.* 183: 6028–6035.

Landrum, P.F. et al. 1987. Reduction in bioavailability of organic contaminants to the amphipod *Pontoporeia hoyi* by dissolved organic matter of sediment interstitial waters. *Environ. Toxicol. Chem.* 6: 11–20.

Lee, B.G. et al. 2000. Influence of acid volatile sulfide and metal concentrations on metal bioavailability to marine invertebrates in contaminated sediments. *Environ. Sci. Technol.* 34: 4517–4523.

Leepipatpiboon, V. 1995. Trace enrichment by solid-phase extraction for the analysis of heavy metals in water. *J. Chromatogr.* A 697: 137–143.

Leloup, J. et al. 2005. Dynamics of sulfate-reducing microorganisms (*dsrAB* genes) in two contrasting mudflats of the Seine estuary (France). *Microb. Ecol.* 50: 307–314.

Leloup, J. et al. 2006. Diversity of the *dsrAB* (dissimilatory sulfite reductase) gene sequences retrieved from two contrasting mudflats of the Seine estuary (France). *FEMS Microbiol. Ecol.* 55: 230–238.

Leonard, E.N., G.T. Ankley, and R.A. Hoke. 1996. Evaluation of metals in marine and freshwater surficial sediments from the Environmental Monitoring and Assessment Program relative to proposed sediment quality criteria for metals, *Environ. Toxicol. Chem.* 15: 2221–2232.

Li, F. et al. 2001. Chemical forms of Pb, Zn, and Cu in the sediment profiles of the Pearl River estuary. *Mar. Pollut. Bull.* 42: 215–223.

Lloyd, J.R. and D.R. Lovley. 2001. Microbial detoxification of metals and radionuclides. *Curr. Opin. Biotech.* 12: 248–253.

Lovley, D.R. 1993. Dissimilatory metal reduction. *Annu. Rev. Microbiol.* 47:263–290.

Luoma, S.N. 1983. Bioavailability of trace metals to aquatic organisms: a review. *Sci. Total Environ.* 28: 1–22.

Luoma, S.N. 1989. Can we determine the biological availability of sediment-bound trace elements? *Hydrobiologia* 176–177: 379–396.

Luoma, S.N. 1995. Prediction of metal toxicity in nature from bioassays: limitations and research needs. In *Metal Speciation and Bioavailability in Aquatic Systems*, Tessier, A. and Turner, D.R., Eds. Chichester: John Wiley & Sons, pp. 609–646.

Luther, G.W. et al. 1998. Simultaneous measurement of O_2, Mn, Fe-I and S-II in marine pore waters with a solid state voltammetric electrode. *Limnol. Oceanogr.* 43: 325–333.

Lyons, B.W. and W.F. Fitzgerald. 1983. Trace metals speciation in estuarine anoxic and suboxic porewaters. In *Trace Metals in Seawater*, Wong, C.S. et al., Eds. New York: Plenum Press, pp. 621–641.

Marin, B. and P. Giresse. 2001. Particulate manganese and iron in recent sediments of the Gulf of Lions continental margin (northwestern Mediterranean Sea): deposition and diagenetic process. *Mar. Geol.* 172: 147–165.

Martin, J.M., P. Nirel, and A.J. Thomas. 1987. Sequential extraction techniques: promises and problems. *Mar. Chem.* 22: 313–341.

Matsunaga, T. et al. 1993. Redox chemistry of iron and manganese minerals in river-recharged aquifers: a model interpretation of a column experiment. *Geochim. Cosmochim. Acta* 57: 1691–1704.

Millero, F J. 1991. The oxidation of H_2S in Chesapeake Bay. *Coast. Shelf Sci.* 33: 521–527.

Mills, G.L., E. McFadden, and J.G. Quinn. 1987. Chromatographic studies of dissolved organic matter and copper organic complexes isolated from estuarine waters. *Mar. Chem.* 20: 313–325.

Miramand, P. et al. 2001. Contamination of the biological compartment in the Seine estuary by Cd, Cu, Pb, and Zn. *Estuaries* 24: 1056–1065.

Morel, F.M.M. 1983. *Principles of Aquatic Chemistry*. New York: John Wiley & Sons.

Morrison, G.M.P. 1989. Trace element speciation and its relationship to bioavailability and toxicity in natural waters. In *Trace Element Speciation: Analytical Methods and Problems*, Batley, G.E., Ed. Boca Raton: CRC Press, pp. 25–41.

Morse, J.W. and T. Arakaki. 1993. Adsorption and coprecipitation of divalent metals with mackinawite (FeS). *Geochim. Cosmochim. Acta* 57: 3635–3640.

Morse, J.W. et al. 1987. The chemistry of the hydrogen sulfide and iron sulfide systems in natural waters. *Earth Sci. Rev.* 24: 1–42.

O'Day, P.A. et al. 2000. Metal speciation and bioavailability in contaminated estuary sediments, Alameda Naval Air Station, California. *Environ. Sci. Technol.* 34: 3665–3673.

Ouddane, B. et al. 2004. Determination of metal partitioning in porewater extracted from the Seine River Estuary sediment (France). *J. Environ. Monit.* 6: 243–253.

Ouddane, B. et al. 2001. The post-depositional reactivity of iron and manganese in the sediments of a macrotidal estuarine system. *Estuaries*, 24: 1015–1028.

Prica, M. et al. 2008. A comparison of sediment quality results with acid volatile sulfide (AVS) and simultaneously extracted metals (SEM) ratio in Vojvodina (Serbia) sediments. *Sci. Total Environ.* 389: 235–244.

Purdy, K.J. et al. 2001. Use of 16S rRNA-targeted oligonucleotide probes to investigate the distribution of sulfate-reducing bacteria in estuarine sediments. *FEMS Microbiol. Ecol.* 36: 165–168.

Quevauvillier, P. et al. 1997. Certification of trace metal extractable contents in a sediment reference material (CRM 601) following a three-step sequential extraction procedure. *Sci. Total Environ.* 205: 223–234.

Rickard, D., M.A.A. Schoonen, and G.W. Luther. 1995. Chemistry of iron sulfides in sedimentary environments. In *Geochemical Transformation of Sedimentary Sulfur*, Vairavamurthy, M.A. and Schoonen, M.A.A. Eds., ACS Symposium, Series 612. Washington: American Chemical Society, pp. 168–193.

Roychoudhury, A.N. 2006. Time-dependent calibration of a sediment extraction scheme. *Mar. Pollut. Bull.* 52: 397–403.

Sakata, M. 1985. Diagenetic of manganese, iron, copper and lead in anoxic sediment of a freshwater pond. *Water Res.* 19: 1033–1038.

Salomons, W. and U. Förstner. 1980. Trace metals analysis on polluted sediment. II: Evaluation of environmental impact. *Environ. Technol. Lett.* 1: 506–517.

Salomons, W. and U. Förstner. 1984. *Metals in the Hydrocycle*. Berlin: Springer.

Salomons, W. et al. 1987. Sediments as a source for contaminants? *Hydrobiologia* 149: 13–30.

Saulnier, I. and A. Mucci. 2000. Trace metal remobilisation following the resuspension of estuarine sediments: Saguenay Fjord, Canada. *Appl. Geochem.* 15: 203–222.

Shaw, T.J., J.M. Gieskes, and R.A. Jahnke. 1990. Early diagenesis in differing depositional environments: the response of transition metals in porewater. *Geochim. Cosmochim. Acta* 54: 1233–1246.

Shuttleworth, S.M., W. Davison, and J. Hamilton-Taylor. 1999. Two-dimensional and fine structure in the concentrations of iron and manganese in sediments pore-waters. *Environ. Sci. Technol.* 33: 4169–4175.

Simpson, S.L., S.C. Apte, and G.E. Batley. 1998. Effect of short-term resuspension events on trace metal speciation in polluted anoxic sediments. *Environ. Sci. Technol.* 32: 620–625.

Sutherland, R.A. et al. 2004. Metal extraction from road-deposited sediments using nine partial decomposition procedures. *Appl. Geochem.* 19: 947–955.

Taillefert, M., A.B. Bono, and G.W. Luther. 2000. Reactivity of freshly formed Fe(III) in synthetic solutions and (pore) waters: voltammetric evidence of an aging process. *Environ. Sci. Technol.* 34: 2169–2177.

Tercier, M.L. and J. Buffle. 1993. *In situ* voltammetric measurements in natural waters: future prospects and challenges. *Electroanalysis* 5: 187–200.

Tessier, A., P.G.C. Campbell, and M. Bisson. 1979. Sequential extraction procedure for the speciation of particulate trace elements. *Anal. Chem.* 51: 844–851.

Twiss, M.R. and J.W. Moffet. 2002. A comparison of copper speciation in coastal marine waters estimated using analytical voltammetry and diffusion gradient in thin film (DGT) techniques. *Environ. Sci. Technol.* 36, 1061–1068.

Ure, A. and Davidson C. 2002. *Chemical Speciation in the Environment*, 2nd ed. New York: John Wiley & Sons.

USEPA. 1995. http://www.epa.gov/sab/pdf/epe95020.pdf

USEPA. 2004. The incidence and severity of sediment contamination in surface waters of the United States (National Sediment Quality Survey, 2nd Ed., EPA 823-R-04-007. Washington: U.S. Environmental Protection Agency.

Van den Berg, C.M.G. and E.P. Achterberg. 1994. Automated in line sampling and analysis of trace elements in surface waters with voltammetric detection. *Trends Anal. Chem.* 13: 348–352.

Wagner, M. et al. 1998. Phylogeny of dissimilatory reductases supports an early origin of sulfate respiration. *J. Bacteriol.* 180: 2975–2982.

Zhang, H. and W. Davison. 1995. Performance characteristics of diffusion gradients in thin films for the *in situ* measurement of trace metals in aqueous solution. *Anal. Chem.* 67: 3391–3400.

Zhang, H. et al. 1998. *In situ* measurements of solution concentrations and fluxes of trace metals in soils using DGT. *Environ. Sci. Technol.* 32: 704–710.

Zhang, H. and W. Davison. 1999. Diffusional characteristics of hydrogels used in DGT and DET techniques. *Anal. Chim. Acta* 398: 329–340.

Zhang, H. and W. Davison. 2000. Direct *in situ* measurements of labile inorganic and organically bound metal species in synthetic solutions and natural water using diffusive gradient in thin films. *Anal. Chem.* 72: 4447–4457.

Zhang, H. et al. 2002. Localised remobilisation of metals in a marine sediment. *Sci. Total Environ.* 296: 175–187.

6 Organic Contaminants in Coastal and Estuarine Food Webs

Alain Abarnou

CONTENTS

6.1 INTRODUCTION

High levels of chemical compounds in organisms have been observed for many years and warrant serious concerns about the environment and human health. The first findings of potentially toxic substances in wildlife tissues that led to ecological awareness on the part of the general public occurred in the 1960s. *Silent Spring*, Rachel Carson's famous book published in 1962, is often cited as a milestone leading to the relatively recent consciousness of the degradation of the environment as a result of releases of multitudes of chemicals into surface waters and the atmosphere.

In some cases, the fear of exposure to harmful substances via food consumption greatly increased the amount of available information about chemical contaminants in biota. One well known example was the tragic Minamata disease caused by mercury present in chemical plant wastes that contaminated the surrounding bay and its living resources. High levels of exposure to methyl mercury through consumption of fish resulted in severe injuries and 1784 deaths (Anon. National Institute for Minamata Disease). More recently, the presence of dioxins and other persistent organic pollutants (POPs) in farmed salmon (Hites et al. 2004) received extensive media coverage and led to distrust of the fish and seafood industries and fear of poisoning from fish and seafood consumption.

Progress in environmental analysis has increased the production of data concerning the presence of contaminants in biota. More sensitive and specific analytical techniques now allow measurement of trace contaminants at very low levels and created a network of accurate information about the distribution and fate of organic compounds in the environment, and more particularly in organisms.

At present, our entire environment is contaminated by a multitude of chemicals of various natures and origins that potentially represent hazards to wildlife and humans. Appropriate methodologies now allow the determination in environmental samples of many compounds belonging to the same chemical family, such as PCBs (polychlorinated biphenyls), PAHs (polyaromatic hydrocarbons) or PBDEs (polybrominated diphenyl ethers). Within these chemical groups, the relative distribution of each individual compound is largely dependent on chemical and biological processes acting on its bioaccumulation or conversely on its biotransformation. Thus detailed examinations of these contaminant fingerprints in biota allow us to trace the history of chemicals in the environment and their progress along food chains to higher animals.

Early studies of contaminants in coastal marine environments dealt with the identification and quantification of contaminants in biological tissues without necessarily looking to their origins to the processes leading to their bioaccumulation. In this context, aquatic organisms have been used as natural "bioextractors" for determining very low levels of compounds present in the water column. These measurements serve as the bases for the implementation of networks of pollution monitoring programs such as "Mussel Watch." Sentinel species selected for these programs must satisfy various criteria (Phillips 1990) and sound interpretation of data obtained requires the ability to differentiate among the factors acting on bioaccumulation, particularly biology-dependent factors considered normal and those that have been modified by anthropogenic influences.

In the case of PCBs and many other man-made contaminants, one of the major results of pollution monitoring has been the identification of estuaries as "hot spots." One hot spot consists of the Seine estuary and the eastern part of the Bay of Seine, both of which are heavily contaminated by PCBs and many other chemical substances. Much relevant information has been obtained within the framework of the Research Programme Seine Aval (Anon 2001)—mainly about PCBs considered typical contaminants that biomagnify, and, to a lesser extent, about other organic micropollutants such as PAHs, polychlorodibenzoparadioxins (PCDDs) and polychlorodibenzofurans (PCDFs), and PBDEs (polybromodiphenyl ethers).

The objective of this chapter is to review the main results obtained to date about organic compounds in estuarine food chains and demonstrate how these studies have contributed to the still-growing knowledge base about the process that act on the distribution and fate of these compounds in biota and affecting other organisms further along the food chain such as human consumers. After defining common terms, the discussion will cover the chemical and biological factors affecting the transfer of organic compounds along food chains. Special attention will be given to PCBs and PAHs as typical examples of widespread contaminants that behave oppositely in food webs: they are bioaccumulated or biotransformed and eliminated. More recently studied contaminants such as PBDEs and dioxins in coastal and estuarine environments will also be discussed; their distributions and behaviours in the estuarine food chain will be compared to those of the more typical contaminants.

6.2 PROCESSES LEADING TO BIOACCUMULATION

6.2.1 Basic Definitions

Regardless of the sources of contamination and the detailed mechanisms involved, bioaccumulation in any organism occurs when chemical input exceeds output. Very simply, the accumulation of any compound in biota may be expressed by a ratio between its concentration in the organism and its concentration in the external medium (water, sediment, food). Common concepts have been defined and discussed elsewhere (van der Oest et al. 2003).

For aquatic species, an early idea was to include concentration in the surrounding water in the ratio, leading to the concept of the bioconcentration factor (BCF):

$$BCF = C_B/C_W = \text{concentration in biota/concentration in water} \qquad (6.1)$$

Bioconcentration is the accumulation in an organism of a chemical from water; in all cases, water is the only source of contamination by adsorption in the case of smaller species such as phytoplanktonic cells, and/or by direct uptake of compounds through external surfaces such as the respiratory surfaces of larger organisms.

Conversely, *bioaccumulation* involves consideration of other sources of contamination such as food. The dietary uptake of contaminants supposes that bioconcentration occurred at an earlier stage. In a very general sense, bioaccumulation is also considered to increase with time of exposure of an individual organism.

Biomagnification and *ecological magnification* are recent terms that refer to an increase of contamination from organisms at a low trophic level in a food chain to higher predators. This last process—an enrichment of contaminants along a food chain—is a reiteration of similar processes that depend on the same chemical and biological factors as bioaccumulation. PCBs are typical contaminants that bioaccumulate and biomagnify in food webs.

6.2.2 Bioconcentration: Importance of Chemical Characteristics

Bioconcentration is the basic process leading to bioaccumulation: the entry of contaminants into biological cycles occurs by adsorption and/or absorption of compounds that are initially dissolved in the water column and possess high affinities for solid particles or lipid-rich tissues; this ability is characteristic of hydrophobic compounds that tend to leave the dissolved phase.

Bioconcentration may be followed by measuring concentrations of contaminants in aquatic organisms during exposure experiments. When placed in aquaria filled with contaminated water at a known concentration, fish and other aquatic organisms become contaminated; concentrations in tissues increase until they reach an equilibrium at which they remain fairly constant.

The bioconcentration factor (Equation 6.1) corresponds to this equilibrium concentration in biota divided by the concentration in water. In many cases, depending on the characteristics of the compounds, the equilibrium concentration in biota is reached only after lengthy experiments that may lead to difficulties in determining accurate concentrations. This phenomenon is known as the steady state BCF.

According to several authors, for instance Davies and Dobbs (1984), it is far easier to choose a kinetic approach to evaluate bioconcentration factors. Indeed, if concentrations in biota (C_B) and in the water (C_W) are tracked during the duration of an experiment, the C_B may be described by a simple first-order kinetic law that describes the uptake of contaminants from water and their elimination or clearance:

$$dC_B/dt = K_1C_W - K_2C_B \qquad (6.2)$$

When equilibrium is reached, the concentration in biota remains constant and accordingly, Equation (6.2) becomes:

$$K_1/K_2 = C_B/C_W = BCF \qquad (6.3)$$

The bioconcentration factor is the ratio of rate constants K_1 (uptake) and K_2 (elimination). The constants can be determined at the early stage of accumulation and elimination in short-term experiments.

Correlations between BCF and physico-chemical characteristics were found a long time ago. Among those relationships, the octanol–water partition coefficient (K_{ow}) and the aqueous solubility (Sol.) are useful factors for describing the bioconcentration potentials of compounds. Ernst (1980) provided an early example by establishing correlations of the aqueous solubilities of several organo-chlorinated compounds with their bioconcentrations by various mollusc species.

Veith et al. (1979), Bysshe (1982), Davies and Dobbs (1984), Spacie and Hamelink (1984), and Connell (1992) reviewed the correlations obtained between bioconcentrations factors of different organo-chloro pesticides in different fresh water species and the octanol–water partition coefficient (K_{ow}) in the following general form:

$$\log BCF = a \log(K_{ow}) + b \qquad (6.4)$$

The best correlations are obtained with persistent compounds, mainly halogenated aromatic hydrocarbons, and species at low trophic level. The constant a is generally close to unity, whereas the constant b corresponds to the lipid content in the considered tissue (expressed in percent of the weight).

Similar equilibrium equations have been used to describe other exchange processes at the interfaces that act on the distribution of hydrophobic substances in the various compartments of the environment. For example, the distribution of compounds, between water and sediment or suspended particular matter, is defined by the adsorption coefficient K_D ($K_D = C_{Part}/C_W$) that depends on the chemical substance, its octanol–water partition coefficient, and solid particle characteristics such as granulometry and organic carbon content.

These examples emphasize the importance of the octanol–water partition coefficient that appears to be a key characteristic parameter to explain and predict bioconcentration. K_{OW}s serve as the basis of the concept of fugacity and its use in environmental multimedia models (Mackay and Paterson 1981). These models are useful for assessing the distribution of compounds within the different compartments of the environment: water column, biota, air, sediment, and soil and gaining information about the distribution of chemical substances at equilibrium.

The bioaccumulation of organic compounds depends primarily on the natures of the compounds. Among the main characteristics of aquatic organisms that affect bioaccumulation, we must consider their presence in the surrounding waters, their bioavailability, their hydrophobicity or affinity for lipid-rich tissues, their persistence, and their reduced capacities to be transformed metabolically.

Connell (1992) summarized the main chemical factors governing bioaccumulation (Table 6.1). These specifications determine the fates of substances in the abiotic environment and consequently their availability and capacity to be bioaccumulated. Molecular weights and dimensions are directly related to chemical structures. They

TABLE 6.1
General Physico-Chemical Characteristics of Organic Compounds Exhibiting Bioaccumulation (Modified after Connell 1992)

Characteristic	Features Indicating Bioaccumulation
Chemical structure	High capacity: high proportion of C–C (aliphatic), C–H, and C–halogen bonds; limited capacity; low proportion of bonds above with presence of variety of functional groups
Molecular weight	>100 Da, giving maximum capacity at ~350 Da, then declining to very low capacity at ~600 Da
Molecular dimensions	Cross-section width <9.5Å, molecular surface area 20 to 460Å^2, molecular volume 260 to 760Å^3
Stability	Resistance to degradation reflected in soil persistence in order of years
Log K_{OW}	>2, giving maximum capacity ~4 to 6 and decline to very low capacity ~10 to 12
Water solubility	<18, giving a maximum at ~0.002 mole.dm^{-3} with declining capacity of lower values
Degree of ionization	Very low

determine the basic physico-chemical properties and hence the fates of compounds in the environment. Their octanol–water partition coefficients or (inversely) water solubilities and vapour pressures are useful parameters for describing fates.

Obviously, bioconcentration and subsequent bioaccumulation may occur when compounds are present in the water column in a way that favours their uptake by aquatic organisms. Generally, fish and most other aquatic species take up contaminants from water when chemicals are present in the dissolved phase. This means that bioconcentration occurs if the compounds are present in their dissolved forms for sufficient time. Of course, volatile compounds that are rapidly removed from the water column by evaporation fall into the category of contaminants that are not bioconcentrated.

At the other end of the volatility range are the compounds that possess heavier molecular masses, larger dimensions, and apolar characters. These very stable compounds are nevertheless not bioconcentrated due to their stronger adsorption to solids. Under environmental conditions, these compounds that have very high octanol–water partition coefficients are readily adsorbed onto suspended particulate materials and consequently escape the dissolved phase. The adsorption to solid particles is also favoured by the high organic contents of these particles that thus play against bioconcentration and bioaccumulation by (other) biota. The addition of organic-rich or carbon-rich material should reduce the transfer of contaminants from contaminated sediments to bottom fauna. Special attention is warranted by the adsorption of contaminants onto phytoplanktonic cells and other minute species that limits the presence of compounds in the dissolved phase. This process affects the entry of chemical substances into biota and will be discussed later related to biological aspects of bioaccumulation.

The sizes and stereochemistries of molecules are other factors that may limit bioconcentration. In a group of related compounds, an increase of the number of substitutants on a molecule also means a higher K_{ow} and consequently a higher bioconcentration potential. However, in spite of higher octanol–water partition coefficients, such compounds are less bioconcentrated than expected, which may be due to the size of the molecules and their steric hindrance that limit their mobility across biological membranes and their transport within organisms.

The presence of chemical substances in the water compartment is also related to their resistance to degradation processes. The more chemically unreactive a molecule is, the longer it will stay in solution and be more easily bioconcentrated. On the other hand, it is generally accepted that dissolved compounds are more easily degraded by chemical or biological processes. Again, lower solubility favours bioaccumulation by preventing rapid degradation. This means that the simplest molecules that lack reactive functional groups are potentially bioconcentrated compounds. The hydrocarbons and the organo-halogenated hydrocarbons are the most typical examples of bioconcentrated compounds that may subsequently be bioaccumulated.

The presence of a polar group in a molecule limits bioconcentration. Polar groups enhance the degradability of a compound. Hence their presence on a molecule greatly reduces its affinity for lipid-rich tissues. For similar reasons, ionisable compounds exhibit little or no bioconcentration. For these compounds, the equilibrium between dissociated and undissociated forms decreases the possibility of bioconcentration.

In summary, bioconcentration and subsequent bioaccumulation depend on several chemical characteristics of compounds. A primary condition for bioconcentration is the proper form of the contaminant, usually the dissolved phase, for a sufficient period of time. Accordingly, bioconcentration and bioaccumulation of volatile compounds and very hydrophobic compounds that show stronger affinities for solid material are expected to be limited. Characteristics related to the structure of a molecule such as its size or the presence of functional groups are also believed to exert effects on bioconcentration.

Concerning bioaccumulation, the concentrations in biota at equilibrium are calculated from the concentrations in water and from the known or estimated octanol–water partition coefficients using the more appropriate evaluation of BCFs from empirical relationships. This thermodynamic approach does not consider the rates of these passive exchanges or the importance of the main biological functions that for all animals act on the extents and kinetics of these exchanges.

6.2.3 BIOCONCENTRATION AND BIOACCUMULATION: IMPORTANCE OF BIOLOGICAL FUNCTIONS

Bioaccumulation arising from the consumption of both food and water means that bioconcentration occurred earlier at a lower trophic level from contaminated food and water. Biological factors are of great concern related to bioaccumulation. Obviously the rate and the extent of bioaccumulation represent dynamic processes that depend on physico-chemical characteristics discussed above for bioconcentration and also on the rates of exchanges and the efficiency of biological functions that contribute to the enrichment of contaminants in biota. Many common observations, such as the presence of contaminants in biota, the increase of concentrations with age, and increased concentrations up the food chain implicitly establish a link between the accumulation of compounds and biological processes. Table 6.2 summarizes the main biological processes acting on bioaccumulation to increase or decrease concentrations within organisms. Unlike chemical factors, these processes are very specific to each type of organism; they depend on the trophic levels of the organisms and are more or less subjected to variation according to ecological and environmental conditions like temperature that can influence food requirements and the availability of prey.

6.2.3.1 Uptake of Contaminants: Bioconcentration and Adsorption

The uptake of contaminants from water (bioconcentration) represents the first and most important step that determines chemical behaviours in biota. The entrance of contaminants into the biological cycle is governed by the adsorption of contaminants onto solid inert particles or living materials. This partition process is passive, rapid, and surface-dependent and becomes very important for smaller particles and organisms that serve as foods of higher trophic level organisms. Several authors (Brown et al. 1982; Harding 1986) reviewed the dynamics of organic contaminants in planktonic species. The distribution of contaminants between the water column and the phytoplanktonic cells or even the smallest zooplanktonic species may be related to

TABLE 6.2
Main Biological Processes Acting on Bioaccumulation

Factor	Organ or Process	Measurement Parameter
Factors Increasing Bioaccumulation		
Water uptake	Adjust with direct absorption: passive process	
	Diffusion: more important for small species	
	Adsorption: more important for low tropic level species	
	Respiration	Active process: respiration rate
	O₂ consumption: assimilation coefficient	
Dietary uptake	Food: increasing importance for higher consumers	Feeding rate, diet composition
	Sediment: increasing importance for more hydrophobic compounds	Assimilation coefficient, lipid content, organic carbon content
Factors Decreasing Bioaccumulation		
Excretion	Adjust via faeces: faster for relatively soluble compounds	Excretion rate
	Via urine	Assimilation coefficient
Biotransformation	Persistent metabolites (e.g., DDE) may be bioaccumulated	
	Liver: presence of specified enzymes	Metabolic rate, metabolites, assimilation coefficient
Growth	Redistribution, dilution within body	Allometric relationships, lipid content
Reproduction (source of contamination for next generation; important for mammals	Gonads: maturation and spawning; transfer *in utero* and via milk	GSI/RHI, seasonal variation, reproduction status

Source: After Connell 1992.

the octanol–water partition coefficient and the lipid content of the material. Brown et al. (1982) proposed such an experimental relation:

$$\log[PCB_{phyto}] = \log[PCB_{water}] + 0.46 \log K_{OW} + \log[lip] + 0.714 \qquad (6.5)$$

that allows the estimation of the PCB concentration in planktonic cells from their concentration in water and the lipid content of the plankton. This relation is very similar to those used to calculate contaminant concentrations in suspended particulate matter. It seems very important to point out the rapidity of this partitioning process compared to most of the other biological processes leading to bioaccumulation.

Respiration — For aquatic animal species, the uptake of contaminants occurs primarily during respiration. Large amounts of water pass across gills where exchanges of oxygen and contaminants take place. Several studies emphasized the role of respiration and consequently the importance of water as a major source of contamination. Bioconcentration is a rapid process that does not seem to be influenced by the structural characteristics of compounds. In bioaccumulation models, the uptake of contaminants from water is related to the respiration rate of the animal and an assimilation coefficient. Temperature and dissolved oxygen that characterize the surrounding waters are environmental parameters that act indirectly on the extent of bioaccumulation.

Dietary uptake — The ingestion of contaminated food contributes to bioaccumulation. Whatever the type of food (living prey, suspended particulate material [SPM], or sediment), it is assumed that contamination has occurred by bioconcentration at a previous step according to the processes described above. Bivalve molluscs and other filter feeding organisms absorb chemicals by ingestion of SPM that carries contaminants. Sediment represents an important source of contamination for endobenthic species. For these organisms, uptake occurs from the interstitial water or from ingested particulate material. The PCB distribution in various benthic invertebrates shows that feeding behaviour has a strong influence on the distribution of PCBs in these species. In filter-feeding molluscs, the low-chlorinated components are present at slightly higher concentrations than in other species, whereas in worms and crustaceans that feed on detritus, highly chlorinated compounds become more frequent.

The relative importance of both sources of contamination, water and food, is still frequently discussed. Bioconcentration remains very important in organisms from lower trophic levels; its influence on bioaccumulation decreases toward upper trophic levels. The more hydrophobic a compound, the more predominant dietary uptake becomes (over water) as a source of contamination.

6.2.3.2 Processes Limiting Bioaccumulation

After ingestion into the gastrointestinal tract, accumulated contaminants are subjected to several processes that contribute to decreasing accumulated concentrations. They may be eliminated directly, with or without biotransformation, redistributed within the body, remobilized during reproduction, and eventually transferred to new generations.

6.2.3.2.1 Elimination and Biotransformation

Direct elimination may occur without modification of the structures of compounds. This direct excretion of foreign compounds occurs via faeces and urine. On the basis of bioconcentration experiments, the excretion of contaminants is included in a clearance process that is slower than uptake. In the case of PCBs, the depuration rate is generally slower than the uptake rate and depends on the structure of the molecule; low-chlorinated soluble compounds are eliminated more rapidly. In bioaccumulation models, the term related to elimination is the *assimilation coefficient* that expresses the portion of the contamination available in the prey that remains in the organism.

Biotransformation contributes to the reduction of contamination. It constitutes part of the detoxifying mechanism that takes place in the liver or equivalent organ in which parent contaminants are transformed via enzymatically mediated reactions into more polar metabolites that are more easily excreted. However, some metabolites are found in the tissues of organisms and are then subjected to bioaccumulation, for example, hydroxylated and methyl-sulfonated metabolites of PCBs. DDE is another well known example of a bioaccumulated compound. Note that the bioconcentration concept does not apply to these compounds because in theory they are absent from the water compartment. The identification of these metabolites and the kinetics of their formation represent difficult tasks. In many cases, a single parent compound may lead to several derivatives at very much lower concentrations in biological tissues. Several derivatives of PBDEs fall into this category of bioaccumulated metabolites.

In the case of PCBs, the structural requirements for these biotransformation reactions were covered briefly above. The biotransformation reactions depend on the capacities of organisms to metabolize contaminants. Generally, vertebrates are more efficient than invertebrates. Observations of the PCB accumulation patterns in organisms from various trophic levels have revealed transformation capacities even in invertebrates at low trophic levels. This capacity—which is minimal or absent in molluscs—increases in organisms like fish and decapod crustaceans and predators like sea birds and marine mammals.

Biotransformation reactions are controlled by structural conditions like the relative positions of the substitutions in a molecule. Stereochemistry is an important feature that may affect enzymatic processes, leading to formation of derivatives, and thus acting directly on bioaccumulation, biomagnifications, and biological activities. The coplanarity of a molecule seems to be an important factor. Studies have pointed out the optically selective character of biotransformation reactions of certain contaminants like α-hexachlorocyclohexane (α-HCH) and a few optically active PCB congeners (Hühnerfuss et al. 1993; Hühnerfuss et al. 1995). These biotransformation reactions differ from most chemical degradation reactions that, in contrast, are not selective. Consequently a different enantiomer enrichment may occur in organisms and in the food web and this chemical signal may be used to decipher a possible effect of trophic relation on the distribution of contaminants in organisms.

6.2.3.2.2   Growth and Tissue Distribution

Parent non-metabolized contaminants and persistent metabolites are redistributed throughout the body. The more lipid-rich biological tissues are, the more they are contaminated. The contaminants are stored in fatty reserves and may be remobilized according to energy requirements.

Growth is a function that obviously affects the concentration of contaminants within organisms. The variations of contamination that occur with age seem to be similar to variations on the growth curve, with an apparently steady state reached at adulthood. In bioaccumulation models, growth is usually considered a dilution of the same amount of contaminant in a larger body.

6.2.3.2.3   Reproduction

For all organisms, reproductive function has a very marked effect on the distribution of contaminants. This process, well known in mammals, has also been observed in other vertebrates and invertebrates. In fish, sexual maturity creates new energy requirements that are satisfied by the remobilization of fatty reserves. Lipids and the associated hydrophobic contaminants are used for the formation of eggs. During spawning, a substantial portion of accumulated contaminants may be rapidly eliminated from the body and delivered into millions of eggs.

For marine mammals, contaminants are stored mainly in blubber and transferred to the foetus, then after birth, to the young animal that feeds on its mother's milk. Obviously, while reproduction, including gestation and particularly lactation, is an important factor that reduces contamination in the mother's body, the same process also contributes to the contamination of the offspring until it can feed independently.

6.2.4 DYNAMIC APPROACH TO BIOACCUMULATION

The above discussion of the various chemical and biological factors that act together to produce bioconcentration and bioaccumulation may provide a complex picture of the situation. Several bioaccumulation models have been proposed in attempts to produce a simple description of the interactions of these various counteracting processes. In such models, for instance those of Norstrom et al. (1976), Thomann and Connolly (1984), Connolly (1991) and Loizeau et al. (2001a,b), the basic equation of bioaccumulation establishes a mass balance of contaminants within an organism: the inputs from water and from food equal their output due to the elimination, transformation, and redistribution of compounds within the body. This equation can be written as follows:

$$dC_i/dt = R_i \cdot a_{iw} \cdot C_w + \Sigma_{ij} \cdot F_i \cdot P_{ij} \cdot a_{ij} \cdot C_j - (E_i + G_i + Rep.i + Metab.i) C_i \quad (6.6)$$

where C_w, C_i, and C_j are the concentrations of the bioaccumulated contaminant in water, in the organism and in its prey, respectively; R_i, F_i, E_i, G_i, $Rep.i$, and $Metab.i$ are rates of respiration, feeding, elimination, growth, reproduction, and metabolisation, respectively; P_{ij} represents the contributions of various prey j to the diet of i; a_{iw} and a_{ij} are the uptake efficiency terms for the component by the organism i when it is absorbed from the water or from the prey j.

This equation indicates that the inputs of contaminants occur from water, probably during respiration, and from feeding on contaminated prey. Several processes contribute to decreases of contaminant concentrations in biota: growth (which may be considered a dilution), excretion of compounds, their biotransformation, their redistribution within the organism, and their transfer to the next generation via reproduction.

Based on Equation 6.6, the dietary uptake of contaminants may become very complicated when several types of prey from different trophic levels are considered; that is the case for biomagnification. The importance of biomagnification has been discussed elsewhere (Gray 2002). Only a few groups of contaminants are biomagnified in food webs which means that their concentration in biological tissue increases from one trophic level to the next upper one. Biomagnification occurs when the ratio of lipid-normalized concentrations of contaminants in predator to lipid-normalized concentrations of contaminants in prey exceeds unity.

If only one source of food is considered (one source is sufficient for discussing the biological factors acting on bioaccumulation), Equation 6.6 becomes simpler:

$$dCi/dt = Ri \cdot aiw \cdot C_w + Fi \cdot aiF \cdot CF - (Ei + Gi + Rep.i + Metab.i) \, Ci \quad (6.7)$$

where F represents the feeding rate without specifying the type of food.

Equations 6.6 and 6.7, which are very similar to Equation 6.2, clearly show that the efficiencies of the processes are determined by physiological functions such as respiration rate, feeding rate, biotransformation capability, and reproduction. If, in Equations 6.6 and 6.7, the inputs of contaminants are restricted only to the water, Equation 6.8 looks like Equation 6.2 describing bioconcentration:

$$dCi/dt = Ri \cdot aiw \cdot Cw - (Ei + Gi) \, Ci \quad (6.8)$$

Under equilibrium conditions, the concentration in the organism remains constant. Based on a comparison of Equations 6.8 and 6.2, uptake is related to respiration only, whereas the elimination rate depends on several processes including excretion, biotransformation, and reproduction.

6.3 POTENTIALLY BIOACCUMULATED COMPOUNDS

Most of the commonly studied compounds possess octanol–water partition coefficients in the range 104 to 109, demonstrating their capacities to be bioconcentrated and, depending on their persistency, to be bioaccumulated and biomagnified in food webs. The maximum bioaccumulation potentials were observed for compounds with log K_{ow} values between 5.5 and 7.5. Octanol–water partition coefficients (log K_{ow}) published for PCBs (Rapaport and Eisenreich 1984; Hansen et al. 1999), for PAHs (Güsten et al. 1991), for PCDDs and PCDFs (Govers and Krop 1998), and for PBDEs (Braekevelt et al. 2003) show that these organic contaminants are prone to be bioconcentrated and bioaccumulated. K_{ow} determination is not always easy and some values may be uncertain: a systematic literature review would report an uncertainty of about 0.5 log unit on log K_{ow} which would indicate a need for careful use of these coefficients and on the extensive use of empirical correlations of BCF and K_{ow}.

6.3.1 PCBs

Polychlorinated biphenyls are generally considered as the best examples of contaminants that bioaccumulate. They are distributed worldwide in all compartments of the environment. Despite their presence at very low levels in oceanic waters, typically at the pg.dm^{-3} level for individual components (Schulz et al. 1988), the highest concentrations were determined in top predators. Concentrations of 1 to 100 mg.kg^{-1} were determined in the fatty tissues of marine mammals (Boon et al. 1997), corresponding to an overall bioaccumulation factor of about 10^8.

Much information exists on PCB contamination in biota. The interest in assessing PCBs in bioaccumulation studies relates to their membership of a group of 209 congeners with varying physicochemical properties, depending on the numbers and positions of the chlorine atoms in the molecules. These features act directly on the octanol–water partition coefficient of a compound and consequently on its bioavailability, its capacity to be bioconcentrated and biomagnified, and its biological effects on higher consumers.

Muir and Sverko (2006) reviewed analyses of PCBs in environmental samples and in biota. The current analytical methodology, based on appropriate extraction and clean-up procedures followed by low resolution gas chromatography and electron capture detection (GC-ECD) or gas chromatography and low resolution mass spectrometry (GC-MS) determination, allows the selective measurement of single congeners at very low levels in biological tissues, typically in the 0.1 to 1 μg.kg^{-1} range, at least in the case of indicator PCBs. At this point, the differences between indicator PCBs and dioxin-like (DL) PCBs are worth mentioning. Marker or indicator PCBs are the seven congeners currently measured in monitoring programs by commonly used analytical methodology (GC-ECD). These congeners (CB28, 52, 101, 118, 138, 153, 180 according to the usual PCB designation; see Ballschmiter and Zell 1980) are predominant compounds in formerly used technical mixtures and environmental samples.

The DL-PCBs are non-ortho-substituted (CB77, 81, 126, 169) and mono-ortho-substituted congeners (CB105, 114, 118, 123, 156, 157, 167, 189). Due to the position of the chlorine atom, a DL-PCB has a coplanar configuration, is structurally related to 2378-TCDD, and possesses the same toxicological properties as the 17 other 2378-substituted PCDDs and PCDFs. Like other 2378-substituted PCDD and PCDF congeners, DL-PCBs were assigned toxicity equivalent factors (van den Berg et al., 1998 and 2006). DL-PCBs are present in the environment at much lower levels than indicator PCBs. The exception is CB118, which belongs to both groups.

As an example, CB169 concentrations in fish and molluscs (2 to 2000 ng.kg^{-1} wet weight [ww]) are about three orders of magnitude lower than those of CB153 (1 to 100 μg.kg^{-1} ww). Like dioxins, their analysis requires gas chromatography-high resolution mass spectrometry (HRGC-HRMS) and is much more expensive than analyses of indicator PCBs; this explains the limited information available about DL-PCBs. However, recent data indicate that in molluscs and fish from the Baie de Seine, DL-PCBs contributed 60 to 70% of the total toxic equivalent quantity (TEQ) corresponding to DL-PCBs, PCDDs, and PCDFs (Abarnou and Duchemin 2008).

Bivalve molluscs are commonly used in pollution monitoring programs. Typical PCB concentrations found in mussels sampled along the French shoreline vary over a very large range. For CB153, one of the main congeners in environmental matrices that is refractory to degradation (22'44'55' substitution scheme) and representative of biomagnified congeners, the concentrations varied from 0.5 μg.kg^{-1} in the flesh of mussels from Northern Brittany to 100 to 1500 μg.kg^{-1} in similar samples from the Seine estuary. Very variable concentrations are also reported in fish, particularly flatfish that are more or less sedentary, feed on benthic species, and are presumably more contaminated than other pelagic species. Due to the persistence of PCBs, long-living species are particularly exposed to them. PCB distributions and accumulation patterns are subjected to important modifications based on the position of the exposed animal in the food web (Muir et al. 1988). Discrepancies from the expected linear BCF and K_{ow} correlations have been also observed. The correlations are better described by a polynomial regression including maximum bioconcentration potentials for certain K_{ow}s and decreases of bioconcentration for elevated K_{ow}s.

The uptake of PCBs from water follows rapid kinetics independently of the characteristics of the compounds. The lower BCF values observed for low-chlorinated CBs (<4 Cl/molecule) are explained by more rapid elimination. On the other hand, reductions of bioconcentration potentials for highly chlorinated compounds arise from their higher hydrophobicity that decreases their bioavailability because of more pronounced adsorption onto solid material. Moreover, the steric hindrance due to the increase of chlorine substitution, particularly in the ortho position, may partly reduce passage through biological membranes and consequently limit the bioaccumulation of larger molecules.

In the very persistent PCB group, several components may have disappeared, depending on the species of concern. Certain structural requirements are needed to allow biotransformation of CBs and are related to the presence of two vicinal H atoms in the biphenyl molecule. This means that, depending on the number of unsubstituted pairs in o–m or in m–p positions, the compounds will be partially biotransformed (Kanan et al. 1995; Boon et al. 1997; Sijm et al. 1996). On the other hand, compounds that have similar K_{ow} coefficients, such as CB101 and CB118 (log K_{ow} = 7.07 and 7.12, respectively; Rapaport and Eisenreich 1984) are differently bioaccumulated. Biotransformation is an important process that limits bioaccumulation (Sijm et al. 1997; Boon et al. 1997); this detoxifying mechanism facilitates the elimination of PCBs by formation of more soluble hydroxylated metabolites.

6.3.2 PAHs

Polycyclic aromatic hydrocarbons (PAHs) are widely distributed in the coastal marine environment. The major sources of PAHs arise from the transport and the use of fossil fuels (petrogenic hydrocarbons) and other combustion processes (pyrolytic hydrocarbons). Specific molecular ratios have been proposed to assess the sources of PAH contamination in the environment (Baumard et al. 1998). Despite their presence in superficial sediments near estuaries and their elevated K_{ow} levels, PAHs have not been systematically found at high levels in biota, particularly in fish.

These contaminants have been found in zooplankton (Chapter 10), in mussels (RNO 2006; Minier et al. 2006; Abarnou and Duchemin 2008), and in sediment-dwelling worms. In mussels from the Seine estuary, PAH concentrations (sum of 15 PAHs) ranged from 20 to 50 $\mu g.kg^{-1}$ ww; much lower levels, typically around 10 $\mu g.kg^{-1}$ ww, were found in similar samples from "clean" areas. PAHs are virtually absent in fish muscle due to their rapid and efficient biotransformation.

Several characteristics may partly limit the bioaccumulation of PAHs. In coastal marine waters, the occurrence of these substances is related to local atmospheric inputs from many combustion processes like fuel- or coal-fired power stations, solid waste incinerators, oil refineries, steel factories, and vehicle exhausts. The substances emitted by these activities are strongly associated with carbon-rich solid particles (fly ashes are made of particles containing graphite-like carbon, a planar adsorbent with a similar structure to structures of PAHs, that possess strong affinity for this material) and are less bioavailable. The size and the planarity of the molecule are other factors that reduce PAH access and transport into biological materials.

Finally, probably more importantly, PAHs are rapidly biotransformed into more polar metabolites, favouring their elimination: hydroxylated metabolites of PAHs have been found in the bile and the faeces of fish. Hydroxy-PAHs are commonly measured in the bile of aquatic organisms as biomarkers of exposure to PAHs. In conclusion, PAHs are present in bivalves through ingestion of contaminated suspended particulate material but they are probably not biomagnified along food chains due to their limited availability and rapid biotransformation. The presence of PAHs in the marine coastal environment is therefore a problem as a result of their toxicological properties; several compounds exert mutagenic and carcinogenic activities.

6.3.3   PCBs and PAHs in Estuarine Food Webs

PCBs and PAHs were measured in plankton: these compounds have comparable octanol–water partition coefficients, (K_{ow} 10^3 to 10^7) and are assumed to enter the food chain in a similar way, by passive exchanges of contaminants from the water column onto the surfaces of phytoplanktonic cells; this step corresponds to bioconcentration.

Abarnou et al. (2002) performed a study in the eastern part of the Baie de Seine. Plankton was collected by filtering surface water using successive nets with decreasing mesh sizes (from 500 to 20 μm), allowing the separation of zooplankton from phytoplankton (<80 μm) and detritus. The results (Figure 6.1) show the presence of both PCBs and PAHs in the smaller size group mainly consisting of phytoplankton; both classes of compounds are bioconcentrated by phytoplankton via adsorption processes. At higher trophic levels, mainly larger zooplanktonic species (copepods), PCBs are bioaccumulated whereas PAH concentrations decrease. PAHs are very rapidly biotransformed even by the smallest zooplanktonic species.

These two groups of compounds were analyzed in the flounder (*Platichtys flesus*) and sea bass (*Dicentrarchus labrax*) food webs in the estuary as part of the Seine Aval research project (Abarnou et al. 2002). PCBs were bioaccumulated from phytoplankton to higher trophic levels; in contrast, PAHs entered the food chain and were

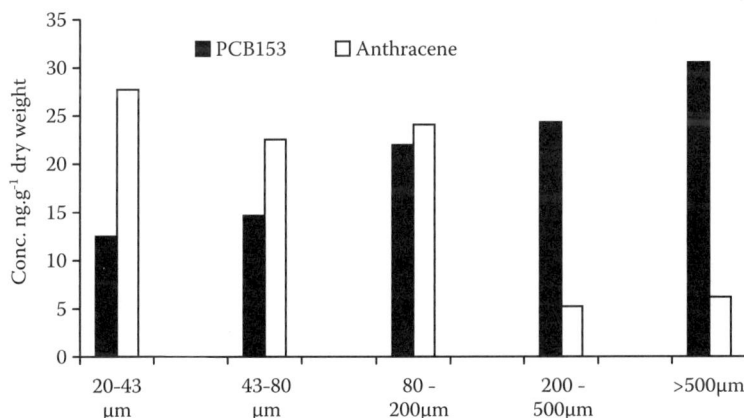

FIGURE 6.1   PCBs (PCB153) and PAHs (anthracene) in plankton of increasing size classes.

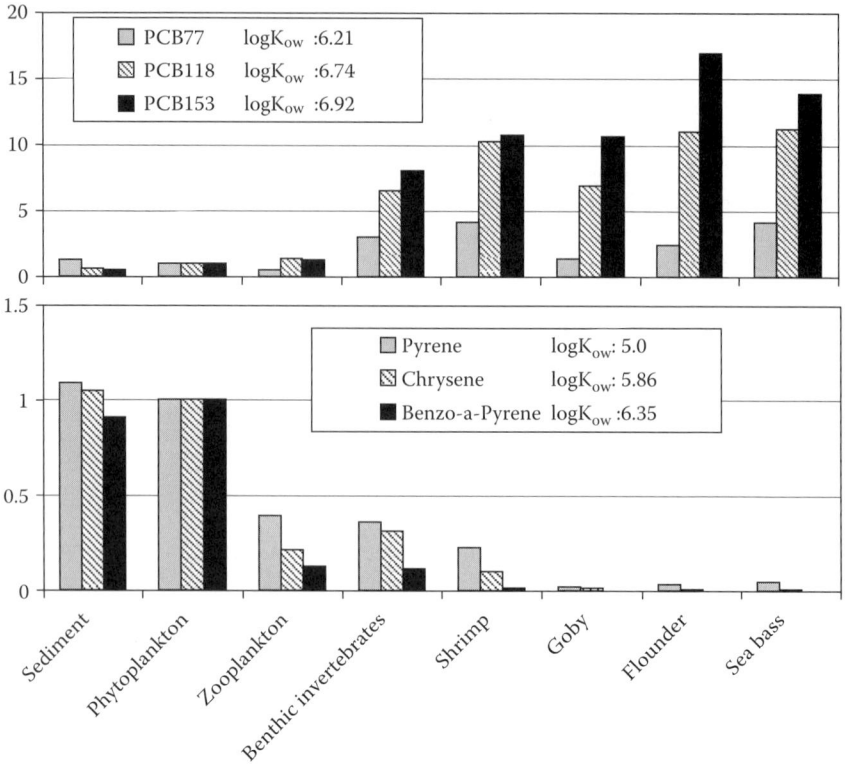

FIGURE 6.2 Biomagnification factors in the sea bass *(Dicentrarchus labrax)* and flounder *(Platichthys flesus)* food webs of the Seine estuary. BF = concentration in organisms/ concentration in phytoplankton) for a few PCBs and PAHs. Benthic invertebrate species include molluscs *(Abra alba, Ensis ensis)*; ophiuroids *(Acrocnida brachiata)*; worms *(Owenia fusiformis, Pectinaria koreni)*; natantian crustaceans *(Crangon crangon* shrimp); and goby fish *(Pomatoschistus microps)*.

rapidly biotransformed, depending on compound structure and the metabolising capacities of the various species.

The apparent biomagnification factors in Figure 6.2 are estimated relative to phytoplankton contamination—the basis of the trophic web. This method facilitates comparisons of the fates of compounds between trophic levels and shows the biomagnification of PCBs by an approximate factor of 1 to 15, in contrast to the very rapid biotransformation of PAHs by estuarine species (BF <<1). PCBs are known to behave more or less in the same manner; little difference appears, for example, in a comparison of PCB77 and PCB110, both of which are partially biotransformed by shrimp. In the case of PAHs, all compounds present the same general behaviour but biotransformation depends on chemical structure and on species capacity to biotransform the compounds.

The influence of chemical structure on the trophic transport of chemical substances is demonstrated by comparing the behaviours of PCBs and PAHs to the

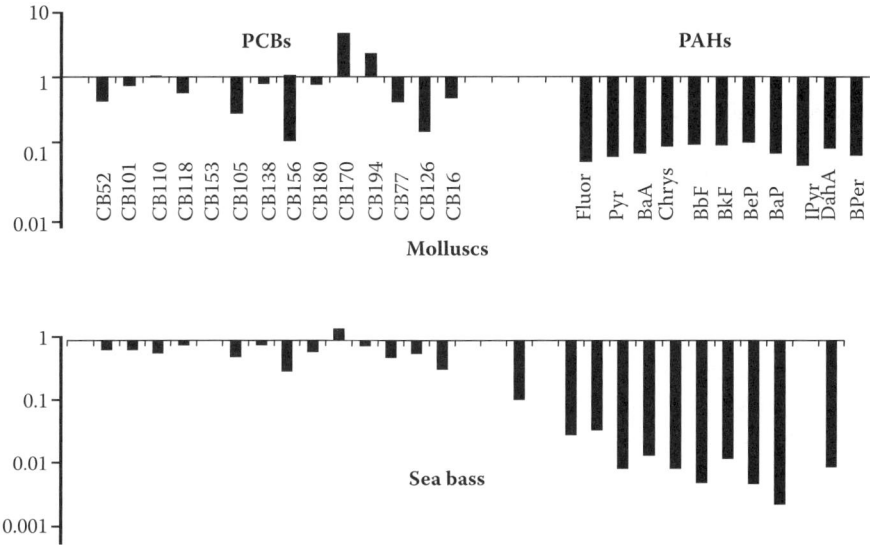

FIGURE 6.3 Trophic transport indices of PCBs and PAHs in two very simplified food chains (phytoplankton to filter feeding molluscs and phytoplankton to sea bass).

behaviour of PCB153—the main PCB congener in all organisms that is very hydrophobic and persistent, and thus a typical example of a biomagnified contaminant. A trophic transport index (TTI) is proposed:

$$TTI = (X \text{ predator}/CB153 \text{ predator})/(X \text{ prey}/CB153 \text{ prey}) \qquad (6.9)$$

The comparison is proposed for PCBs and PAHs and mussels and sea bass, leading to different TTI values based on the capacities of these species to biotransform contaminants (Figure. 6.3). In both cases, phytoplankton is the original food organism at the base of the food chain. Mussels receive and accumulate contaminants directly from phytoplankton, while sea bass receive compounds that were biotransformed or bioaccumulated earlier in lower trophic levels. This representation highlights the well known bioaccumulation of PCBs with TTI levels of 1 or higher (PCB194), and the biotransformation of PAHs—very low in mussels but much more efficient in sea bass with indices <0.01 (99% biotransformation of accumulated contaminant).

6.3.4 PBDEs

Polybromodiphenylethers (PBDEs) constitute a large family of environmental contaminants that have been studied relatively recently. PBDEs and PCBs possess several similarities: chemical structure, common nomenclature, and widespread occurrence in the global environment. In a PBDE molecule, the two phenyl rings are linked by an oxygen atom (diphenylether) and substituted by 1 to 10 bromine atoms leading to 209 congeners theoretically. Technical mixtures of various bromination levels (penta-, octa-, and deca- brominated products) were used as fire retardants in electronic equipment and textile materials. The deca-bromo mixture is the only one that continues to be produced since the banning of certain brominated products in 2001.

Concentrations of PBDEs are present at the microgram per kilogram level in most organisms, approximately 100 to 1000 times lower than PCB concentrations. PBDEs are known to be endocrine disruptors, but the main concern about their omnipresence in the environment is the rapid increase of contamination over recent years. As an example, during the past 2 to 30 years, PBDE concentrations in human milk doubled every 5 years in Sweden (Noren and Meironyte 2000) and later confirmed in the U.S. (Hites 2004). However, retrospective analyses of banked samples of guillemot eggs in Sweden (Sellstrom et al. 2003) and mussels in France (Johansson et al. 2006) demonstrated a stability and even a possible decrease of PBDE contamination since the mid-1980s. Data on levels in the marine environment have greatly improved. A few results for estuarine organisms are given in Table 6.3 and include data from supposed clean areas.

PBDE levels from the Seine estuary were around 1 ng.g^{-1} ww in mussels and 1 to 3 ng.g^{-1} in fishes—about 50 to 100 times less than PCB contamination in similar samples. It seems that PBDEs are mostly transported by SPMs into coastal waters and these particulate materials become the contamination sources for filter feeders like bivalves. Comparing the distribution of these two groups of contaminants in the Bay of Seine, on the basis of the PCB/PBDE ratios found in mussels from Villerville near the estuary and in mussels from Montfarville-Le Moulard (northwest Bay of Seine, Pointe de Barfleur), PBDEs have more diffuse origins than PCBs.

Looking at the relative distributions of PBDE congeners in biota, fewer compounds are observed in the case of brominated contaminants compared to PCBs. BDE47, a tetrabrominated compound, is by far the most common and represents approximately 50 to 60% of the sum of the measured congeners in mussels and 50 to 80% of the sum in fish. Because the deca-bromo mixture is the main source of contamination, the result is that the PBDE contamination results from debromination processes of highly brominated compounds and these biotransformation processes are still active within the food web. The degradation processes leading to the formation of BDE47 may also change the relative importance of intermediary congeners BDE99 and BDE100 found in biota when considering the relative importance of these two compounds in mussels and fish.

Because of structural similarity with PCBs, the potential of PBDEs to be biomagnified is of great concern, as already highlighted by measurements in mammals and other top predators. Their presence and fate in food webs were investigated in two major French estuaries, those of the rivers Loire and Seine (Bragigand et al. 2006). BDE47 was the predominant PBDE congener, measured in all samples to be approximately 50 to 70% of the sum of the measured components (PBDE28, 47, 99, 100, 153, and 154). Levels three to five times higher were found in species collected from the Seine than from the Loire estuary, in agreement with the more important urbanisation and industrial activities of the Seine basin. These results (Table 6.4) also show the biomagnification of PBDEs in the estuarine food web in relation with the trophic levels of the organisms and their lipid contents.

TABLE 6.3
PBDE Contamination Levels in Estuarine Organisms

Organism	Location and Duration	BDE47 ($\mu g.kg^{-1}$ ww)	S PBDE ($\mu g.kg^{-1}$ ww)	Reference
Molluscs				
Mussel *Mytilus edulis*	Villerville (2001–2004) Seine estuary	0.04–0.36	0.075–0.710	Seine Aval Program
	Le Moulard (2001–2004) Western Bay of Seine	0.07–0.1	0.13–0.18	
	Villerville (1981–2003) Retrospective analyses	0.35–2	1–5	Johansson et al. 2006
	Villerville	0.40–0.60	0.8–1.1	Abarnou and Duchemin 2008
	Bay of Seine	0.08–0.11	0.15–0.20	
	West Cotentin Peninsula	0.07–0.11	0.12–0.22	
Mussel *Mytilus edulis*	Scheldt and Ems estuaries, Wadden Sea, Rotterdam Harbour (1999)	0.18–0.86	1–2	de Boer et al. 2003
Mussel *Mytilus californicus*	San Francisco Bay	1.8–5.5 (3.1)	2.7–9.4 (4.7)	
Crustaceans (Concentrations in Claw Meat)				
Spider crab *Maja brachydactyla*	Bay of Seine (Octeville)	0.02	0.032	Bodin et al. 2007a
	Western Brittany (Roscoff, Le Conquet, Le Guilvinec)	0.004–0.014	0.008–0.24	
Crab *Cancer pagurus*	Bay of Seine (Octeville)	0.017	0.031	
	West Cotentin (Granville)	0.011	0.021	
	Western Brittany (Roscoff, Le Conquet, Le Guilvinec)	0.04	0.007	
Flatfishes (Concentrations in Muscle Tissues)				
Dab *Limanda limanda*	Scheldt estuary		0.13	Vorspoels et al. 2003
	Southern North Sea		0.03–0.08	
Plaice *Pleuronectes platessa*	Scheldt estuary		1.2	
	Southern North Sea		0.1–0.19	
Sole *Solea solea*	Escaut estuary		0.016–1.4	
	Southern North Sea		0.016–0.15	

TABLE 6.3 (continued)
PBDE Contamination Levels in Estuarine Organisms

Organism	Location and Duration	BDE47 (µg.kg⁻¹ ww)	S PBDE (µg.kg⁻¹ ww)	Reference
Flounder	Seine		3.4 ± 2.6	Tapie 2006
Platichthys flesus	Ster (Southern Brittany)		0.6	
	Loire		0.8 ± 0.4	
	Gironde		2 ± 1.2	
Sole	Eastern Bay of Seine	0.20–0.35	0.41–0.70	Abarnou
Solea solea	Western Bay of Seine	0.02–0.05	0.04–0.1	and Duchemin 2008
Plaice	Eastern Bay of Seine	0.14–0.92	0.23–1.35	
Pleuronectes platessa	Western Bay of Seine	0.04–0.1	0.05–0.15	

Source: After Bragigand et al. 2006.

TABLE 6.4
PBDEs in Estuarine Food Webs

Organism	Trophic Level	Loire Estuary Lipids % ww	Loire Estuary PBDE47* ng.g⁻¹ ww	Seine Estuary Lipids % ww	Seine Estuary PBDE47* ng.g⁻¹ ww
Bivalve	Primary consumer				
Scrobicularia plana		0.35	0.01–0.03 (0.01)	0.44	0.01–0.06 (0.02)
Worm					
Nereis diversicolor	Omnivore	1.01	0.03–0.07 (0.05)	1.05	0.08–0.12 (0.1)
Shrimp	Omnivore				
Crangon crangon		0.72	0.10–0.16 (0.13)		
Palaemon elegans		0.44	0.10–0.30 (0.16)		
Palaemon serratus		0.32	0.08–0.25 (0.15)		
Sole	Omnivore				
Solea solea		0.58	0.09–0.39 (0.28)		
Flounder	Omnivore				
Platichthys flesus		0.62	0.06–0.5 (0.15)	0.60	0.42–4.15 (0.61)
Eel	Supercarnivore				
Anguilla anguilla		29.2	0.13–0.57 (0.31)	8.0	2.67–7.84 (3.37)

* Concentration range and median value in parentheses.

6.3.5 PCDDs AND PCDFs

Polychlorinated dibenzo-p-dioxins (PCDDs) and dibenzofurans (PCDFs), commonly referred to as dioxins, are among the most toxic compounds present in the environment. Based on their chemical properties, these compounds are potentially accumulated by aquatic organisms. They come mainly from combustion processes and from chemical industries using chlorination processes. PCDFs are unwanted by-products synthesized during the production of the formerly used PCB mixtures. Limited information about their presence in marine biota exists, mainly due to high analysis costs. Moreover, most of the existing data are restricted to the seventeen 2378-substituted congeners—those having the highest toxicity properties.

PCDD and PCDF concentrations are commonly expressed in toxic equivalent quantities (TEQs) using appropriate Toxic Equivalent Factors (TEFs; van den Berg et al. 1998, 2006). In most cases, TEQs are always <2 pg.g^{-1} ww and <4 pg.g^{-1} ww, the maximum level for dioxins in fish and fishery products according to the European Commission regulation 199 (2006). Higher concentrations may be found in mussels from estuarine areas. PCDD and PCDF contamination decreases in the Baie de Seine and fish are generally not contaminated at risk levels. Dioxins behave intermediately between PCBs and PAHs; some of the PCDD and PCDF congeners are partly biotransformed.

PCDFs generally appear more persistent than PCDDs. These biotransformation reactions are species-dependent and lead to a distribution of PCDDs and PCDFs in fish that differs from the distribution observed in bivalves that have low capacities to metabolize organic contaminants. PCDD and PCDF concentrations are cited in TEQs. Therefore it is worth noting that in estuarine organisms, because of the great influence of anthropogenic activities, the contributions of DL-PCBs to the total TEQ is important and exceeds by a factor of three to five the contributions of dioxins.

6.4 BIOACCUMULATION MODELS

As noted, bioaccumulation results from several interacting mechanisms that depend on the characteristics of the compounds and on biological factors. Various attempts have been made to model bioaccumulation in order to describe and possibly to predict the fates of organic contaminants in food webs. BCF and its prediction from a BCF-K$_{ow}$ correlations are basic and very efficient tools for determining the bioaccumulation of a substance. Multimedia models are based on physico-chemical properties, thus allowing the prediction of the distribution of chemicals in the various environmental compartments (air, water, sediment, or biota). Koelmans et al. (2001) reviewed more sophisticated and integrated models built to take into account the interactions between primary production and bioaccumulation.

Statistical models have been also proposed to interpret the contaminant concentrations measured in biota and their relations to explanatory factors. In estuarine ecosystems Bragigand (2005) followed such an approach to compare the distribution of PBDEs in two estuarine food chains from the Loire and Seine estuaries and determined that lipid content was the main factor acting on the bioaccumulation of PBDE47.

In the Seine estuary, food web bioaccumulation models have been built to describe the fate of PCBs in the sea bass food chain (Loizeau et al. 2001a and b). Following earlier approaches (Norstrom et al. 1976; Thomann and Connolly 1984; Connolly 1991), the contaminant–sea bass model is based on the bioaccumulation equation (6.6) described above. In this bioenergetic model, the uptake and clearance of contaminants rely upon trophic relations and on the biological functions specific to each organism that participates in the sea bass food chain. Thus the structure of the food web and its functioning are key points of these models. In the case of the sea bass from the Seine estuary, the information was mainly obtained from investigations of stomach contents.

Recently, the use of the C and N isotopic signatures has allowed the allocation of a precise trophic level to each organism. Bodin et al. (2007b) used the enrichment of the heavier stable N-15 isotope to trace the biomagnification of PCBs in spider crab food webs from differently contaminated coastal sites. However, data about the nature of the prey and feeding rate are essential when devising bioaccumulation models. The same holds true for all other biological functions: respiration, growth, elimination, action of metabolisation on bioaccumulation. These functions are described by empirical numerical equations found in the literature; obviously the equations are very specific and cannot be generalized to other organisms (Loizeau et al. 2001a and 2001b). PCBs are entered into a model in the terms corresponding to respiration and feeding. The first step in the food chain remains the association between contaminants and suspended material, described by empirical equations using the octanol–water partition coefficient K_{ow} that still serves as the major descriptor of bioaccumulation.

After a model is set up, it must be validated at the stationary state by PCB measurements in sea bass and their different prey. The model has demonstrated the importance of dietary intake in the contamination of higher predators, particularly in the case of highly chlorinated congeners. The subsequent dynamic model takes aging into account and differentiates males and females when getting older, and thus demonstrates reproduction to be a decontamination process for females. Coupling such bioaccumulation models with hydrobiological models constitutes a very promising tool for revealing the distribution of contaminants in estuaries, particularly if the models can include the distributions of suspended matter and phytoplankton. The distribution may be very complex in estuaries where environmental conditions change continuously. Results obtained by the models also depend on the appropriate definition of assimilation coefficients of contaminants that are taken up from water through respiration or from food. These parameters can be determined by successive readjustments or, even better, from bioaccumulation experiments that are, however, very time consuming.

Simplified approaches are needed to evaluate exposure to contaminants in estuarine food chains. The GEMCO model (Le Gall et al. 2003) is a degraded version of the food web model that considers simplified food chains and aims to evaluate the transport, fate, and impact of contaminants in estuaries. Bragigand (2005) used this model to assess PBDEs and observed a good agreement between measured concentrations and those calculated by the model. When models do not take into account the biotransformation of contaminants, the calculated concentrations greatly exceed the true measured concentrations and thus lead to a situation considered "the worst case." An empirical approach to more realistic conditions was used in the GEMCO model: the concentrations of contaminants were first calculated in the same manner as persistent contaminants (i.e., PCBs) and the over-estimated values were then corrected by a biotransformation coefficient deduced from experimental studies. The accumulation of partially metabolized PCB congener was compared to accumulations of PCB153 and BaP (benzo[a]pyrene), respectively, the reference compounds of bioaccumulated or biotransformed contaminants. This attention to biotransformation should be encouraged if we seek more precise models able to predict the behaviours of partially metabolizable compounds corresponding as much as possible to real conditions.

6.5   CONCLUSIONS

Estuaries may be exposed to many chemical wastes from various activities in the basins they drain. Many potentially toxic substances of various natures and origins may reach the marine coastal zone and enter and climb the food chains up to higher consumers and thus create a risk to aquatic life, the quality of the environment, and human health. Persistent bioaccumulative and toxic (PBT) compounds warrant serious environmental and health concern, and therefore should be measured. Their distribution and fate in estuarine ecosystems should be better known and their risks should be adequately assessed and minimized.

Several PBT contaminants such as PCBs, PAHs, and other POPs (persistent organic pollutants), are regularly measured within the framework of pollution monitoring and in food safety control programs. From an environmental view, the situation has greatly improved in Europe and in other developed countries over the past two to three decades. Decreasing trends have been observed for many monitored substances as a result of public awareness and appropriate regulations. This progress indicates that the chemical contamination of our environment can be efficiently managed and reduced. However, marine coastal and estuarine zones are still at risk because of historical contamination by very persistent compounds, such as PCBs in the Seine estuary, and also because of the presence of new potentially toxic substances downstream in rivers that drain large inhabited and industrialized areas.

The presence of organic contaminants in estuarine food chain represents an important challenge concerning food safety. Fish and shellfish (crustaceans and molluscs) contribute importantly to human exposure to contaminants and the role of seafood in dietary intake is larger for certain subpopulations such as fishermen. In the context of this chapter, humans are the highest consumers of the estuarine food web. Estimations of human exposure to contaminants are commonly obtained by measuring concentrations in food and daily food intake, after which the estimated dietary contaminant exposures are compared to toxicological reference values fixed for each group of substances by regulations or recommended by food agencies.

Generally, the dietary intakes of contaminants estimated by only the consumption of fish and seafood would be well below maximum values, partly because of the limited importance of coastal fish in the daily diet. However, consumers of large amounts of fish may be at risk because of excessive exposure. That may be the case for a few local consumers who may be at risk because they go fishing and frequently eat large amounts of their own harvests.

The Seine estuary remains heavily contaminated by PCBs, dioxins, PAHs, and PBDEs. The main concern about the quality and safety of seafood remains PCB contamination: maximum admissible daily intake (ADI) may be exceeded regularly. In the case of PAHs, a similar situation would occur much less frequently and only through the consumption of large amounts of shellfish. PBDEs are detectable at reasonable low levels which probably allow maintenance of safe exposure levels. However, regulatory reference values have not been yet fixed. These issues, particularly the presence of persistent and bioaccumulative contaminants in fish and seafood from estuarine areas, plead for regulation of fishing activities in certain exposed areas and for appropriate accompanying social and economic measures in favour of professional fishermen who may have limited livelihood options.

For all these reasons, we must promote further investigations of organic compounds in estuarine organisms to allow better assessment and management of environmental and human health. The impacts of biological and chemical factors on bioaccumulation and their importance have been discussed in this chapter. Some possible directions can be proposed for future work.

Bioaccumulation obviously depends on the organism and trophic interactions, its predator–prey relations, feeding rate, contaminant assimilation coefficient, and capacity to store contaminants in lipids. Investigation of some of these aspects may require more advanced experimental approaches. The metabolisation of contaminants is important for explaining the distribution in biota: detoxification mechanisms should be studied with special attention to identifying persistent metabolites that may remain in organisms. Such studies would also provide information on the disappearance rate of parent contaminants from biota. PBDEs appear as a typical example to be studied because PBDE47, the major compound found in biota, is virtually absent in the technical mixture originally released into the environment.

This point takes us back to the chemical factors governing bioaccumulation. The identification of chemical substances in biota should always be supported, and such studies in estuaries should merit high priority because these coastal regions are continuously exposed to quantities of chemicals.

Among the compounds recently studied, a few so-called emerging substances may provide new perceptions and knowledge about accumulation processes. Perfluorinated alkylated substances represent such a class. They are water- and stain-repellent and are constituents of many well known products (e.g., Teflon® and Goretex®); their residues have been identified in surface waters, in biota and in human serum; they are both hydrophilic and hydrophobic and thus can be associated to proteins in biota. A growing information base about these compounds may lead to a partial revision of the lipid partition theory that explains the bioaccumulation of organic compounds and the associated use of the octanol–water partition coefficient in most studies on bioaccumulation. Whatever new substances of interest arise, they must be categorized based on their bioaccumulation or biotransformation capacities. One possibility would involve measuring their accumulated concentrations both in bivalves and fish, and then comparing their relative distributions to those of typical persistent (PCBs) or conversely biotransformed ones (PAHs).

The predictive values of bioaccumulation models have greatly improved because of adequate formulations of the processes involved. They allow the prioritisation of the various mechanisms and are probably useful tools for chemical risk assessment and environmental management, but using them is a matter of choice. The more specific they are to defined conditions, the more precise and complex they are.

New questions may soon arise. Global change has become a major concern. How will new environmental conditions affect bioaccumulation? The changing climate should probably exert little effect on bioaccumulation because of the slight effects of temperature change on the structures of food webs and/or rates of biological functions. Further questions may arise about chemicals in the environment. Can climate change indirectly lead to increased use of pesticides against new invasive species? Can intensification of rainfall increase the release of persistent contaminant in estuarine areas?

Finally, and to finish on a more positive note, the presence of organic compounds in biota will decrease only if humans have the will to modify their modern ways of life by limiting their use of chemicals. Monitoring programs have revealed several situations in which contamination trends are decreasing so chemical pollution is not inescapable. This is an important message to remember.

REFERENCES

Abarnou, A. and J. Duchemin. 2008. Distribution et devenir de contaminants persistants dans les écosystèmes littoraux. Comparaison Manche Ouest–Manche Est. IFREMER-AESN www.eau-seine-normandie.fr/.../Expert/Etudes_et_Syntheses/etude_2008/Etude_AESN_IFREMER_corr_2008.pdf

Abarnou, A. et al. 2002. Contaminants in marine foodwebs. *Rev. Med. Vet. Toulouse* 153: 425–432.

Anon. 2001. The Seine estuary: a man-altered macrotidal system. *Estuaries* 24(6) Part B (dedicated issue).

Anon. National Institute for Minamata disease, http://www.nimd.go.jp/archives/english/

Ballschmiter, K.H. and M. Zell. 1980. Analysis of PCBs by glass capillary column–gas chromatography: composition of technical Arochlor and Clophen PCB mixtures. *Fresenius Z. Anal. Chem.* 302: 20–31.

Baumard, P. et al. 1998. Concentrations of PAHs (polycyclic aromatic hydrocarbons) in various marine organisms in relation to those in sediments and to trophic level. *Mar. Pollut. Bull.* 36: 951–960.

Bodin, N. et al. 2007a. PCB, PCDD/F and PBDE levels and profiles in crustaceans from the coastal waters of Brittany and Normandy (France) *Mar. Pollut. Bull.* 54: 657–668.

Bodin, N. et al. 2007b. Variability of stable isotope (delta C13 and delta N15) signatures in two spider crab populations (*Maja brachydactyla*) in Western Europe. *J. Exp. Mar. Biol. Ecol.* 343: 149–157.

Boon, J.P. et al. 1997. Concentration-dependent changes of PCB patterns in fish-eating mammals: structural evidence for induction of cytochrome P450. *Arch. Environ. Contam. Toxicol.* 33: 298–311.

Boon, J. et al. 2002. Level of polybrominated diphenyl ether (PBDE) flame retardants in animals representing different trophic levels on the North Sea food web. *Environ. Sci. Technol.* 36: 4025–4032.

Braekevelt, E., S.A. Tittlemier, and G.T. Tomy. 2003. Direct measurement of octanol–water partition coefficients of some environmentally relevant brominated diphenyl ether congeners. *Chemosphere* 51: 563–567.

Bragigand, V. 2005. Recherches écotoxicologiques sur les retardateurs de flamme bromés dans les écosystèmes estuariens (Loire et Seine). Ph.D. Thesis, University of Nantes, France.

Bragigand, V. et al. 2006. Influence of biological and ecological factors on the bioaccumulation of polybrominated diphenyl ethers in aquatic food webs from French estuaries. *Sci. Total Environ.* 368: 615–626.

Brown, M.P. et al. 1982. A mathematical model of PCB accumulation in plankton. *Ecol. Model.* 15: 29–47.

Bysshe, S.E. 1982. Bioconcentration factor in aquatic organisms. In *Handbook of Chemical Property Estimation Methods: Environmental Behavior of Organic Compounds*, Lyman, W.J. et al., Eds. New York: McGraw-Hill, pp. 5–30.

Christensen, J.H. et al. 2002. Polybrominated diphenyl ethers (PBDEs) in marine fish and blue mussels from southern Greenland *Chemosphere* 47: 631–638.

Commission Regulation EC 199. 2006 *Off. J. Europe. Union* L32.

Connell, D.W. 1992. Quantitative structure activity relationships (QSARs) for modelling bioconcentration of lipophilic compounds. In *Proceedings of a Bioaccumulation Workshop*, Miskiewicz, A.G., Ed. Water Board and Australian Marine SC. Assoc. 332 pp.

Connolly, J.P. 1991. Application of a food chain model to polychlorinated biphenyl contamination of the lobster and winter flounder food chains in New Bedford harbor. *Environ. Sci. Technol.* 25: 760–770.

Davies, R.P. and A.J. Dobbs. 1984. The prediction of bioconcentration in fish. *Water Res.* 18:1253–1262.

de Boer, J. et al. 2003. Polybrominated diphenyl ethers in influents, suspended particulate matter, sediments, sewage treatment plant and effluents and biota from the Netherlands. *Environ. Pollut.* 122: 63–74.

Ernst, W. 1980. Effects of pesticides and related organic compounds in the sea. *Helgolander Meeresun.* 3: 301–312.

Govers, H.A.J. and H.B. Krop. 1998. Partition constants of chlorinated dibenzo-furans and dibenzo-p-dioxins. *Chemosphere* 37: 2139–2152.

Gray, J.S. 2002. Biomagnification in marine systems: the perspective of an ecologist. *Mar. Pollut. Bull.* 45: 46–52.

Güsten, H., D. Horvatic, and A. Sabljic. 1991. Modelling N-octanol/water partition coefficients by molecular topology: polycyclic aromatic hydrocarbons and their alkyl derivatives. *Chemosphere* 23: 199–213.

Hansen, B.G. et al. 1999. QSARs for K_{ow} and K_{oc} of PCB congeners: a critical examination of data, assumptions and statistical approaches. *Chemosphere* 39: 2209–2228.

Harding, G.C. 1986. Organochlorine dynamics between zooplankton and their environment: a reassessment. *Mar. Ecol. Prog. Ser.* 33: 167–191.

Hites, R. 2004. PBDEs in the environment and in people: a meta-analysis of concentrations. *Environ. Sci. Technol.* 38: 945–956.

Hites, R.A. et al. 2004. Global assessment of organic contaminants in farmed salmon. *Science* 303: 226–229.

Hühnerfuss, H. et al. 1993. Enantioselective and non-enantioselective degradation of organic pollutants in the marine ecosystem. *Chirality* 5: 393–399.

Hünhnerfuss, H. et al. 1995. Stereochemical effects of PCBs in the marine environment: seasonal variation of coplanar and atropisomeric PCBs in blue mussels (*Mytilus edulis* L.) of the German Bight. *Mar. Pollut. Bull.* 30: 332–340.

Johansson, I. et al. 2006. Polybrominated diphenyl ethers (PBDEs) in mussels from selected French coastal sites: 1981–2003. *Chemosphere* 64: 296–305.

Kannan, N. et al. 1995. Chlorobiphenyls: model compounds for metabolism in food chain organisms and their potential use as ecotoxicological stress indicators by application of the metabolic slope concept. *Environ. Sci. Technol.* 29: 1851–1859.

Koelmans, A.A. et al. 2001. Integrated modelling of eutrophication and organic contaminant fate and effects in aquatic ecosystems: a review. *Water Res.* 35: 3517–3536.

Law, R.J. et al. 2006. Levels and trends of brominated flame retardants in the European environment. *Chemosphere* 64: 187–208.

Law, R.J. et al. 2003. Levels and trends of polybrominated diphenyl ethers and other brominated flame retardants in wildlife. *Environ. Int.* 29: 757–770.

Le Gall, A.C. et al. 2003. The food web model. Final report of GEMCO Project: generic estuary modelling system to evaluate transport, fate and impact of contaminants. IFREMER EL/EC/03.01, 159 pp.

Loizeau, V. et al. 2001a. A model of PCB bioaccumulation in the sea bass food web from the Seine estuary (Eastern English Channel). *Mar. Pollut. Bull.* 43: 242–255.

Loizeau, V., A. Abarnou, and A. Menesguen. 2001b. A steady-state model of PCB bioaccumulation in the sea bass (*Dicentrarchus labrax*) food web from the Seine estuary, France. *Estuaries*, 24: 1074–1087.

Mackay, D. and S. Paterson. 1981. Calculating fugacity. *Environ. Sci. Technol.* 15: 1006–1014.

Minier, C. et al. 2006. A pollution-monitoring pilot study involving contaminant and biomarker measurements in the Seine Estuary, France, using zebra mussels (*Dreissena polymorpha*). *Environ. Toxicol. Chem.* 25: 112–119.

Muir, D. and E. Sverko. 2006. Analytical methods for PCBs and organochlorine pesticides in environmental monitoring and surveillance: a critical appraisal. *Anal. Bioanal. Chem.* 386: 769–789.

Muir, D.C.G., R.J. Norstrom, and M. Simon. 1988. Organochlorine contaminants in Arctic marine food chain: accumulation of specific polychlorobiphenyls and chlordane-related compounds. *Environ. Sci. Technol.* 22: 1071–1079.

Norén, K. and D. Meironyté. 2000. Certain organochlorine and organobromine contaminants in Swedish human milk in perspective past 20–30 years. *Chemosphere* 40: 1111–1123.

Norstrom, R.J., A.E. McKinnon, and A.S.W. de Freitas. 1976. Bioenergetics based model for pollutant accumulation by fish: simulation of PCB and methyl mercury residue level in Ottawa river yellow perch (*Perca flavescens*). *J. Fish Res. Board Can.* 33: 248–267.

Opperhuizen, A. and T.H.M. Sijm. 1990. Bioaccumulation and biotransformation of polychlorinated dibenzo-p-dioxins and dibenzofurans in fish. *Environ. Toxicol. Chem.* 9: 175–186.

Phillips, D.J.H. 1980. *Quantitative Aquatic Biological Indicators: Their Use to Monitor Trace Metal and Organochlorine Pollution.* London: Applied Science, 488 pp.

Rapaport, R.A. and S.J. Eisenreich. 1984. Chromatographic determination of octanol–water partition coefficients (K_{ow}) for 58 PCB congeners. *Environ. Sci. Technol.* 18: 1071–1079.

RNO. 2006. Ligne de base: les contaminants chimiques dans les huîtres et les moules du littoral français. *Bull. RNO* 27–51.

Schultz, D.E., G. Petrick, and J.C. Duinker. 1988. Chlorinated biphenyls in North Atlantic surface and deep waters. *Mar. Pollut. Bull.* 19: 526–531.

Segstro, M.D. et al. 1995. Long term-fate and bioavailability of sediment-associated polychlorinated dibenzo-p-dioxins in aquatic mesocosms. *Environ. Toxicol. Chem.* 14: 1799–1807.

Sellström, U. et al. 2003. Temporal trend studies on tetra- and pentabrominated diphenyl ethers and hexabromocyclododecane in guillemot egg from the Baltic sea. *Environ. Sci. Technol.* 37: 5496–5501.

Sijm, D.T.H. et al. 1997. Biotransformation in environmental risk assessment: SETAC Workshop. Noordwijkerhout, 28 April–1 May 1996.

Spacie, A. and J.L. Hamelink. 1984. Bioaccumulation. In *Fundamentals of Aquatic Toxicology: Methods and Applications*, Rand, G.M. and Petrocelli, S.R., Eds. New York: Hemisphere.

Tapie, N. 2006. Contamination des écosystèmes aquatiques par les PCB et PBDE: application à l'estuaire de la Gironde. Ph.D. Thesis, University of Bordeaux, France.

Thomann, R.V. and J.P. Connolly. 1984. Model of PCB in the Lake Michigan lake trout food chain. *Environ. Sci. Technol.* 18: 65–71.

van den Berg, M. et al. 1998. Toxic equivalency factors (TEFs) for PCBs, PCDDs, PCDFs for humans and wildlife. *Environ. Health Perspect.* 106: 775–792.

van den Berg, M. et al. 2006. The 2005 World Health Organization reevaluation of human and mammalian toxic equivalency factors for dioxins and dioxin-like compounds. *Toxicol. Sci.* 93: 223–241.

van der Oest, R., J. Beyer, and N.P.E. Vermeulen. 2003. Fish bioaccumulation and biomarkers in environmental risk assessment. *Environ. Toxicol. Pharmacol.*13: 57–149.

Veith, G.D., D.L. Defoe, and B.J. Bergstedt. 1979. Measuring and estimating the bioconcentration factor of chemicals in fish. *J. Fish Res. Board Can.* 36: 1040–1048.

Voorspoels, S., A. Covaci, and P. Schepens. 2003. Polybrominated diphenyl ethers in marine species from the Belgian North Sea and the Western Scheldt estuary: levels, profiles, and distribution. *Environ. Sci. Technol.* 37: 4348–4357.

7 Tolerance in Organisms Chronically Exposed to Estuarine Pollution

Claude Amiard-Triquet, Thierry Berthe,
Anne Créach, Françoise Denis, Cyril Durou,
François Gévaert, Catherine Mouneyrac,
Jean-Baptiste Ramond, and Fabienne Petit

CONTENTS

Tolerance may be defined as the ability of organisms to cope with stress to which they are exposed in their environment. Stressors may be natural factors such as dramatic changes in temperature, salinity, or oxygenation experienced by estuarine species or anthropogenic disturbances including contamination by numerous classes of chemicals.

The importance of the variability of sensitivity to stress has been documented in numerous taxa. In freshwater microorganisms, the relative sensitivity based on the inhibition of photosynthesis (IC_{50}) has shown that cyanobacteria are sensitive to Cd, Cu, and Zn whereas Chlorophyceae tend to have high tolerance. Within each of these taxa, inter-specific differences of IC_{50} can reach one (Cd, Zn) to two (Cu) orders of magnitude (Takamura et al. 1989). Among diatoms exposed to PCBs, determination of the growth rate showed that *Asterionella japonica* was more resistant than

Ditylum brightwellii or *Thalassiosira nordenskioldii* (Cosper et al. 1984, 1988). In invertebrates, it is generally accepted that inter-specific variability in tolerance is linked to very important variability in adaptive strategies (limiting the influx of contaminants, storage in detoxified physicochemical form, increased elimination) (Mason and Jenkins 1995).

Phylogenetic differences have been reported (Rainbow 1998; De Pirro and Marshall 2005), and even within relatively restricted zoological groups with similar modes of feeding, striking inter-specific differences may be observed such as those reported in filter-feeding bivalves exposed to Ag (Berthet et al. 1992). Numerous studies carried out after oil spills have clearly established that crustaceans, particularly amphipods, are very sensitive to petroleum hydrocarbons, whereas the impact on polychaetes is much more limited (Gómez Gesteira and Dauvin 2000 and literature cited therein).

In addition to the tolerance characteristics of different species, many studies have reported that within the same species, populations chronically exposed to chemicals in their environment have been able to cope more efficiently than "naïve" individuals. The best known examples are the resistance of bacteria to antibiotics and other chemicals (Sardessai and Bhosle 2002; Stepanauskas et al. 2005; Baker-Austin et al. 2006; Sprocati et al. 2006; Stepanauskas et al. 2006), of terrestrial plants to metals (Clemens 2006), and of insects to pesticides (ffrench-Constant et al. 2004; Hemingway et al. 2004). In a recent review, Amiard-Triquet et al. (2008) drew attention to acquired tolerance in microalgae and cyanobacteria after exposure to metals, PCBs, and different classes of pesticides; in different taxa of annelids exposed to several metals; in crustaceans exposed to metals and pesticides; and in fish exposed to metals, PCBs, PCDDs, and PAHs.

Tolerance may be acquired by physiological acclimation during exposure to sublethal concentrations at some period; in this case, it is not transferable to the next generation. The acquisition of tolerance may also be the consequence of genetic adaptation in populations that have been exposed for generations to the selection pressure resulting from chemical stress. Because organisms may become tolerant to a pollutant for two reasons, tolerance must be examined using physiological and genetic approaches. A classical way to demonstrate tolerance in an exposed population is to perform comparative toxicity tests with a population originating from a clean site as similar as possible in natural factors. The endpoints may be lethality, biochemical changes, disturbance of physiological functions, behavioural disturbances, etc. Another way is to compare the performance or the genetic patterns of populations or communities living in comparatively clean and impacted sites. In the present study, the Authie estuary, located in an area far from any local pollution source, was used as the control site for comparison against the multipolluted Seine estuary.

7.1   METAL-RESISTANT BACTERIA IN ESTUARINE MUDFLATS

Bacterial resistance to mercury is one of the numerous examples of the genetic and physiological adaptations of microbial communities exposed to high available concentrations of contaminants in their environment (Nies 1999; Rasmussen and Sorensen 2001; Barkay et al. 2003). The specific bacterial mercury resistance mechanism relies on the reduction of toxic mercuric ion Hg^{2+} into the volatile and less toxic elemental mercury Hg^0 by mercuric reductase MerA (Barkay et al. 2003). Thus, in the mercury-contaminated mudflats of estuaries, Hg^R (*merA*) bacteria may play a significant role in the biogeochemical cycle of mercury by contributing to the emission of elemental mercury to the atmosphere, thus leading to the detoxification of the ecosystem (Barkay et al. 2003).

The *mer*-mediated mercury resistance mechanism is one of the most widely observed in bacteria, and even in Archaea (Vetriani et al. 2005; Schelert et al. 2006), whereas for other toxic metals, many different genetic systems encountered can be transferred only between phylogenetically related bacterial species (Nies 1999; Barkay et al. 2003). Thus, in water, soil, and sediment, the *merA* gene is a relevant model in ecological studies for assessing relationships between microbial mercury resistance and bioavailable mercury contamination of the environment (Nazaret et al. 1994; Barkay et al. 2003; Nì-Chadhain et al. 2006).

In the present study, the relative abundance of the *merA* gene was monitored by cPCR and corresponds to the proportion of Gram-negative *merA* bacteria in the total microbial community (Oger et al. 2001; Leloup et al. 2004). The main advantage of cPCR is that it overcomes the effects of contaminants (trace metals) and humic acids that inhibit equally the amplification of target and competitor sequences. Thus, cPCR is well adapted to the comparative study of contrasting contaminated estuarine mudflats.

The occurrence of the Gram-negative *merA* gene was investigated in the Authie reference site and compared with three contrasting mudflats of the Seine estuary: the Oissel mudflat located in the upstream estuary in the freshwater zone; the northern intertidal mudflat located in the middle estuary and subjected to the influence of both tides and marine waters; and a subtidal mudflat located in the Bay of Seine (downstream estuary, see Figure 1.3 in Chapter 1) (Deloffre et al. 2005). The higher abundance of the Gram-negative *merA* gene from Seine estuary mudflat sediments (203.9 ± 67.9 to 430 ± 59.2 gene copies.ng^{-1} of total extracted DNA) when compared to the Authie estuary reference site (mean value of 6.7 ± 5.1 gene copies.ng^{-1} of total extracted DNA along the core depth) indicates a relationship between this abundance and the degree of anthropogenic mercury contamination of the estuary watershed (Figures 7.1 and 7.2). Thus, except for the September 2001 core from the Oissel mudflat (Figure 7.2A) and the September 2002 core from the subtidal mudflat (Figure 7.2C), the abundance of the *merA* gene was significantly higher in the Seine estuary mudflats than in any of the cores from the reference site (Figure 7.1) despite a low and constant presence of the Gram-negative *merA* gene in the Authie mudflats.

Nevertheless, in the three mudflats of the Seine estuary (Figure 7.2), the maxima of Gram-negative *merA* abundance were always located in fresh sediment deposits. In contrast, cores sampled after an erosion period displayed a low abundance profile of the *merA* gene along the core. Thus, in dynamic environments such as estuarine mudflats, *merA* gene abundance is greatly influenced by hydrosedimentary processes, with a high abundance of *merA* bacteria in fresh sediment deposits, and should not be directly compared only with mercury concentrations.

The low *merA* gene abundance in the deep sediments could result from the limited bioavailability of inorganic mercury, as shown by Cardona-Marek et al. (2007).

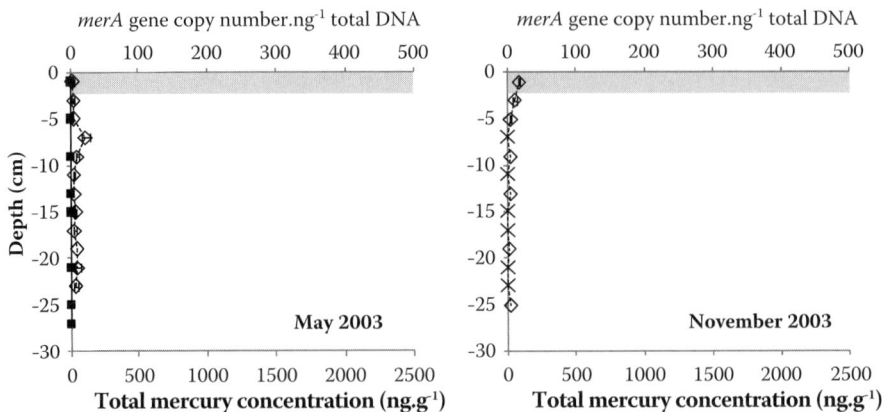

FIGURE 7.1 Vertical distribution of the relative abundance of the Gram-negative *merA* gene (copy numbers per nanogram total DNA), total Hg concentrations (nanograms per gram), and hydrosedimentary processes in the Authie mudflat. The *merA* gene was quantified as described by Ramond et al. (2008) (◇) along 30-cm-long cores collected in May and February 2003. Filled symbols ◆ = *merA* gene amplified but not quantified (detection limit: 278 *merA* copies). ✕ = no *merA* gene detected. Values shown are means ± standard deviations of triplicates. ■ = total Hg concentration. Fresh sediment deposits indicated by grey areas.

Thus, Gram-negative *merA* bacteria are deposited on the Seine estuary mudflats via hydrosedimentary processes. These bacteria have probably been selected elsewhere in another reservoir of the mercury-contaminated Seine watershed and associated with the particulate matter deposited on mudflat surfaces. Nevertheless, the intensive release of trace metals occurring through early diagenesis during the decomposition of organic substances in the surface sediment pore water cannot be excluded (Ouddane et al. 2004).

7.2 PHOTOSYNTHETIC ACTIVITY OF MICROPHYTOBENTHOS EXPOSED TO CHEMICALS IN THE FIELD

Microphytobenthos (mainly diatoms and cyanobacteria) is the major contributor to the primary productivity in intertidal mudflats and also a key food resource for many benthic invertebrates. Extracellular polymeric substances (EPS) of algae and bacteria such as polysaccharides form a substantial biofilm that can integrate the effects of environmental conditions over extended periods (Underwood and Paterson 1993; Sabater et al. 2007). By physical adsorption, the biofilm can remove toxicants (e.g., trace metals or pesticides) from the water or sediment that can be later assimilated by the organisms feeding at the sediment surface. The binding capacity can also act as a mechanism of protection of microorganisms against chemicals and has been found to depend on the age and succession status of the community. Biofilms in an early colonization stage are more vulnerable than mature biofilms to metal exposure, and exposure history is also a determinant parameter (Ivorra et al. 2000). Several other abiotic factors (temperature, light, pH, nutrient availability) also affect the responses of biofilms (Sabater et al. 2007).

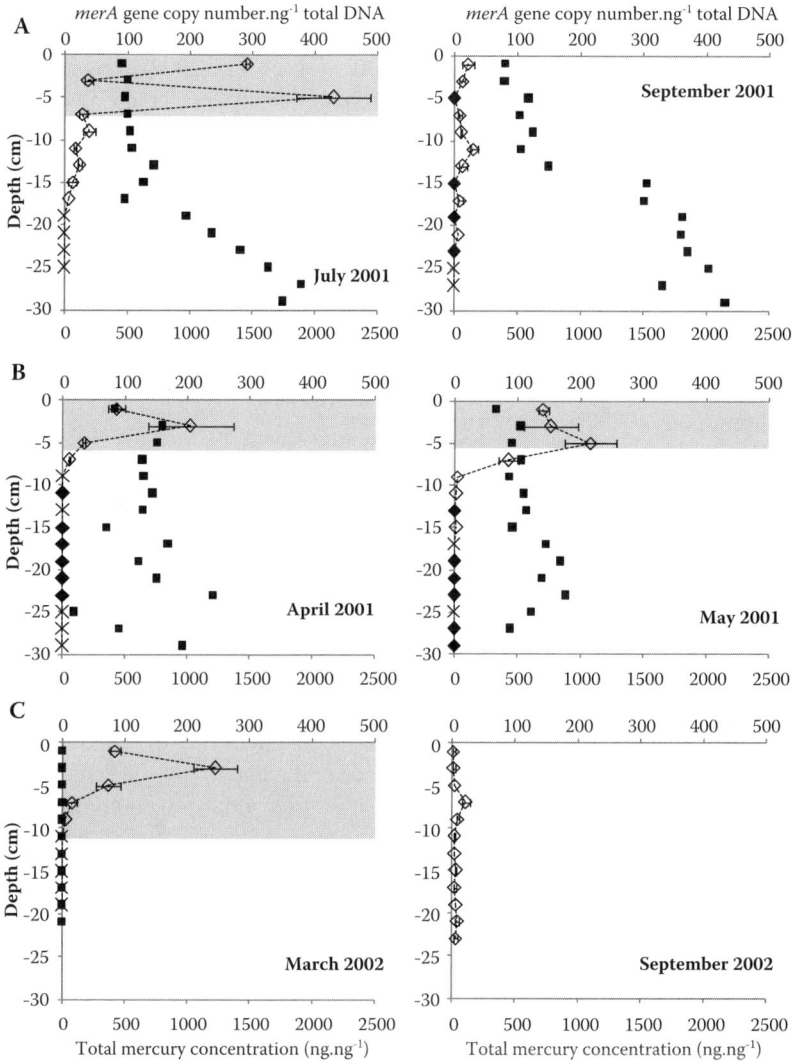

FIGURE 7.2   Vertical distribution of the relative abundance of the Gram-negative *merA* gene (copy numbers per nanogram total DNA), total Hg concentrations (nanograms per gram), and hydrosedimentary processes in the Seine estuary: (A) Oissel intertidal mudflat, (B) northern intertidal mudflat, and (C) subtidal mudflat. The *merA* gene was quantified as described by Ramond et al. 2008 (◇) along 30-cm-long cores collected in May and February 2003. Filled symbols ◆ = *merA* gene amplified but not quantified (detection limit: 278 *merA* copies). ✕ = no *merA* gene detected. Values shown are means ± standard deviations of triplicates. ■ = total Hg concentration. Fresh sediment deposits indicated by grey areas.

The effects of pollutants on biofilms can be expressed as short-term variations of physiological properties (for example, photosynthesis or respiration) and long-term change of the community structure. The photosynthetic activity of microphytobenthic communities has been measured by monitoring variations in amounts of dissolved oxygen (Glud et al. 1992), by ^{14}C-based fixation rates of natural microphytobenthos populations (Barranguet and Kromkamp 2000; Perkins et al. 2002) or by using pulse amplitude-modulated (PAM) fluorescence. The application of chlorophyll fluorescence as a non-invasive technique has increased in recent years (Hartig et al. 1998; Barranguet and Kromkamp 2000; Brotas et al. 2003; Migné et al. 2007). It is used to estimate the photosystem II quantum efficiency of microalgae (Φ_{PSII}) and then the linear electron transport rate (generally expressed in a relative form, rETR, as far as the optical absorption cross-section is not measured with the biofilm).

Photosynthetic performances of microphytobenthos (rETR) were measured directly *in situ* in the intertidal Authie and Seine mudflats during the periods of emersion in the summer. Three home-made supports were placed randomly in a plane surface of the sediment allowing the measurements of Φ_{PSII} periodically at the same three spots. rETR was then calculated as:

$$rETR = \Phi_{PSII} \times PAR \times 0.5$$

where PAR (in micromol photons per square meter per second) is the photosynthetically active radiation measured at the sediment surface. Changes in rETR were followed as a function of variation in ambient light to establish composite PI curves (Figure 7.3) as described in Migné et al. (2007). The relationship was described by the equation of Eilers and Peeters (1988) and the characteristic photosynthetic parameters of the microphytobenthic communities were then assessed: the initial slope of the non-saturated photosynthetic rate (α), the light saturation parameter (I_k), and the saturating light intensity (I_m) obtained for the maximum rETR ($rETR_m$).

Figure 7.3 shows that photosynthetic performances of the two microphytobenthic communities measured under the same environmental conditions (light and temperature) were similar. Moreover, rates of photosynthesis estimated *in situ* from PAM fluorometry and CO_2 fluxes (Migné et al. 2007) were compared to propose a way to calculate carbon fixation rates from PAM measurements. Despite the differences in sediment features and sediment chlorophyll-*a* content between sites, a strong correlation between the fluorescence parameter (rETR) and the community level carbon fixation rate (gross community primary production in milligrams of carbon per square meter per hour) was obtained. Our results suggest that the Seine estuary has developed microphytobenthos communities tolerant to chemicals.

7.3 METAL TOLERANCE OF WORMS *NEREIS DIVERSICOLOR* FROM CLEAN AND CONTAMINATED ESTUARIES

The selection of a metal-tolerant population of the polychaete annelid *N. diversicolor* was demonstrated in Restronguet Creek, Cornwall, a metal-contaminated estuary (Bryan and Gibbs 1983), and the heritability of tolerance was established, indicating a genetic basis of the phenomenon in this area with a long history of metal contamination linked to mining (Grant et al. 1989). Decades later, comparable metal bioavailabilities are still present (Berthet et al. 2003) and thus it is not surprising that

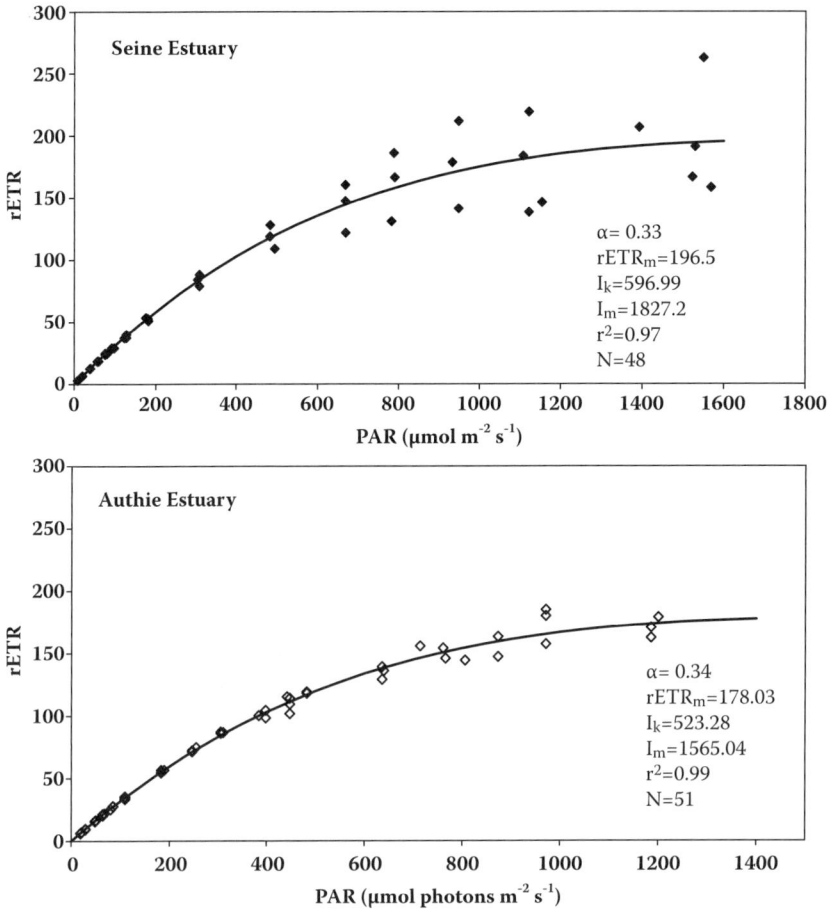

The Seine Estuary plot shows:
$\alpha= 0.33$
$rETR_m=196.5$
$I_k=596.99$
$I_m=1827.2$
$r^2=0.97$
$N=48$

The Authie Estuary plot shows:
$\alpha= 0.34$
$rETR_m=178.03$
$I_k=523.28$
$I_m=1565.04$
$r^2=0.99$
$N=51$

FIGURE 7.3 Photosynthetic characteristics in response to irradiance in microphytobenthos of the Authie and Seine mudflats. Comparison of the relative electron transport rate (rETR) over the whole period of emersion in summer (low tide at 12:13 in the Authie on 26 August 2003 and at 11:36 in the Seine on 27 August 2003).

local populations of ragworms exhibit metal tolerance based on sublethal (Burlinson and Lawrence 2007) or lethal parameters (Mouneyrac et al. 2003). However, metal concentrations in sediments and worms from Restronguet Creek were extremely high compared to many other estuarine and coastal areas (Berthet et al. 2003). Thus we aimed to determine whether populations from the Seine estuary exposed to less dramatic metal contamination have adapted to it.

7.3.1 COMPARATIVE TOXICITY TESTS

Classical toxicity tests were performed to determine LT_{50} values (lethal time for 50% of the experimental population) in worms exposed up to 21 days to a range of

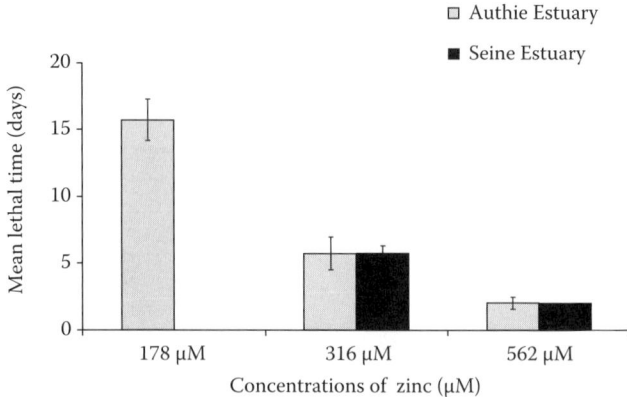

FIGURE 7.4   Lethal time (mean and standard deviation) determined in worms originating from a multipolluted estuary (Seine) and a reference site (Authie) and exposed to a gradient of Zn concentrations in water at salinities nearing those at their sites of origin (15 and 25‰, respectively). No death registered at day 21 among Seine specimens.

Cd, Cu, and Zn concentrations (for details, see Durou et al. 2005). In addition to being differently impacted by anthropogenic stressors, the sites of collection in the Seine and Authie estuaries also showed different salinities (see Chapter 1). Several reports indicate increased tolerance to metals in brackish water compared to normal seawater (Bryan 1976; Fernandez and Jones 1990b; Ozoh 1992). Thus to avoid any interaction of salinity, the bioassays were carried out at salinities of their site of origin for each population.

The Seine specimens had not developed tolerance to Cd or Cu but increased tolerance to Zn was shown (Figure 7.4). At the lowest concentrations tested (31.6 and 100 μM), no intersite differences were shown. At a Zn concentration of 178 μM, all the worms originating from the Seine estuary survived after 21 days of exposure whereas the LT_{50} was only 16 days in worms from the Authie estuary. At the highest concentrations, no intersite differences were observed, most likely because defence mechanisms were overwhelmed, even in tolerant organisms, under such acute stress.

The potential influence of salinity on tolerance to Zn was investigated by determining LT_{50} values in specimens from each site at 15 and 25‰. Independently of the geographical origins of worms, a better survival was observed at 15‰ in agreement with the literature mentioned above. Thus, whether the higher tolerance of worms from the Seine estuary was due to a more favourable salinity (15‰) remains questionable. However, even when bioassays were carried out at 25‰, nearing the salinity usually experienced in the Authie, the LT_{50} in the Seine estuary (23 d) was higher than the LT_{50} in the Authie estuary (16 d).

Bioassays were carried out on experimental samples representative of the natural populations, with small individuals from the Seine estuary and a much larger range of sizes of specimens from the Authie estuary (see Chapter 8). Fernandez and Jones (1990a) have not shown a clear difference in sensitivity according to the size of *N. diversicolor* exposed to zinc. However, the potential influence of this natural factor on tolerance to Zn was investigated independently of any effect of contamination by comparing specimens from the Authie, the mean individual wet weight of which was 317 or 101 mg. A lower sensitivity of the larger specimens was revealed at the clean site. These results agree with most of the literature findings, showing that small and/or young individuals are more sensitive to chemicals than bigger and/or older ones. It is generally admitted that such differences depend on the surface-to-volume ratio and a higher basal metabolism in small and/or young individuals. Despite their larger size, worms of the Authie estuary were more sensitive to Zn than small worms from the Seine estuary, a finding that reinforces the hypothesis of adaptation by a population chronically exposed to chemicals including Zn in the Seine estuary.

7.3.2 ORIGIN OF TOLERANCE IN *NEREIS DIVERSICOLOR*

The mechanisms of defence used by invertebrates to cope with high levels of metals in their environment have been reviewed by several authors (Mason and Jenkins 1995; Marigomez et al. 2002; Amiard et al. 2006; Amiard-Triquet et al. 2008). Limiting the penetration of metals and other chemical stressors may be achieved by mucus production. In ragworms experimentally exposed to Ag or Cu, greater mucus production was observed in specimens adapted in the field to metal pollution (Restronguet Creek, Cornwall) than in specimens from a control site (Mouneyrac et al. 2003). However, this mechanism has only a transitory effect since higher bioaccumulated metal concentrations were observed in contaminated sites compared to controls (Poirier et al. 2006 and literature cited therein).

After metals have entered an organism, other mechanisms of defence are activated. The main methods of detoxification are binding to metallothioneins (MTs) and biomineralization. Metallothionein-like proteins (MTLPs) or MTs are induced by metal exposure in the field or in the laboratory in annelid worm species (see review by Amiard et al. 2006). MTLPs are present in *N. diversicolor* (Berthet et al. 2003), but no significant relationship is indicated between MTLP and metal concentrations. This may be due at least partly to the importance of metal-containing granules in ragworms (see Chapter 4).

Increased excretion of an accumulated contaminant is another potential mechanism of tolerance. In fact, little evidence indicated significant differences in rates of excretion of accumulated Zn, Cd, or Cu between a population of *N. diversicolor* known to be tolerant to Zn and Cu (Restronguet Creek) and a reference population (Geffard et al. 2005). In addition to specific mechanisms allowing the worms to cope with metals, *N. diversicolor* and closely related species respond to pro-oxidant agents (transition metals, PCBs, some pesticides concomitantly present in the Seine estuary; see Chapter 3) by involving antioxidant processes like superoxide dismutase, catalase, or glutathione-*S*-transferase (GST) activities (Geraticino et al. 2004a; Ait Alla et al. 2006). No intersite differences were observed in catalase activities for ragworms from the Seine estuary or the Authie reference site, but higher GST activity was registered in specimens from the Seine. Individuals from the Seine did not suffer the effects of pro-oxidant agents as shown by the absence of intersite differences in TBARS levels (including malondialdehyde, which reflects the state of lipid peroxidation of membranes).

7.3.3 LIMITS OF TOLERANCE

Considering a pollution gradient related to the development of a stress syndrome, the response profiles of biomarkers such as catalase and GST activities or MT content increase at low and intermediate exposures, then decrease at very high exposure. Even at intermediate levels, these defence mechanisms are not totally efficient and changes may be observed for biomarkers of damage (See Figure 4.9 after Dagnino et al. 2007). In agreement with this pattern, higher GST activity was determined in *N. diversicolor* originating from a contaminated estuary in Morocco (Oued Souss) than in a comparatively clean site (oyster farming area of Oualidia), but this did not prevent higher TBARS levels. However, no consequences were observed at the individual level and even after an improvement of water quality in Oued Souss, Cu and Zn tolerances were maintained in the population (Ait Alla et al. 2006).

In certain freshwater invertebrates, mucus secretion represents 13 to 32% of absorbed energy (Wicklum and Davies 1996). In the marine gastropod *Patella vulgata*, mucus production represents 23% of the absorbed energy (Davies et al. 1990). In the marine gastropod *Nucella lapillus* exposed to Cd, decreased oxygen consumption rate and glycogen concentration could be linked to both the production of mucus and MT synthesis (Leung et al. 2000). In oysters exposed to Ag, tolerance was mainly due to the deposition of the toxicant as crystals of Ag_2S in basal membranes and hemocytes. Biomineralization was accompanied by a strong depletion of glycogen, the concentrations of which were rapidly restored when oysters were transferred into clean water (Berthet et al. 1990).

In mussels living in an area impacted by metals (Ni, Cr, Fe), significant correlations were shown between the response of an anti-oxidant system (glutathione peroxidase) and scope for growth (Tsangaris et al. 2007). These findings are convincing examples of the physiological costs of combating chemical toxicants, the ecological consequences of which have been conceptualized by Calow (1991). In agreement again with this concept, *N. diversicolor* from the Seine, despite being tolerant to at least one metal and able to use a number of defence mechanisms, exhibited lower physiological and populational status than in the reference site (Chapter 8).

7.4 GENETIC DIVERSITY IN *NEREIS DIVERSICOLOR* POPULATIONS

Natural populations can adapt to environmental constraints genetically as a result of selection (Klerks and Weis 1987). Four important processes are (1) mutation, which is responsible for genetic diversity by creating new allelic forms, (2) genetic drift, which leads to a decrease in intra-populational genetic diversity and an increase in inter-populational genetic diversity, (3) selection of the effects of which on genetic structure differ according to the origin of the forces and the sensitivity of the organisms, and (4) migration, which homogenises intra-specific genetic structure via gene flow. Analysis of the genetic structures of populations offers a valuable tool to identify divergences of populations that are or are not under environmental stress and to assess ecotoxicological impact (Belfiore and Anderson 2001).

Such analysis takes place at intra- and inter-populational levels. Intra-populational variability affects the adaptability of each population, ultimately affecting its ability to survive a stress. Inter-populational diversity reveals the extent of gene flow limiting the degree of differentiation between populations and the subsequent ability of a species to recolonize a site after a local extinction due to disturbance of the environment. Gene flow may also influence the degree of tolerance of each population via improvement of the available genetic pool. Among the mechanisms that modify genetic diversity, some are random processes whereas others are selective and largely result from ambient environmental conditions including anthropogenic contamination.

7.4.1 ECOLOGICAL CONTEXT AND BIOLOGICAL SPECIFICITY

Based on the extremely fragmented distribution of estuaries, each estuary can be considered a specific environment able to influence the genetic structure of its local populations. Low levels of genetic exchange may be expected among populations in different estuaries, potentially leading to high genetic inter-populational diversity. Estuarine populations are arguably isolated but their degree of isolation is related to their reproductive strategy and their position along the estuarine cline. Estuaries are notorious for accumulating contaminants that may interact synergistically. In these stressful conditions, populations of typical species can only be considered to be control populations if they possess sensitive phenotypes.

The simultaneous application of ecotoxicological and genetic studies, here for the polychaete *Nereis diversicolor*, presents several relevant advantages. On the one hand, since this species is ubiquitous, one can investigate genetic homogeneity of populations between sampling areas. This homogeneity is associated with the degree of inter-populational mixing arising from genetic flow. Conversely, the presence of the annelids at polluted and unpolluted sites allows us to investigate the role of tolerance that is genetically based in one of the most representative species of estuarine fauna.

N. diversicolor is an errant polychaete that digs burrows where it lays eggs. Its behaviour presents two characteristics that may influence inter-populational gene flows. First, although cross-fertilisation occurs externally, the eggs laid by the females are kept in the burrows until the juvenile stage, limiting the planktonic larval phase generally considered important for inter-site dissemination. Second, in contrast to observations of some errant bristle worms, the adults of *N. diversicolor* have no epitoke stage during reproduction. This may also limit genetic exchanges between populations. Nevertheless many ragworms can be caught in the water column at the time of reproduction, and the resuspension of deposited mud in the mouths of estuaries during periods of the strong tidal currents may support the transfer of individual worms and thus genes between sites.

The study of genetic variability can proceed on two fronts. First, genetic diversity can be analysed with noncoding markers. They are not involved in metabolic processes, so they are prone to selective mechanisms. This approach enables the determination of genetic variability in and between populations submitted to evolutionary drivers in the absence of selection pressures (mutation, genetic drift, and migration). On the other hand, genetic diversity can be analysed with coding sequences or with the products of the expression of the functional genome as isoenzymes that potentially are submitted to the evolutionary force of selection. These processes are sources of inter-populational variations detectable by biochemical genetic studies and are generally associated with the adaptability of natural populations to environmental conditions (Nevo et al. 1986). In these two families of markers (coding and noncoding), we found markers that were divergent enough to allow detection of modifications of the genetic structures between close geographic populations and to analyse fine-scale genetic structures.

7.4.2 Nonfunctional Genome

To discover the genetic diversity not arising from selective forces, we analysed the genetic variations of sequences that were not under selective processes (nonfunctional genome). We used the genetic nuclear marker ITS1 that belongs to the rDNA cluster of tandemly repeated units. It differs from the coding region in its rate of evolution (Vogler and DeSalle 1994) and can reveal genetic relationships at the population level (Miller et al. 1996).

The total genomic DNA was extracted from individual adults using a standard proteinase K extraction with a 2% CTAB solution (Winnepenninckx et al. 1993) and the sequences were amplified using specific primers. Direct sequencing of PCR products allowed us to evaluate the degree of polymorphism of the sequences using the ABI PRISM® 310 Genetic Analyzer. The sequences were verified visually and aligned using CLUSTAL W (Thomson et al. 1994). The amplified ITS1 sequences were 495 bp long. Their variability was high enough to allow study of the intra- and inter-genetic variations of the populations.

Among the 55 analysed sequences, 36 haplotypes characteristic of the populations were identified with a divergence of 4.6% (number of divergent nucleotides: 23) and <7% of indels (insertions or deletions). The higher intra-specific variability of the polychaete *Perinereis aibuhitensis* (10%, Chen et al. 2002) was largely explained by indels (95%). The inter-population variations (Table 7.1) were estimated with the genetic distance calculated with Tajima and Nei's index (1984). The PN population was the most different. Although it is geographically close to the HON population, genetically it was as far from the HON (0.58) as from the Authie populations (0.62). In the Authie estuary, the most divergent population was the one with a genetic distance of 0.47, while the mean intra-estuary genetic distance was 0.39.

The genetic results were illustrated with a median-joining network (Bandelt et al. 1999; Figure 7.5). It confirmed the previous analysis. The genetic variability selects two major clusters mainly composed from

TABLE 7.1
Inter-Populational Distances between ITS1 Sequences

	AN	AP	AS	HON	PN
AN	0.00	0.39	0.23	0.35	0.58
AP		0.00	0.55	0.57	0.67
AS			0.00	0.51	0.63
Hn				0.00	0.58
PN					0.00

Note: AN = Authie Nord. AP = Authie Port. AS = Authie Sud. HON = Honfleur. PN = Pont de Normandie. See Chapter 1 and Figure 1.3.

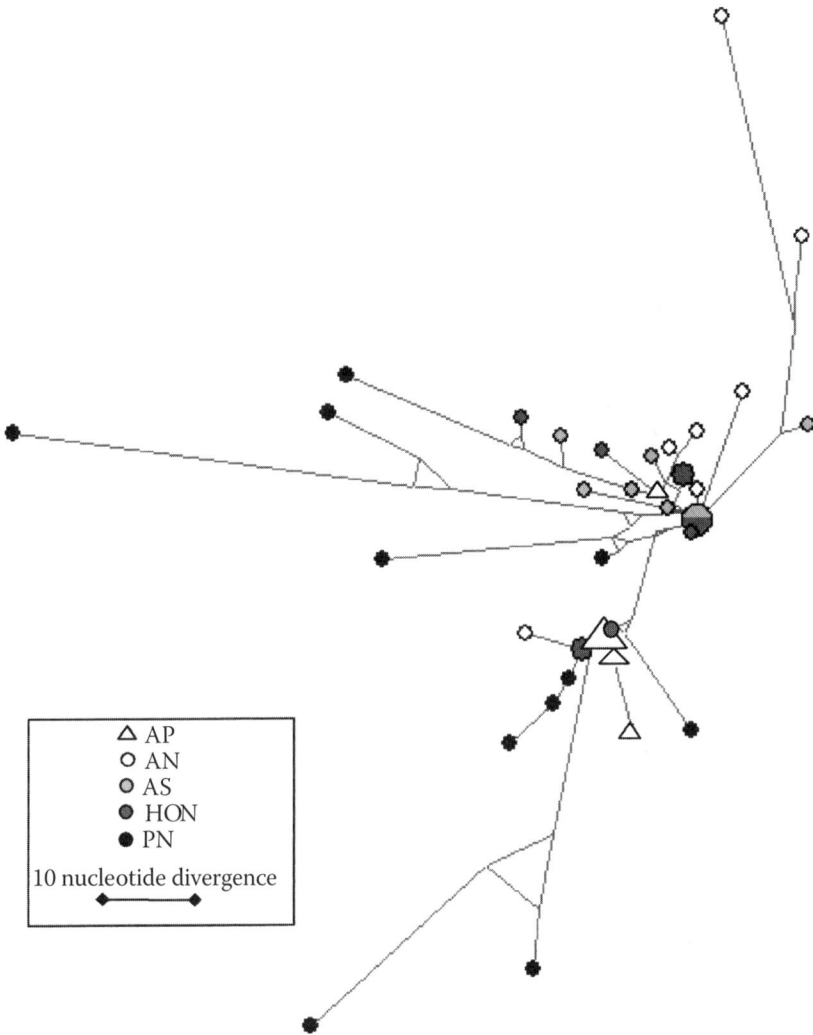

FIGURE 7.5 Genetic relationships of the ITS1 sequences from the studied populations illustrated with a median-joining network (Bandelt et al. 1999). The symbol size relates to the number of identical sequences they represent. The distance between two haplotypes is proportional to the number of mutations between them. AN = Authie Nord. AP = Authie Port. AS = Authie Sud. HON = Honfleur. PN = Pont de Normandie.

three populations (AS and HON in part, AP and HON in another part). The remaining population of the Authie estuary (AN) showed a higher diversity but all these sequences were related to each other and to the first cluster (AS–HON). Thus, the Authie populations showed a star organisation that underlined genetic relationships in this geographical area. No direction of genetic structure could be determined in the Authie estuary. The genetic proximity of the HON population to the Authie populations showed that genetic divergence between populations could not be attributed to isolation among estuaries. The most variable population (PN) appeared separated in two groups of haplotypes excluding bottleneck or founder effects due to contaminant concentrations; one was close to the AS–HON cluster, the other group was close to the AP–HON cluster.

This underlined the weak but real relationships between the PN and the HON populations without any selective processes, based on the noncoding status of the ITS1 sequence. The weakness may be attributed to intra-estuary fragmentation as observed by Virgilio et al. (2006). The microscale structuring appears to be an important reason for the genetic divergence. The results emphasized the absence of continuance of the populations because of the low capacities of dispersion of the larvae and adult forms. Even though the microscale structuring may be promoted by bottleneck processes, the double organisation of the PN population did not seem to support this hypothesis. These results for noncoding sequences showed that genetic diversity can be explained without directional evolutionary forces.

7.4.3 Functional Genome

To study the genetic variability of the part of the genome under selective processes (functional genome), we used two kinds of genetic markers: we estimated polymorphism of the products of gene expression (isoenzymes) involved in basal metabolism, and we analysed the level of divergence of the COI sequences from the mitochondrial genome.

In order to estimate the polymorphism of the isoenzymes, horizontal electrophoresis was performed using starch gels according to the procedures of Pasteur et al. (1987). The studied isoenzymes function in two pathways of basal metabolism that may be affected by anthropogenic contaminants. An energy pathway was addressed with four isoenzymes: glucose-phosphate isomerase (GPI: EC 5-3-1-9), phosphoglucomutase (PGM: EC 5-1-2-2), malate dehydrogenase (MDH: EC 1-1-1-37), and mannose-phosphate isomerase (MPI: EC 3-5-1-8). These isoenzymes revealed six loci. Osmoregulatory metabolism was analysed with two isoenzymes: cytosolic leucine amino acid I (CAP-I: EC3-4-11-1) and glutamate oxaloacetate transferase (GOT: EC 2-6-1-1). These enzymes revealed two loci.

Statistical analyses were carried out using GENEPOP 3.3 software (Raymond and Rousset 1995): allele frequencies, observed and expected heterozygosities (Nei 1978), genotypic disequilibrium, level of fixation index F_{is} (Wright, 1965), and levels of genetic differentiations among populations F_{st} (Wright 1965). Among the eight loci, six were polymorphic and were kept to estimate the inter- and intra-populational variation (GPI, PGM, MDH-2, MPI-2, CAP-I, GOT). Except for PGM and GOT that were monomorphic at one population each, the others were polymorphic at the 95% level in all populations. Thus the percentage of polymorphic loci was high (70%). The intra-populational heterozygosity varied from 52 to 57%. This is higher than the observations of Virgilio et al. (2003) for Mediterranean populations. Nevertheless, the allelic richness was weak and the mean number of alleles per locus was stable (1.58 to 1.63).

No linkage disequilibrium was detected, highlighting the independence of the information. The index of fixation showed low heterozygote deficiencies that may be

TABLE 7.2
Inter-Populational Variation Estimation (Fst) for Each Locus and in Total for All Populations and Both Estuaries

	GPI	MDH 2	MDH 1	MPI 2	MPI 1	CAP-I	Total
Total	0.111[a]	0.042	0.047	0.105*	0.025	0.007	0.053
Authie	0.035	0.010	0.026	0.061	0.032	0.008	0.029
Seine	0.046	0.027	0.051	0.099*	0.026	0.033	0.047

[a] Values significantly different from 0 (P <0.05).

associated to two loci, MDH 2 and CAP-I, for which the indices of fixation were statistically significant. Previous studies have not shown a correlation between multilocus heterozygosity and tolerance to metal pollutants (Virgilio et al. 2005). It has been hypothesised that heterozygotes may be better buffered against environmental stress because they produce different forms of the same enzyme (Diamond et al. 1989) that may exhibit different kinetic properties (Geracitano et al. 2004b). On the other hand, it has been suggested that significant deficits in heterozygotes are associated with the structuring processes between and within the populations by inbreeding depression (Maltagliati 1999). The enzymatic markers underlined a low genetic variability. This confirmed earlier studies (Fong and Garthwaithe 1994; Röhner et al. 1997) but was in opposition to the observations of Scaps et al. (2001). This may be due to the selection of alleles conferring adaptive advantages in a wide range of environmental conditions (Cognetti 1994). No genetic structure was observed between the estuaries by the level of genetic differentiation estimator for the loci all together, but specific analysis underlined genetic differentiations for two loci from energy metabolism (GPI and MPI 2; see Table 7.2). This observation indicated that the energy pathway may be under selective pressure but the osmoregulatory one is not.

In other studies, Virgilio et al. (2003) and Virgilio and Abbiati (2004) showed that the patterns of genotype tolerance responses in laboratory experiments implicated the PGI locus. Nevertheless, in our study, the intra-estuary level results indicated that the MPI 2 locus was the only one that significantly identified a population (PN) as different (Table 7.1). The MPI 2 locus may act as a nonspecific marker of tolerance to environmental toxicity. Among the special characteristics of this population, higher polymorphism was noted. Thus, the polymorphic patterns may be associated with higher tolerance in a stressful environment and the MPI 2 locus may serve as an indicator of a differentiation between populations from polluted and nonpolluted environments.

Another study of a coding gene focused on the mitochondrial genome. Cytochrome oxydase I was analysed because its product is involved in energy metabolism. The experimental protocol was similar to that of ITS1 analysis. The specific primers of Folmer et al. (1994) were used. The sequences were 602 bp in length. The genetic variability was low (<1%) but not absolutely absent as observed by Breton et al. (2003). Nevertheless no inter-populational variations were detected. Although mitochondrial DNA is known to have a high level of mutation, the COI marker cannot be used for monitoring intra-populational genetic variation. Other mitochondrial markers should be tested such as the *Cytb* gene whose genetic divergence is actually known to be higher.

7.5 CONCLUSIONS

From the literature, it can be concluded that most populations in polluted areas have increased tolerance. At a first glance, this seems to be good news for the conservation of biodiversity, particularly in the case of keynote species in communities such as those examined in the Seine estuary, and a good outcome for the conservation of estuarine ecosystem structure and function. However, depending on the endpoints chosen to compare the sensitivities of organisms chronically exposed or not to stress, the conclusions can change dramatically.

In the Seine and Authie estuaries, no intersite differences in the photosynthetic activity of microphytobenthic communities were demonstrated, suggesting that the Seine community had evolved tolerance. However, considering the abundance and structure of the benthic diatom assemblages (an important constituent of microphytobenthos), the situation appeared clearly contrasted in the polluted and reference sites (Chapter 12).

This apparent discrepancy may be highlighted with the help of the concept of pollution-induced community tolerance (PICT); proposed by Blanck and Wängberg (1988). PICT is based on the exclusion of sensitive organisms from communities when exposed to toxicants in high enough concentrations for sufficient time. The resulting restructured communities become more tolerant (Molander and Blanck 1992; Ivorra et al. 2000; Paulsson et al. 2000). Several mechanisms have been proposed to explain algal tolerance to elevated levels of heavy metals or herbicides (Gorbatynk and Arsan 2008), such as reduced uptake (Levy et al. 2007), cellular biotransformation (Yang et al. 2002; Karadjova et al. 2008), binding to glutathione (Kawakami et al. 2006) or phytochelatins or class III MTs (Perales-Vela et al. 2006), vacuolar compartmentalisation (Pawlik-Skowronska 2003), and development of defence systems protecting cells from oxidative stress that generally results at elevated levels of metals. Numerous cases of cross-resistance have been reported between metals or between organic chemicals such as PCBs and DDT (Cosper et al. 1987 and literature cited therein; Takamura et al. 1989). This phenomenon may clearly have ecological importance since many sites are submitted to mixtures of pollutants.

At the individual (survival) level and at the infra-individual (anti-oxidant activity, detoxification ligands as MTLP and biomineralization granules) level, the infaunal worm *Nereis diversicolor* seems well adapted in the Seine estuary. However, considering acute toxicity tests on populations of these worms from various metal-contaminated sites, worms from the Seine have evolved increased tolerance to Zn only. Conspecifics from Oued Souss are tolerant to both Cu and Zn (Ait Alla et al. 2006) and populations from Restronguet Creek are tolerant to Cd, Cu, and Zn (Bryan and Gibbs 1983; Mouneyrac et al. 2003; Burlinson and Lawrence 2007).

These interpopulation differences demonstrate that living in a contaminated environment is alone not sufficient to induce a large spectrum of tolerance: the pollution pressure must be high enough. This is in agreement with our results showing that at the lowest concentrations tested, no intersite differences in mortality were shown based on geographical origin (Seine or Authie), suggesting that the stress was under a threshold acceptable by populations whether adapted or not. Despite exhibiting a number of traits indicating a potential for tolerance to metals and other compounds that have been recognized in the Seine estuary (Chapter 3), the population of *N. diversicolor* living in the Seine estuary seems to be at risk, possibly as a consequence of the changes in energy allocation required to meet the physiological cost of surviving in an impacted site (Chapter 8).

This study has revealed that it remains difficult to monitor the impacts of pollutants by analysis of genetic variability across populations because of the lack of knowledge about genotype tolerance responses. The combined effects of adaptation to local environment and isolation processes were strong factors influencing genetic differentiations among geographically close populations, so genetic structure should be carefully analysed at different spatial and temporal scales before it is used as a tool for assessing impacts of pollutants. Nevertheless, even if no correlation between contamination and genetic structure is directly shown, one cannot exclude the possibility that high concentrations of pollutants may promote genetic divergence. The genotype tolerance pattern at the MPI 2 locus suggests that it may be used as marker of genetic responses to pollutant contamination for *N. diversicolor*. This genetic trait concerning an isoenzyme involved in energy metabolism seems consistent with the observations collected in parallel on other physiological parameters linked to the energy budgets of the worms (Chapter 8). This study confirms the importance of analysing both noncoding and coding regions of the genome. In the latter case, isoenzymes may be considered relevant but it seems still difficult to establish a useful genomic marker subject to selection forces.

Bacterial resistance to mercury is one of the numerous examples of the genetic and physiological adaptations of microbial communities exposed to available concentrations of contaminants in their environment. The mer-mediated mercury resistance mechanism is widely disseminated in microbial communities. For this reason, in this chapter we have focused on bacterial resistance to mercury, in interaction with estuarine hydrodynamics, the importance of which in interpreting the distribution of resistant bacteria has been underlined.

Many different genetic systems for other toxic metals have been described but they can be transferred only among phylogenetically related bacterial species (Silver and Phung 1996). Thus, the cadA gene encoding a Cd^{2+} efflux pump involved in a Cd resistance mechanism has been reported only in six Gram-positive bacteria in both plasmid and chromosomal locations (Silver and Phung 1996; Oger et al. 2003). Molecular quantification of the cadA gene in total DNA extracted from sediments of mudflats from both estuaries of interest show that cadA-resistant-bacteria are detected more often in mudflats from the Seine estuary. In this environment, Oger et al. (2003) reported eleven distinct *Staphylococcus* species (including pathogenic *S. aureus*), and also bacteria belonging to the *Micrococcus* and *Halobacillus* genera carrying the cadA gene. This determinant was mostly plasmid-borne in the *Staphylococcus* genus. The IS257 sequences known to participate in antibiotic resistance gene dissemination in *S. aureus* were located near the cadA gene in 16 of 31 cadmium-resistant *Staphylococcus* strains and one *Micrococcus* strain.

Thus IS257 may contribute to the dissemination of resistant genes among microbial communities.

Considering this aspect and the cost of tolerance, the acclimatization or adaptation of organisms to chemical stress in their environment must be carefully evaluated, taking into account both positive and negative outputs.

Finally, practical factors related to tolerance must be considered in the conduct of bioassays. Major regulatory texts recently published in Europe concerning conservation of the aquatic environment imply intensive use of bioassays (Water Framework Directive, REACH). To improve environmental quality and risk assessments, it is necessary to put more *eco* into *eco*toxicological tests by using keystone species from the area assessed (Chapman 2002). This is particularly important for estuarine water bodies because most current toxicological parameters are derived from standard bioassays carried out on freshwater species and a smaller number of marine species, using "safety factors." In addition, Chapman (2002) stresses that we focus on organisms that are easily cultured in a laboratory and genetically stable. If animals from the "real world" are used to improve the realism of toxicological assessment, the implications of resistance must be carefully taken into account to avoid false negatives or positives in our conclusions.

REFERENCES

Ait Alla, A. et al. 2006. Tolerance and biomarkers as useful tools for assessing environmental quality in the Oued Souss estuary (Bay of Agadir, Morocco). *Comp. Biochem. Physiol.* 143C: 23–29.

Amiard, J.C. et al. 2006. Metallothioneins in aquatic invertebrates: their role in metal detoxification and their use as biomarkers. *Aquat. Toxicol.* 76: 160–202.

Amiard-Triquet, C., C. Cossu-Leguille, and C. Mouneyrac. 2008. Les biomarqueurs de défense, la tolérance et ses conséquences écologiques. In *Les biomarqueurs dans l'évaluation de l'état écologique des milieux aquatiques*, Amiard, J.C. and Amiard-Triquet, C., Eds. Paris: Lavoisier, pp. 55–94.

Baker-Austin, C. et al. 2006. Co-selection of antibiotic and metal resistance. *Trends Microbiol.* 14: 176–182.

Bandelt, H.J., P. Forster, and A. Röhl. 1999. Median-joining networks for inferring intraspecific phylogenies. *Mol. Biol. Evol.* 16:37–48.

Barkay, T., S.M. Miller, and A. Summers. 2003. Bacterial mercury resistance from atom to ecosystems. *FEMS Microbiol. Rev.* 27: 355–384.

Barranguet, C. and J. Kromkamp. 2000. Estimating primary production rates from photosynthetic electron transport in estuarine microphytobenthos. *Mar. Ecol. Prog. Ser.* 204: 39–52.

Belfiore, N.M. and S.L. Anderson. 2001. Effects of contaminants on genetic patterns in aquatic organisms: a review. *Mutat. Res.* 489: 97–122.

Berthet, B., C. Amiard-Triquet, and R. Martoja. 1990. Effets chimiques et histologiques de la décontamination de l'huître *Crassostrea gigas* Thunberg préalablement exposée à l'argent. *Water Air Soil Poll.* 50: 349–363.

Berthet, B. et al. 1992. Bioaccumulation, toxicity and physico-chemical speciation of silver in bivalves molluscs: ecotoxicological and health consequences. *Sci. Total Environ.* 125: 97–122.

Berthet, B. et al. 2003. Accumulation and soluble binding of Cd, Cu and Zn in the polychaete *Hediste diversicolor* from coastal sites with different trace metal bioavailability. *Arch. Environ. Contam. Toxicol.* 45: 468–478.

Blanck, H. and S.A. Wängberg. 1988. Pollution-induced community tolerance: a new ecotoxicological tool. In *Functional Testing of Aquatic Biota for Estimating Hazards of Chemicals*, Cairns, J.J. and Pratt, J.R., Eds., ASTM STP 988. Philadelphia: American Society for Testing and Materials, pp. 219–230.

Breton, S. et al. 2003. Population structure of two northern hemisphere polychaetes, *Neanthes virens* and *Hediste diversicolor* (Nereididae), with different life-history traits. *Mar. Biol.* 142: 707–715.

Brotas, V.N. et al. 2003. *In situ* measurements of photosynthetic activity and respiration of intertidal benthic microalgal communities undergoing vertical migration. *Ophelia* 57: 13–26.

Bryan, G.W. 1976. Some aspects of heavy metal tolerance in aquatic organisms. In *Effects of Pollutants on Aquatic Organisms*, Lockwood, A.P.M., Ed. Cambridge: Cambridge University Press, pp. 7–34.

Bryan, G.W. and P.E. Gibbs. 1983. Heavy metals in the Fal estuary, Cornwall: a study of long-term contamination by mining waste and its effects on estuarine organisms. *Occ. Publ. Mar. Biol. Ass. U.K.* 2: 1–112.

Burlinson, F.C. and A.J. Lawrence. 2007. Development and validation of a behavioural assay to measure the tolerance of *Hediste diversicolor* to copper. *Environ. Pollut.* 145: 274–278.

Calow, P. 1991. Physiological costs of combating chemical toxicants: ecological implications. *Comp. Biochem. Physiol.* 100C: 3–6.

Cardona-Marek, T. et al. 2007. Mercury speciation, reactivity and bioavailability in a highly contaminated estuary, Berry's creek, New Jersey meadowlands. *Environ. Sci. Technol.* 41: 8268–8274.

Chapman, P.M. 2002. Integrating toxicology and ecology: putting the "eco" into ecotoxicology. *Mar. Pollut. Bull.* 44: 7–15.

Chen, C.A. et al. 2002. Nucleotide sequences of ribosomal internal transcribed spacers and their utility in distinguishing closely related *Perinereis* polychaetes (Annelida; Polychaeta; Nereididae). *Mar. Biotechnol.* 4: 17–22.

Clemens, S. 2006. Toxic metal accumulation, responses to exposure and mechanisms of tolerance in plants. *Biochimie* 88: 1707–1719.

Cognetti, G. 1994. Colonization of brackish waters. *Mar. Pollut. Bull.* 28: 583–586.

Cosper, E.M. et al. 1987. Induced resistance to polychlorinated biphenyls confers cross-resistance and altered environmental fitness in a marine diatom. *Mar. Environ. Res.* 23: 207–222.

Cosper, E.M., C.F. Wurster, and M.F. Bautista. 1988. PCB-resistant diatoms in the Hudson River estuary. *Estuar. Coast. Shelf Sci.* 26: 215–226.

Cosper, E.M., C.F. Wurster, and R.G. Rowland. 1984. PCB resistance within phytoplankton populations in polluted and unpolluted marine environments. *Mar. Environ. Res.* 12: 209–223.

Dagnino, A. et al. 2007. Development of an expert system for the integration of biomarker responses in mussels into an animal health index. *Biomarkers* 12: 155–172.

Davies, M.S., S.J. Hawkins, and H.D. Jones. 1990. Mucus production and physiological energetics in *Patella vulgata* L. *J. Mollus. Stud.* 56: 499–503.

De Pirro, M. and D.J. Marshall. 2005. Phylogenetic differences in cardiac activity, metal accumulation and mortality of limpets exposed to copper: a prosobranch–pulmonate comparison. *J. Exp. Mar. Biol. Ecol.* 322: 29–37.

Deloffre, J. et al. 2005. Sedimentary processes on an intertidal mudflat in the upper macrotidal Seine estuary, France. *Estuar. Coast. Shelf Sci.* 64: 710–720.

Diamond, S.A. et al. 1989. Allozyme genotype and time to death of mosquitofish *Gambusia affinis* during acute exposure to inorganic mercury. *Environ. Toxicol. Chem.* 8: 613–622.

Durou, C., C. Mouneyrac, and C. Amiard-Triquet. 2005. Tolerance to metals and assessment of energy reserves in the polychaete *Nereis diversicolor* in clean and contaminated estuaries. *Environ. Toxicol.* 20: 23–31.

Eilers, P.H.C. and J.H.C. Peeters. 1988. A model for the relationship between light intensity and the rate of photosynthesis in phytoplankton. *Ecol. Model.* 42: 199–215.

Fernandez, T.V. and N.V. Jones. 1990a. The influence of salinity and temperature on the toxicity of zinc to *Nereis diversicolor*. *Trop. Ecol.* 31: 40–46.

Fernandez, T.V. and N.V. Jones. 1990b. Studies on the toxicity of zinc and copper applied singly or jointly to *Nereis diversicolor* at different salinities and temperatures. *Trop. Ecol.* 31: 47–55.

ffrench-Constant, R., P.J. Daborn, and G. Le Goff. 2004. The genetics and genomics of insecticide resistance. *Trends Genet.* 20: 163–170.

Folmer, O. et al. 1994. DNA primers for amplification of mitochondrial cytochrome c oxidase subunit I from diverse metazoan invertebrates. *Mol. Mar. Biol. Biotech.* 3: 294–299.

Fong, P.P. and R.L. Garthwaithe. 1994. Allozyme electrophoretic analysis of the *Hediste limnicola–H. diversicolor–H. japonica* species complex (Polychaeta: Nereididae). *Mar. Biol.* 118: 463–470.

Geffard, A. et al. 2005. Kinetics of trace metal accumulation and excretion in the polychaete *Nereis diversicolor*. *Mar. Biol.* 147: 1291–1304.

Geracitano, L.A. et al. 2004a. Oxidative stress in *Laeonereis acuta* (Polychaeta, Nereididae): environmental and seasonal effects. *Mar. Environ. Res.* 58: 625–630.

Geracitano, L.A. et al. 2004b. Oxidative stress responses in two populations of *Laetonereis acuta* (Polychaeta, Nereididae) after acute and chronic exposure to copper. *Mar. Environ. Res.* 58: 1–17.

Glud, R.N., N.B. Ramsing, and N.P. Revsbech. 1992. Photosynthesis and photosynthesis-coupled respiration in natural biofilms quantified with oxygen microsensors. *J. Phycol.* 28: 51–60.

Gomez Gesteira, J.L. and J.C. Dauvin. 2000. Amphipods are good bioindicators of the impact of oil spills on soft-bottom macrobenthic communities. *Mar. Pollut. Bull.* 40: 1017–1027.

Gorbatynk, L.O. and O.M. Arsan. 2008. Functioning of aquatic plants under conditions of toxic influence of pesticides on aquatic system (a review). *Hydrobiol. J.* 44: 73–85.

Grant, A., J.G. Hateley, and N.V. Jones. 1989. Mapping the ecological impact of heavy metals on the estuarine polychaete *Nereis diversicolor* using inherited metal tolerance. *Mar. Pollut. Bull.* 20: 235–238.

Hartig, P. et al. 1998. Photosynthetic activity of natural microphytobenthos populations measured by fluorescence (PAM) and ^{14}C tracer methods: a comparison. *Mar. Ecol. Prog. Ser.* 166: 53–62.

Hemingway, J. et al. 2004. The molecular basis of insecticide resistance in mosquitoes. *Insect Biochem. Molec.* 34: 653–665.

Ivorra, N. et al. 2000. Differences in the sensitivity of benthic microalgae to Zn and Cd regarding biofilm development and exposure history. *Environ. Toxicol. Chem.* 19: 1332–1339.

Karadjova, I.B., V.I. Slaveykovo, and D.L. Tsaley. 2008. The biouptake and toxicity of arsenic species on the green microalga *Chlorella salina* in seawater. *Aquat. Toxicol.* 87: 264–271.

Kawakami, S.K., M. Gledhill, and E.P. Achterberg. 2006. Production of phytochelatins and glutathione by marine phytoplankton in response to metal stress. *J. Phycol.* 42: 975–989.

Klerks, P.L. and J.S. Weis. 1987. Genetic adaptation to heavy metals in aquatic organisms: a review. *Environ. Pollut.* 45: 173–205.

Leloup, J. et al. 2004. Molecular quantification of sulfate-reducing microorganisms (carrying *dsr*AB genes) by competitive PCR in estuarine sediments. *FEMS Microbiol. Ecol.* 47: 207–214.

Leung, K.M.Y., A.C. Taylor, and R.W. Furness. 2000. Temperature-dependent physiological responses of the dogwhelk *Nucella lapillus* to cadmium exposure. *J. Mar. Biol. Ass. U.K.* 80: 647–660.

Levy, J.L., J.L. Stauber, and D.F. Joley. 2007. Sensitivity of marine microalgae to copper: the effects of biotic factors on copper adsorption and toxicity. *Sci. Total Environ.* 387: 141–154.

Maltagliati, F. 1999. Genetic divergence in natural populations of the Mediterranean brackish water killifish *Aphanius fasciatus*. *Mar. Ecol. Prog. Ser.* 179: 155–162.

Marigomez, I. et al. 2002. Cellular and subcellular distribution of metals in molluscs. *Microsc. Res. Tech.* 56: 358–392.

Mason, A.Z. and K.D. Jenkins. 1995. Metal detoxication in aquatic organisms. In *Metal Speciation and Bioavailability in Aquatic Systems*, Tessier, A. and Turner, D.R., Eds. Chichester: John Wiley & Sons, pp. 479–608.

Migné, A. et al. 2007. Photosynthetic activity of intertidal microphytobenthic communities during emersion: *in situ* measurements of chlorophyll fluorescence (PAM) and CO_2 flux (IRGA). *J. Phycol.* 43: 864–873.

Miller, B.R., M.B. Crabtree, and H.M. Savage. 1996. Phylogeny of fourteen *Culex* mosquitos species, including the *Culex pipiens* complex, inferred from the internal transcribed spacers of ribosomal DNA. *Insect Mol. Biol.* 5: 93–107.

Molander, S. and H. Blanck. 1992. Detection of pollution-induced community tolerance (PICT) in marine periphyton communities established under diuron exposure. *Aquat. Toxicol.* 22: 129–144.

Mouneyrac, C. et al. 2003. Physico-chemical forms of storage and the tolerance of the estuarine worm *Nereis diversicolor* chronically exposed to trace metals in the environment. *Mar. Biol.* 143: 731–744.

Nazaret, S. et al. 1994. *merA* gene expression in aquatic environments measured by mRNA production and Hg(II) volatilization. *Appl. Environ. Microbiol.* 60: 4059–4065.

Nei, M. 1978. Estimation of average heterozygosity and genetic distance from a small number of individuals. *Genetics* 89: 583–590.

Nevo, E. et al. 1986. Genetic diversity and resistance to marine pollution. *Biol. J. Linn. Soc.* 29: 139–144.

Ni-Chadhain, N. et al. 2006. Analysis of mercuric reductase (*merA*) gene diversity in an anaerobic mercury-contaminated sediment enrichment. *Environ. Microbiol.* 8: 1746–1752.

Nies, D.H. 1999. Microbial heavy metal resistance. *Appl. Microbiol. Biotechnol.* 51: 730–750.

Oger, C. et al. 2001. Estimation of the abundance of the cadmium resistance gene *cadA* in microbial communities in polluted estuary water. *Res. Microbiol.* 152: 671–678.

Oger, C., J. Mahillon, and F. Petit. 2003. Distribution and diversity of a cadmium resistance determinant (*cadA*) and occurrence of IS257 insertion sequences in staphylococcal bacteria isolated from a contaminated estuary (Seine, France). *FEMS Microbiol. Ecol.* 43:173–183.

Ouddane, B. et al. 2004. Determination of metal partitioning in porewater extracted from the Seine estuary sediment (France). *J. Environ. Monit.* 6: 1–12.

Ozoh, P.T.E. 1992. The importance of adult *Hediste (Nereis) diversicolor* in managing heavy metal pollution in shores and estuaries. *Environ. Monit. Assess.* 21: 165–171.

Pasteur, N. et al. 1987. *Manuel technique de génétique par électrophorèse des protéines*. Paris: Lavoisier.

Paulsson, M., B. Nyström, and H. Blanck. 2000. Long-term toxicity of zinc to bacteria and algae in periphyton communities from the river Göta Älv, based on a microcosm study. *Aquat. Toxicol.* 47: 243–257.

Pawlik-Skowronska, B. 2003. Resistance, accumulation and allocation of zinc in two ecotypes of the green alga *Stigeoclonium tenue* Kütz. coming from habitats of different heavy metal concentrations. *Aquat. Bot.* 75: 189–198.

Perales-Vela, H.V. et al. 2006. Heavy metal detoxification in eukaryotic microalgae. *Chemosphere* 64: 1–10.

Perkins, R.G. et al. 2002. Can chlorophyll fluorescence be used to estimate the rate of photosynthetic electron transport within microphytobenthic biofilms? *Mar. Ecol. Prog. Ser.* 223: 47–56.

Poirier, L. et al. 2006. A suitable model for the biomonitoring of trace metal bioavailabilities in estuarine sediments: the annelid polychaete *Nereis diversicolor. J. Mar. Biol. Ass. U.K.* 86: 71–82.

Rainbow, P.S. 1998. Phylogeny of trace metal accumulation in crustaceans. In *Metal Metabolism in Aquatic Environments*, Bebianno, M.J. and Langston, W.J., Eds. London: Chapman & Hall, pp. 285–319.

Ramond, J.B. et al. 2008. Relationships between hydrosedimentary processes and occurrence of mercury-resistant bacteria (*merA*) in estuary mudflats (Seine, France). *Mar. Pollut. Bull.* 56: 1168–1176.

Rasmussen, L.D. and S.J. Sorensen. 2001. Effects of mercury contamination on the culturable heterotrophic functional and genetic diversity of the bacterial community in soil. *FEMS Microbiol. Ecol.* 36: 1–9.

Raymond, M. and F. Rousset. 1995. GENEPOP population genetics software for exact tests and ecumenicism. *J. Hered.* 86: 248–249.

Röhner, M., R. Bastrop, and K. Jürss. 1997. Genetic differentiation in *Hediste diversicolor* (Polychaeta: Nereididae) from the North Sea and the Baltic Sea. *Mar. Biol.* 130: 171–180.

Sabater, S. et al. 2007. Monitoring the effect of chemicals on biological communities: the biofilm as an interface. *Anal. Bioanal. Chem.* 387: 1425–1434.

Sardessai, Y. and S. Bhosle. 2002. Tolerance of bacteria to organic solvents. *Res. Microbiol.* 153: 263–268.

Scaps, P. et al. 2001. Electrophoretic heterogeneity in *Hediste diversicolor* (Annelida: Polychaeta) within and between estuaries in northern France. *Vie Milieu* 52: 99–111.

Schelert, J. et al. 2006. Regulation of mercury resistance in the crenarchaeote *Sulfolobus solfataricus. J. Bacteriol.* 188: 7141–7150.

Silver, S. and L.T. Phung. 1996. Bacterial heavy metal resistance: new surprises. *Annu. Rev. Microbiol.* 50: 753–789.

Sprocati, A.R. et al. 2006. Investigating heavy metal resistance, bioaccumulation and metabolic profile of a metallophile microbial consortium native to an abandoned mine. *Sci. Total Environ.* 366: 649–658.

Stepanauskas, R. et al. 2005. Elevated microbial tolerance to metals and antibiotics in metal-contaminated industrial environments. *Environ. Sci. Technol.* 39: 3671–3678.

Stepanauskas, R. et al. 2006. Coselection for microbial resistance to metals and antibiotics in freshwater microcosms. *Environ. Microbiol.* 8: 1510–1514.

Tajima, F. and M. Nei. 1984. Estimation of the evolutionary distance between nucleotide sequences. *Mol. Biol. Evol.* 1: 269–285.

Takamura, N., F. Kasai, and M.M. Watanabe. 1989. Effects of Cu, Cd and Zn on photosynthesis of freshwater benthic algae. *J. Appl. Phycol.* 1: 39–52.

Thompson, J.D., D.G. Higgins, and T.J. Gibson. 1994. CLUSAL W: improving the sensitivity of progressive multiple sequence alignment through sequence weighting, position-specific gap penalties and weight matrix choice. *Nucl. Acids Res.* 22: 4673–4680.

Tsangaris, C., E. Papathanasiou, and E. Cotou. 2007. Assessment of the impact of heavy metal pollution from a ferro-nickel smelting plant using biomarkers. *Ecotox. Environ. Saf.* 66: 232–243.

Underwood, G.J.C. and D.M. Paterson. 1993. Recovery of intertidal benthic diatoms from biocide treatments and associated sediment dynamics. *J. Mar. Biol. Ass. U.K.* 73: 25–45.

Vetriani, C. et al. 2005. Mercury adaptation among bacteria from a deep-sea hydrothermal vent. *Appl. Environ. Microbiol.* 71: 220–226.

Virgilio, M. and M. Abbiati. 2004. Allozyme genotypes and tolerance to copper stress in *Hediste diversicolor* (Polychaeta: Nereididae). *Mar. Pollut. Bull.* 49: 978–985.

Virgilio, M. et al. 2003. Relationships between sediments and tissue contamination and allozymic patterns in *Hediste diversicolor* (Polychaeta Nereididae) in the Pialassa lagoons (north Adriatic Sea). *Oceanol. Acta* 26: 85–92.

Virgilio, M., S. Maci, and M. Abbiati. 2005. Comparison of genotype-tolerance responses in populations of *Hediste diversicolor* (Polychaeta: Nereididae) exposed to copper stress. *Mar. Biol.* 147: 1305–1312.

Virgilio, M., T. Backeljau, and M. Abbiati. 2006. Mitochondrial DNA and allozyme patterns of *Hediste diversicolor* (Polychaeta: Nereididae): the importance of small scale genetic structuring. *Mar. Ecol. Prog. Ser.* 326: 157–165.

Vogler, A.P. and R. DeSalle. 1994. Evolution and phylogenetic information content of the ITS-1 region in the tiger beetle *Cicindela dorsalis*. *Mol. Biol. Evol.* 11: 393–405.

Wicklum, D. and R.W. Davies. 1996. The effects of chronic cadmium stress on energy acquisition and allocation in a freshwater benthic invertebrate predator. *Aquat. Toxicol.* 35: 237–252.

Winnepenninckx, B., T. Backeljau, and R. Dewachter. 1993. Extraction of high molecular weight DNA from molluscs. *Trends Genet.* 9: 407–422.

Wright, S. 1965. The interpretation of population structure by F-statistics with special regard to systems of mating. *Evolution* 19: 395–420.

Yang, S.Y., R.S.S. Wu, and R.Y.C. Kong. 2002. Biodegradation and enzymatic responses in the marine diatom *Skeletonema costatum* upon exposure to 2, 4-dichlorophenol. *Aquat. Toxicol.* 59: 191–200.

8 Linking Energy Metabolism, Reproduction, Abundance, and Structure of *Nereis diversicolor* Populations

Catherine Mouneyrac, Cyril Durou, Patrick Gillet, Herman Hummel, and Claude Amiard-Triquet

CONTENTS

The ragworm *Nereis diversicolor* is a widespread polychaete of coastal lagoons and estuaries from Scandinavia to Morocco and the Mediterranean Sea, Black Sea (by introduction) and Caspian Sea (Clay 1967; Smith 1977). It is a euryhaline species and a useful pollution indicator for estuarine sediments (Bryan et al. 1980; Saiz-Salinas and Francés-Zubillaga 1997; Scaps 2002; Ait Alla et al. 2006a; Durou et al. 2007a and b). This species has been the subject of many studies focusing on diverse aspects of its biology and ecology, including a range of pollution-related subjects.

N. diversicolor is characterized by a high physiological tolerance to extreme variations of many environmental parameters such as temperature and salinity (Dales 1950; Smith 1977; Bartels-Hardege and Zeeck 1990). It is a major link in food webs and is a principal prey species for crustaceans, fishes, and birds (Heip and Herman 1979; Zwarts and Esselink 1989; Arias and Drake 1995). The ragworm plays a key role in the physical, chemical, and biological properties of the marine water–sediment interface, its influence on biogeochemical processes mainly attributed to its sediment reworking and bioirrigation activities (Chapter 1, Section 1.4).

In environmental assessment, research in ecotoxicology focuses on the improvement of knowledge to allow the implementation of predictive, ecologically relevant, diagnostic tools (Amiard and Amiard-Triquet 2008). Among them, biological and physiological indicators are of great interest because they provide important information about the general health status of individuals. Since the input of individuals within a population in terms of growth (biomass) and reproductive output (persistence) governs the maintenance of a population, energy disturbances may have consequences at higher levels of biological organization. Thus, the final goal is to link effects of environmental impacts at subindividual (e.g., energy reserves), individual (e.g., condition indices, oxygen consumption), and population levels (Calow 1991; Maltby et al. 2001).

Homeostatic maintenance at a level compatible with the normal functioning of an organism is energetically expensive. It is generally considered that combating stress or repairing damage implies energy costs with a consequent re-allocation of resources (Calow 1991). Energy acquired from nutrition and devoted to physiological defence is not available for maintenance, growth, or reproduction—with potential consequences on the health status of individuals and the fate of populations. To examine energy metabolism perturbations, many bioenergetic parameters are available such as adenylate energy charge (Le Gal et al. 1997), levels of energy reserves as sugars, lipids and proteins (Mayer et al. 2002), cellular energy allocation (De Coen and Janssen 2003), scope for growth (Widdows et al. 1997; Widdows et al. 2002), and oxygen consumption (Hopkins et al. 1999; Hummel et al. 2000; Rowe et al. 2001). Alterations in energy metabolism result in individual responses such as alterations of growth, mortality rates, time to reproduction, or reproductive success (Marchand et al. 2004; Xie and Klerks 2004).

From an operational view, the consequences of stress on energy metabolism can be revealed by the use of condition indices. It is necessary to have a good knowledge of the biology and growth of the species considered and identify different morphological and/or biometric characteristics. Multiple condition indices are well developed in different animal species (Mayer et al. 2002). To date, no condition indices are available for polychaetes despite their ecological role and their inclusion in biomonitoring programmes. The major difficulty is that polychaetes have soft bodies and hard parts are scarce.

Among the polychaete annelids, the nereids are the most studied and their reproduction is the best known. *Nereis diversicolor* has been intensively studied in terms of development, gametogenesis, and reproduction for more than 100 years (Andries 2001). As in many invertebrates, reproduction is controlled by both environmental and endocrine factors. In contrast to other nereids that undergo metamorphosis to a typical epitokous heteronereid form, *N. diversicolor* remains atokous throughout its life (Clark 1961). In most populations, sexual maturity occurs 1 to 3 years before spawning (Scaps 2002 and literature cited therein). Most nereid polychaetes are semelparous: a single episode of reproduction followed by death. No gonads are developed in *N. diversicolor* and gamete development proceeds in the coelomic cavity. Females lay eggs inside their burrows before males eject sperm into open water above the burrows. Ventilatory activity of the females is intensified, bringing sperm into the burrows. Then, fertilized eggs are brooded by the females in the maternal burrows.

The sex ratio of *N. diversicolor* is largely in favour of females (about 80%) (Dales 1950; Smith 1976; Olive and Garwood 1981; Mettam et al. 1982). Abundant literature discusses oogenesis (Olive and Garwood 1981; Mettam et al. 1982; Arias and Drake 1995; Abrantes et al. 1999). The freely floating oocytes grow by the uptake of nutrients supplied via the coelomic fluid. A key role in the translocation of nutrients to the germ cells is played by eleocytes, cells that store and secrete lipids, carbohydrates, nucleosides, and the vitellogenin yolk protein (Porchet 1984; Fischer and Dhainault 1985; Hoeger et al. 1999). The energy invested in gamete biomass can be as high as 70% of total energy (Olive 1983; Gremare and Olive 1986), and correspondingly a reduction of somatic biomass occurs and growth ceases (Olive 1983; Fischer and Hoeger 1993).

The parameter most frequently used in oocyte growth analysis is the average diameter of the oocyte, although it has been proposed that oocyte morphology may be a more sensitive indicator, particularly in relation with hormonal activity (Schroeder et al. 1977 in Olive 1983). During oogenesis, four phases can be recognized: oogonia (active mitosis), premeiotic, previtellogenic, and vitellogenic (Olive and Clark 1978). When primary oocytes appear, they accumulate nutritive resources (lipids, glycoprotein, and finally lipovitellin) and grow to reach maximum oocyte diameter. Since oogenesis is a long process involving transfer of energy reserves, the mechanisms that control it should be understood.

It is well known that physiological reproductive processes in organisms are modulated by endocrine control. Generally, steroids play a critical role in sexual development. The hormonal regulation of gametogenesis in nereids remains largely unstudied; most work relates to neuroendocrinological aspects (Andries 2001). Growth and gametogenesis are controlled by a neurohormone produced by the cerebral neuroendocrine system. Neuroendocrine activity is high during the early stages of life, decreasing during the accumulation of maturing gametes, which is mediated by a chemical factor (Golding and Yuwono 1994).

The aim of this work was to propose an operational approach to assess the health status of *N. diversicolor* by comparing specimens from populations in the multi-polluted Seine estuary and the comparatively clean Authie estuary. The main sampling stations were designated PN in the Seine and AN in the Authie (see Chapter 1, Figure 1.3). The condition of *N. diversicolor* was assessed by different parameters representative of size and weight. Parameters of energy metabolism (glycogen, lipid and protein levels, oxygen consumption) were also investigated. In parallel with these physiological studies, different aspects related to reproduction (sex steroids, sexual maturity, fecundity) in both the Authie and Seine populations were examined. The objective was to answer two main questions: (1) What are the influences of environmental factors, particularly those related to estuarine quality? (2) Which physiological indicators document the health status of populations? To improve the potential ecological value of these measurements, investigations were carried out to examine links between condition, energy metabolism, reproduction, and population effects according to the sites of origin (Seine or Authie estuaries). Differences in abundance of ragworms were also considered from the view of potential consequences on ecosystem functioning through effects of bioturbation.

8.1 BIOLOGICAL AND PHYSIOLOGICAL INDICATORS

8.1.1 ENERGY RESERVES

In animals, energy reserves are predominantly stored as glycogen and lipids. Glycogen is the form of carbohydrate stored by most animals and can be rapidly mobilized to produce high quantities of adenosine triphosphatase (ATP). Lipids are comparatively more energy-rich than glycogen (Mayer et al. 2002). When energy reserves are depleted to a critical threshold, proteins can be used as energy sources (Le Gal et al. 1997). Thus the status of different energy reserves serves as a potential biomarker of the general health status of an organism.

Knowledge of the biology of the relevant organism is necessary to analyze results. Estuarine invertebrates are subjected to many ecological factors (temperature, salinity, oxygen level) responsible for fluctuations in energy metabolism. Moreover, endogenous factors such as weight and/or age, activity, and reproductive state can also influence metabolism (Olive et al. 1985; Cammen 1987). Thus, the biochemical compositions of invertebrates vary greatly as a consequence of cycles of accumulation and mobilization of energy reserves. Thus before using energy reserves as pollution biomarkers in *N. diversicolor*, the influence of intrinsic and/or environmental factors must be carefully considered.

During the first sampling year (2002), analyses were carried out on specimens of the most representative size from each population, i.e., larger worms for the reference site (Authie estuary) and smaller worms for the impacted site (Seine). Under these conditions, protein, glycogen, and lipid concentrations were significantly higher in specimens from the clean site. However, because a potential influence of weight was suspected, both small and large worms were collected from the reference site between February 2003 and November 2004. When worms of comparable weight were selected, the influence of site on energy reserves was not so clear (for details see Durou et al. 2007a). However, from an ecological view, it was more interesting to compare the whole population at each site. Thus, the 3-year results from different-sized polychaetes collected over four seasons from the two estuaries were investigated using principal component analysis as reported in Chapter 4 (Table 4.2 and Figure 4.7). The first axis (PC1) was negatively correlated with energy reserves and positively with contaminant concentrations. The plots of scores in the coordinates of principal components PC1 versus PC2 opposed the results from the clean and contaminated sites: the data cloud of the Authie estuary was located in the negative part of PC1 (correlated with energy reserves); that of the Seine estuary was located in the positive part of PC1 (positively correlated with contaminant concentrations).

Because of differences in salinity observed between the two sampling sites (see Chapter 1), in September 2002 we examined the influence of salinity on worm energy reserves. To avoid any interference with contaminants, three stations were chosen in the reference estuary along a salinity gradient (Authie Port: 18.9 ± 2.0; Authie Nord: 29.5 ± 2.1; Authie Sud: 33.0 ± 1.0) thus covering the salinities encountered at both sites over the duration of the study (Chapter 1, Table 1.1). No significant differences were observed in glycogen (12 ± 3; 13 ± 3; 14 ± 4 mg.g^{-1} ww), lipid (8 ± 2; 7 ± 2; 8 ± 2 mg.g^{-1} ww), and protein (7 ± 1; 8 ± 2; 10 ± 2 mg.mL^{-1}) concentrations with increasing salinity. In the laboratory, worms from the Authie estuary were maintained for 21 days under two salinity conditions: 25 (salinity most often experienced in the Authie estuary) or 15 (equivalent salinity in the Seine estuary). When worms were exposed to lower salinity than the one experienced in the field, no differences were observed in glycogen and lipid levels (for details, see Durou et al. 2005). When a closely related species, *N. virens*, was exposed to hypo-osmotic conditions (salinity of 15), glycogen concentrations were significantly reduced on a wet weight basis, explained by the inability of *N. virens* to osmoregulate at constant body volume (Carr 1981). Conversely, in *N. diversicolor* (well known for being highly euryhaline) the onset of regulation including the regulation of body volume was at 25 to 35% seawater (8.75 to 12.25 psu) and the critical low salinity was 1 to 2% seawater (Oglesby 1978).

8.1.2 RESPIRATORY ECOPHYSIOLOGY

The measure of an organism's respiration rate provides a record of total metabolic activity because respiration is representative of the intensity of physiological processes at the time of measurement. Increases in standard metabolic rate (SMR) estimated as oxygen consumption were observed in different species (crustaceans, amphibians, reptiles) exposed *in situ* and/or in the laboratory to metals (Rowe 1998; Hopkins et al. 1999; Rowe et al. 2001). In the present study, oxygen consumption was measured in worms from both sites at each sampling period (from February 2003 to February 2004) for 1 to 2 hours at a controlled temperature (4, 10, 16, and 22°C). Figure 8.1 illustrates the results. In agreement with previous literature data (e.g., Kristensen 1983b), an increase of respiration with experimental temperatures was observed. The relationship of respiration and experimental temperature remained similar during the entire sampling period except in February 2003 when oxygen consumption was about double other measurements.

Nevertheless, at temperatures experienced by worms in the field at any time of the year, respiration was relatively constant and close to 1 ml O$_2$.h.g dw (Figure 8.2), as Nithart et al. (1999) observed earlier in acclimatized worms, except at ambient temperatures below 3°C at which respiration decreased. Differences based on the sites of origin were not conspicuous. This constant respiration at ambient temperatures at all places contrasted with several earlier studies in which environmental differences were shown to influence respiration strongly—positively with salinity or oxygen tension and negatively with starvation (Kristensen 1983a and b; Fritzsche and von Oertzen 1995; Nielsen et al. 1995; Julian et al. 2001). Our study results differed from most other study results because the other studies were performed mostly

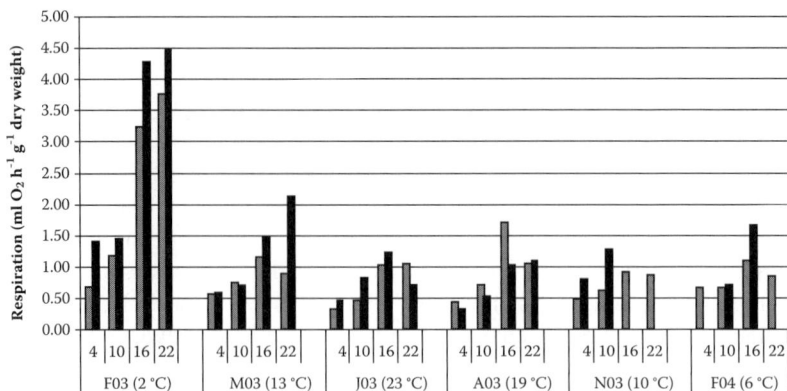

FIGURE 8.1   Variations of respiration in the ragworm *Nereis diversicolor* depending on geographical origin (Authie estuary: grey; Seine estuary: black), sampling date (average temperature measured in field at time of sampling *[in parentheses]*), and range of temperatures (4 to 22°C) at which worms were exposed during respiration tests. F = February. M = May. J = July. A = August. N = November. 03 = 2003. 04 = 2004.

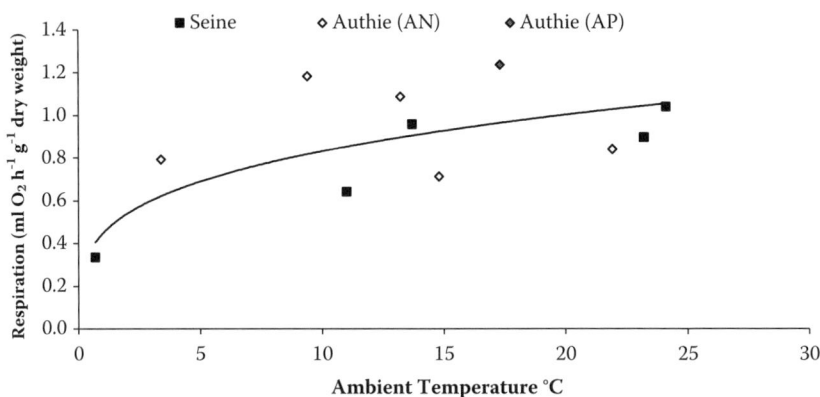

FIGURE 8.2   Variations in respiration rates of *Nereis diversicolor* ragworms at ambient temperatures (trend line: r = 0.77; p <0.01).

under experimental conditions, whereas we performed measurements under ambient conditions with only temperature as a variable.

In our study, respiration at ambient temperatures dropped considerably only in winter at temperatures near freezing (Figure 8.2). At such low ambient temperatures near the freezing point, the sensitivity of the respiration to experimental higher temperatures became two times higher than normal (Figure 8.1). This higher sensitivity of the respiration rate near freezing temperatures may have been caused by cold adaptation related to mitochondrial proliferation as observed in other studies on worms as a measure to compensate for lowered activities at low temperatures (Tschichka et al. 2000).

In clams (*Macoma balthica*) experimentally exposed to polluted sediments (from the Mundaka estuary in southern France), respiration was much higher than in clams from clean sediments (Hummel et al. 2000). In oysters (*Crassostrea virginica*) exposed in the laboratory to Cd (50 µg.L^{-1}) and acclimatized at different temperatures (20, 24, and 28°C), SMR increases were higher at 20°C than at 24°C, whereas no further increase was observed at 28°C in comparison with controls. Thus, in oysters the capacity of oxygen supplying mechanisms may be neutralized by a combination of elevated temperature and Cd exposure (Lannig et al., 2006).

At the cellular level, significant respiratory costs were observed in trout hepatocytes during metabolic processing of pyrene (Bains and Kennedy 2004). However, these findings may hardly be compared with a field situation because contaminant levels used were much higher than those encountered even in highly polluted environments. Interpretation of results must also take into account that observed responses do not consider anaerobic energy production (Abele et al. 1998). Moreover, fluctuations in energy available must also consider the tolerance limits of benthic invertebrates that vary greatly by species. In this sense, *N. diversicolor* can tolerate large variations in temperature and salinity and survive hypoxia. As a consequence, this species can live in naturally fluctuating environments such as upper estuaries (Scaps 2002). As shown in our studies, the ragworm can indeed compensate for a large range of environmental fluctuations (temperature, salinity) and keep its respiration stable even under stress by pollutants as found in the Seine. To our knowledge this has not been observed to this extent earlier.

8.1.3 BIOMETRIC MEASUREMENTS

The ability to cope with environmental stress is expected to have a negative cost to the energy budget (see above), leading potentially to reduced growth. Thus biometric parameters aiming to reveal such an impact were studied in *N. diversicolor*.

Annelids are characterized by segmented bodies. In young animals, the rate of segment proliferation is high; in older individuals, growth is increasingly due to segment enlargement with gain of weight (Barnes et al. 1988). In a natural population, individuals belong to one or more cohorts of different sizes, with the implication that they are of different ages. Several variables are used to assess population size and hence age structure.

The jaws of *N. diversicolor* constitute the only hard body parts (Desrosiers et al. 1988). In another polychaete worm, *Nephtys hombergii*, jaws show annual growth lines (Kirkegaard 1970; Olive 1980) allowing an approximate assessment of the ages of specimens after dissection and microscopic examination. In the case of *N. diversicolor*, jaws do not show such rings but jaw length is regularly employed to assess the size structures of populations (Chambers and Milne 1975; Olive and Garwood 1981; Möller, 1985). The total length and/or weight and the total number of segments of complete worms have also been proposed as relevant measures (Desrosiers et al. 1988). However, these worms can accidentally lose parts of their bodies and also have the ability to autotomize parts in response to stresses such as handling. Thus, some authors have proposed the use of a partial length called L3, corresponding to the sum of the lengths of the prostomium, peristomium, and first chaetigerous segment in *N. diversicolor* (Gillet 1990). In the present study, the biometric variables determined were wet weight, length of jaw, total length of body, L3 and number of segments.

For examining the relationship of size and mass parameters, the general recommendation is to use dry weight instead of wet weight to avoid effects of fluctuations in water content. Many factors such as physiological condition (reproduction, osmoregulation), salinity, and chemical stress can influence the water content of an organism (Lucas and Beninger 1984; Ivankovic et al. 2005; Bianchini et al. 2005). *N. virens* can cope with salinity changes but the response is accompanied by a change of body volume and water content (Carr and Neff 1984).

Osmoregulatory processes in the more euryhaline *N. diversicolor* are very efficient (Oglesby 1969; Fletcher 1974; Smith 1976; Scaps 2002) and so it is not surprising that the water content was relatively constant with regard to the sites of origin of the worms or the sampling period (83 to 85% and 82 to 86% for Authie and Seine specimens, respectively). Significant positive correlations between wet weight and dry weight were observed for worms from the Authie estuary ($R^2 = 0.9683$, n = 90) and the Seine estuary ($R^2 = 0.9553$, n = 70). Thus the wet weight was subsequently used.

The relationship between mass and size parameters was studied in about 550 worms collected at random in February, May, August, November 2003, and February 2004 from both sites (after elimination of specimens that had autotomized segments). An exponential relationship was observed between L3 length and wet weight for worms collected from August 2003 to November 2004 from both sites (Figure 8.3). This pattern corresponds to allometric growth with a first stage characterized by segment proliferation, with segment enlargement taking place during the second stage.

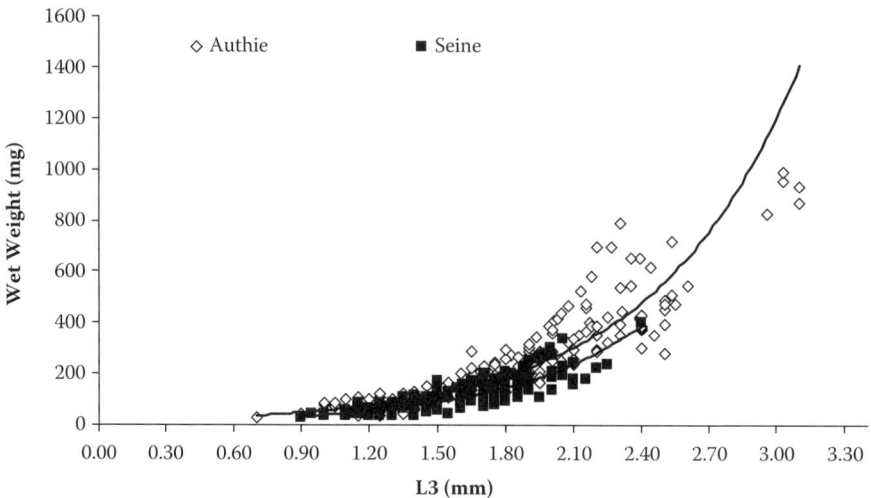

FIGURE 8.3 Relationship between size (as L3 length, see text for definition) and wet weight of *Nereis diversicolor* collected from August 2003 to November 2004 in a multi-polluted estuary (Seine) and a reference estuary (Authie).

TABLE 8.1
Linear Relationships of Log-Transformed Data

Date	Site	N	A	b	R²	p-value
08/03	Authie	48	3.186	3.394	0.899	**0.002**
	Seine	47	2.770	3.437	0.667	
11/03	Authie	38	2.289	3.822	0.944	**< 0.0001**
	Seine	43	2.091	3.759	0.818	
02/04	Authie	38	2.485	3.934	0.935	**< 0.0001**
	Seine	35	2.420	3.748	0.810	
05/04	Authie	63	2.842	3.913	0.909	**< 0.0001**
	Seine	47	3.264	3.297	0.920	
08/04	Authie	49	3.391	3.347	0.960	**< 0.0001**
	Seine	48	2.495	3.579	0.896	
11/04	Authie	41	3.358	3.025	0.972	**0.001**
	Seine	41	2.726	3.067	0.807	

Note: Wet weight and L3 length of *Nereis diversicolor* specimens collected from August 2003 to November 2004 in the multi-polluted Seine estuary and in a reference site (Authie estuary).

The Seine worms clearly showed reduced growth/development in comparison with those from the Authie estuary.

After linear transformation, linear regressions between L3 length and wet weight were obtained. Parameters of these equations are shown in Table 8.1, together with *p* values obtained by analyses of covariance using the site of origin as covariate. These relationships were highly significant at all sampling dates and significant intersite differences were revealed. In the cases of other size variables (length of jaw, total body length) or number of segments, exponential relationships and intersite differences were also observed. For an identical size assessed by any of the measured biometric parameters, the slopes of the regression curves indicated that body weights were generally lower in Seine worms than in Authie worms (for details, see Durou et al. 2008), revealing an impact of local environmental conditions.

8.2 REPRODUCTION

8.2.1 SEXUAL MATURITY

In agreement with literature data, the sex ratio greatly favoured females (Authie: 92 ± 5%; Seine: 89 ± 6%). Consequently, reproduction processes were investigated mostly in female worms. At each sampling date, 30 to 45 females were examined from each site. Five sexual maturity stages (0 to 4) were defined in female worms after microscopic examination of aliquots of coelomic fluid.

Worms in stage 1 correspond to sexually undifferentiated specimens (absence of germ cells). Females from stage 2 (development) contain some primary oocytes

TABLE 8.2
Intersite and Temporal Variations of Sexual Maturity
Indices (SMIs)

	<100 mg		100–200 mg		200–300 mg	
	Authie	Seine	Authie	Seine	Authie	Seine
August 2003	1.46	1.70	1.57	2.56	2.57	—
November 2003	1.60	2.05	1.93	2.68	2.43	3.00
February 2004	1.00	2.13	1.90	3.0	2.67	—
May 2004	1.00	1.17	1.50	2.86	2.75	3.45
August 2004	1.00	1.80	1.38	1.78	1.85	2.29
November 2004	1.13	1.81	1.80	2.86	2.83	—

Note: Mass classes: (<100 mg; 100 to 200 mg; 200 to 300 mg) in a clean (Authie) and a multi-polluted estuary (Seine).

(oocyte diameter ranges from 15 to 65 µm) in germinal cell mass. Then, oocytes initiate a phase of growth (stage 3: vitellogenesis), taking up resources from the coelomic fluid; oocytes are very heterogeneous in size (65 to 190 µm). Stage 4 (mature) is defined by maturation of the oocytes, the maximum diameter of which reached ≥200 µm for gametes ready to be spawned. Stage 0 corresponds to spent females with scarce spherical oocytes or degenerative oocytes. The sexual maturity index (SMI) is calculated as:

$$SMI = [(1*n_1) + (2*n_2) + (3*n_3) + (4*n_4)]/N$$

where n_x is the number of individuals in each stage and N the sum of n_x.

N. diversicolor worms from both sites (Authie and Seine) were assigned into different weight classes, and the distribution among the categories of sexual maturity stages was carried out taking into account the weight class, sampling period, and site of origin (Table 8.2). Regardless of the sites of origin of the female worms, the higher the mass, the greater the degree of sexual maturity. The most striking feature of both populations of worms, whatever the class mass considered (<100 mg, 100 to 200 mg, 200 to 300 mg), is that specimens from the Seine estuary exhibited sexual precocity throughout the sampling period in comparison with those from the Authie estuary.

Considering the influence of the sampling period on sexual maturity, worms in undifferentiated, development, and vitellogenesis stages from both sites were present throughout the year. Females that were well ahead in vitellogenesis (mean oocyte diameter >120 µm) and sexually mature were observed in February and May 2004 in the Authie estuary and only in May 2004 in the Seine estuary. The reproduction periods of worms from the clean (Authie) site seemed larger by comparison with those from the Seine estuary. These observations are in agreement with literature data that indicate one or two spawning periods per year (Dales 1950; Olive and Garwood 1981; Möller 1985; Gillet 1993; Sprung 1994; Fidalgo e Costa et al. 1998; Abrantes et al. 1999).

Environmental factors modulating reproduction in *N. diversicolor* are temperature (Dales 1950), food (Olive and Garwood 1981), and lunar cycle (Dales 1950, Bartels-Hardege and Zeeck 1990). Modifications in temperature regimes are usually proposed to explain differences in the reproduction periods of geographically separated populations. In this study, the differences in spawning periods between the two populations cannot be attributed to photoperiod or lunar cycle because the sites are located at nearly the same latitude, nor to climatic factors since records from meteorological stations near the sites showed very similar changes over the year and a difference of only 0.8°C in annual means.

8.2.2 FECUNDITY

In monitoring studies of populations, fecundity estimation (for technical details, see Durou et al. 2008) represents a useful tool; the number of oocytes is a determining factor in reproductive success.

Temporal variations in the total number of oocytes per female, representative of total fecundity, are shown in Figure 8.4a. A significant influence of the sexual maturity stages on the number of oocytes was observed: when females were in development, the mean number of oocytes was minimal (Authie: 3.9×10^3; Seine: 3.6×10^3) and maximal in the mature stage (Authie: 13.9×10^3; Seine: 8.0×10^3). Both sites exhibited raised production of oocytes in May 2004 corresponding to the period when females were sexually mature. Notably, throughout the study, the mean number of oocytes per female was always higher in specimens from the Authie estuary (7.6×10^3) by comparison with the Seine estuary (5.1×10^3). The difference was statistically significant in August and November 2003 and May 2004.

The relative fecundity was determined as the ratio between the total number of oocytes and the weights of females (Figure 8.4b). This ratio varied from 18 ± 5 oocytes/mg to 76 ± 8 oocytes/mg in Seine females and from 16 ± 4 oocytes/mg to 34 ± 2 oocytes/mg for Authie females. The ratio was significantly higher for Seine females than for Authie females on three occasions (August 2004 and November 2003 and 2004).

8.2.3 SEXUAL STEROIDS

The presence of contaminants in the environment may affect reproduction of aquatic organisms by disrupting the many hormonally controlled processes (Lafont and Mathieu 2007). Numerous studies have focused on the functional roles of polychaete reproductive hormones (Andries 2001 and literature cited therein; Fischer and Dorresteijn 2004) but knowledge about their nature, primary targets, and action mechanisms remains scarce. In different polychaete species, cholesterol (from which steroid sexual hormones may be derived) was found (Voogt 1974; Zeeck et al. 1994; Lee et al. 2005). Recently, 17β-estradiol in *N. virens* (Garcia-Alonso and Rebscher 2005) and progesterone, testosterone, and 17β-estradiol in *N. diversicolor* have been quantified (Mouneyrac et al. 2006).

In vertebrates as in invertebrates, extensive literature indicates that steroid levels depend on sexual maturity stage. In *N. diversicolor*, the links among energy

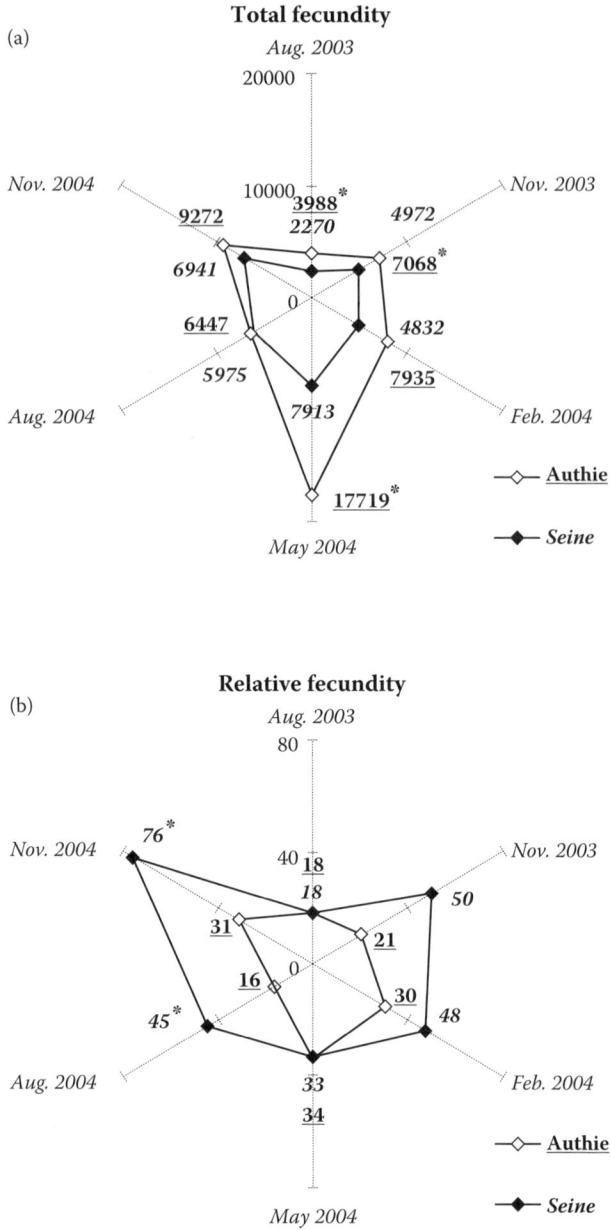

FIGURE 8.4 Oocyte production in *Nereis diversicolor* from the Authie and Seine estuaries: (a) total number of oocytes per female, (b) ratio between total number of oocytes and weight of female. Asterisk indicates intersite difference significant at 95% level. After Durou, C. et al. 2008, *Water Res.* (in press). With permission.

reserves, sexual maturity index, and steroid hormone levels have been studied in detail (Durou and Mouneyrac 2007). Despite the presence of endocrine disruptors in the Seine estuary (Chapter 3), no intersite differences in sexual steroid (progesterone, 17β-estradiol, testosterone) levels in worms were observed.

8.3 INTERSITE COMPARISONS OF *N. DIVERSICOLOR* POPULATIONS

The survey of *N. diversicolor* populations was carried out in parallel with the physiological studies reported above, with quarterly sampling from February 2002 to November 2004 in both estuaries (stations AN and PN shown in Chapter 1, Figure 1.3). Specimens of *N. diversicolor* were sampled down to a depth of approximately 25 cm. Four replicates of 625 cm^2 were taken in the intertidal zone.

8.3.1 DENSITY AND BIOMASS

Temporal variations of the density are depicted in Figure 8.5. At the beginning of the programme in February 2002, the density of *N. diversicolor* was comparable in both estuaries of interest (≈1000 individuals per m^2 at minimum in winter); then it decreased in the Seine estuary to fewer than 300 individuals. In 2002, the density ranged from 1080 to 3580 individuals per m^2 in the Authie estuary and from 140 to 920 in the Seine estuary. In 2003, the densities varied from 910 to 3330 individuals per m^2 in the Authie estuary and from 80 to 290 in the Seine estuary. In 2004, the density varied from 672 to 3510 individuals per m^2 in the Authie estuary and from 120 to 190 in the Seine estuary.

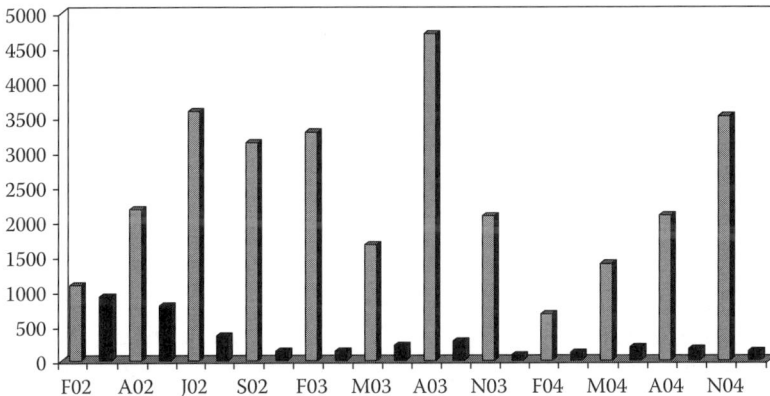

FIGURE 8.5 Temporal variations of the density (number of individuals per m^2) of *Nereis diversicolor* in the Authie estuary (grey) and Seine estuary (black) in February (F), August (A), July (J), September (S), May (M), and November (N) in 2002 (02), 2003 (03), and 2004 (04). After Gillet, P. et al. 2008, *Estuar. Coast Shelf Sci.* 76: 201–210. With permission.

Intersite differences were enhanced when biomasses were compared because the Seine worms were less numerous and were also smaller than those from the Authie estuary (see above). In 2002, the biomass (estimated in dry weight) of *N. diversicolor* showed temporal variations with a maximum of 32.6 g.m^{-2} in July and lower values in February with a biomass of 21.5 g.m^{-2} in the Authie estuary, and varied from 3.4 g.m^{-2} in February to 0.7 g.m^{-2} in September in the Seine estuary. In 2003, the biomass varied from 3.9 g.m^{-2} in February to 26.8 g.m^{-2} in August in the Authie estuary, and from 0.6 g.m^{-2} in February to 1.2 g.m^{-2} in August in the Seine estuary.

8.3.2 COHORTS OF *N. DIVERSICOLOR* POPULATIONS

Size frequency was studied by measuring L3 length which avoided bias resulting from broken specimens (Gillet 1990). The size frequency histograms were treated by the simulated annealing estimating mixture (SAEM) method (for details, see Gillet et al. 2008).

Most of the size frequency histograms were plurimodal with two or three cohorts (see Gillet et al. 2008). In 2002, we observed three cohorts (C1, C2, and C3) in February in the Authie estuary. From April to September, we observed cohorts C2 and C3, then cohorts C3 and C4 in the Authie estuary. In the Seine estuary we observed cohorts C2 and C3 in February, but only cohort C3 in April and July, and two (C3 and C4) in September 2002. In 2003, we observed four cohorts (C3, C4, C5, and C6) in the Authie estuary and only two (C4 and C6) in the Seine estuary. In 2004, we observed four (C5, C6, C7, and C8) in the Authie estuary and three (C6, C7, and C8) in the Seine estuary.

The population of *N. diversicolor* from the Authie estuary belonged to eight cohorts as opposed to six for the population of the Seine estuary (Figure 8.6). The mean size L3 of the cohorts varied from 1.0 to 2.65 mm in the Authie estuary and from 1.1 to 1.94 mm in the Seine estuary. The lifetime for a cohort was 13 months in the Authie estuary and varied from 6 to 11 months in the Seine estuary. The structure of the *N. diversicolor* population in the Seine estuary is simple: one cohort, rarely two (February and September 2002, August and November 2004) per year. In the Authie estuary, the structure of the population is more complex, with generally two or three cohorts per year (never only one).

8.4 CONCLUSIONS

As an endobenthic species, the ragworm *N. diversicolor* is particularly exposed to sediment-bound contaminants (Chapter 3). In the multi-polluted Seine estuary, body sizes (Figure 8.3, and Table 8.1) and concentrations of energy reserves (Figure 4.7 and Table 4.2, Chapter 4) were consistently affected when compared to the situation in the reference Authie estuary.

Biomarkers linked with bioenergetic metabolism have been proposed as potentially valuable biomarkers in environmental monitoring programs (Verslycke et al. 2004).

However, in order to use biochemical, cellular, physiological, and behavioural variations as pollution biomarkers, the change due to contamination must clearly

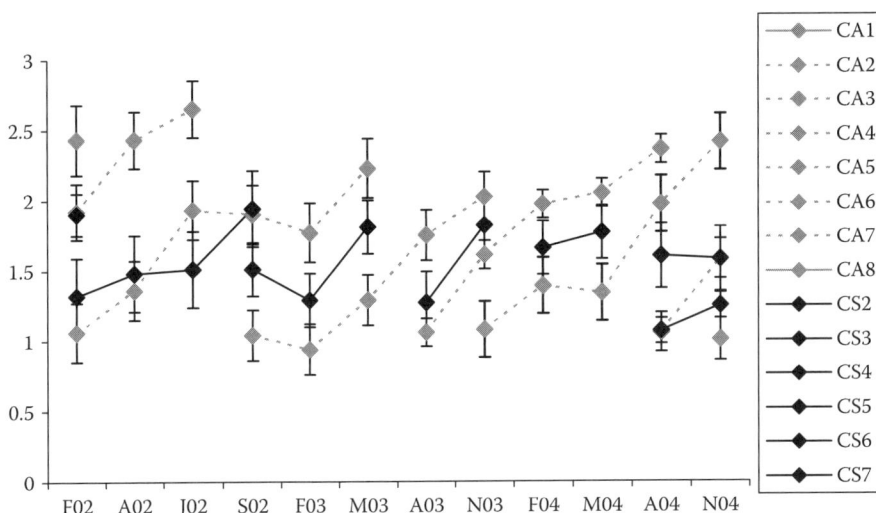

FIGURE 8.6 Growth curves (L3 in mm with standard deviations) of cohorts of *Nereis diversicolor* in the Authie (CA, grey diamonds) and Seine (CS, black diamonds) in 2002, 2003, and 2004. After Gillet, P. et al. 2008, *Estuar. Coast Shelf Sci.* 76: 201–210. With permission.

exceed the natural variability (Cairns 1992). In the present study, the potential influences of different confounding factors on energy reserves were assessed. No significant influence of salinity on energy reserves was shown in accordance with the well documented euryhalinity of *N. diversicolor*. In agreement again with this euryhaline ability, *N. diversicolor* living at different salinities in the Seine and Authie estuaries exhibited identical water contents. Thus the intersite differences in energy reserves shown could not be attributed to salinity differences in the compared habitats.

On the other hand, the influence of weight must be taken into account. Temporal fluctuations of energy reserves in worms from both sites were shown but no consistent trend (annual, seasonal) was clearly established. Many endogenous and environmental factors such as temperature and available food can intervene. Finally, the energy invested in the production of gamete biomass also influences levels of energy reserves (Olive 1983; Gremare and Olive 1986; Durou and Mouneyrac 2007).

In good agreement with the data obtained for energy reserves at a given size (evaluated using L3 length, total length, jaw length, or number of segments), body weights were lower in worms from the contaminated Seine estuary (Figure 8.3 and Table 8.1; for details see Durou et al. 2008). From an operational view, we recommend the use of the L3 length and wet weight because they require less time to determine. In addition, the use of wet weight allows the collection of other measurements from the same specimen, such as energy reserve levels or characterisation of oocyte production in females.

Despite their small size, the worms from the contaminated site were able to reach sexual maturity and even exhibited sexual precocity in comparison with worms from the reference site (Table 8.2). This may correspond to a specific strategy devoted to reproduction under conditions of stress. Despite the presence of endocrine disruptors in the Seine estuary (Chapter 3), no intersite differences in worm steroid hormone levels were apparent.

The numbers of oocytes in females represent determining factors in reproductive success of a species and subsequent maintenance of the population in the field (Figure 8.3). Considering relative fecundity, females from the Seine estuary showed equal or even higher relative production rates than females from the Authie estuary. Concomitantly, lower glycogen levels, generally determined in specimens from the Seine estuary suggest a strategy of considerable energy investment in gametogenesis. However, this situation may also arise from the physiological cost of tolerance to toxicants in the Seine estuary.

Considering total fecundity (Figure 8.4), large females from the clean site generally exhibited higher total numbers of oocytes in comparison with the small specimens from the contaminated site. These results are not surprising because oocytes are freely suspended in the body cavity and so their total production depends strongly upon body volume (Olive 1983). Similarly, in two species from the closely related genus *Perinereis*, a linear and strong relationship was found between oocyte numbers and jaw length or wet weight (Cassai and Prevedelli 1998). In agreement with these findings, highly significant positive correlations between biometric variables (wet weight, number of segments, L3 length, total length) and the number of oocytes in female worms from both sites were observed (Durou et al. 2008).

The mudflats of the Authie estuary support worm population densities and biomasses that are lower than those reported in some areas from Belgium to Portugal, whereas data for Sweden are difficult to compare because only juveniles were considered (Table 8.3). On the other hand, worm densities and biomass in the Authie reference estuary were consistently higher than many values published for estuaries from Northern Europe to Morocco. Low densities and biomasses observed in the Seine estuary during most of the present study were among the lowest registered in the literature.

In the Seine estuary, densities and biomasses decreased from 2002 to 2004 and seemed to show a downward trend in population of *N. diversicolor* (Figure 8.5). The lower densities were comparable to densities from polluted parts of the Tees estuary (Gray 1976). In the Ems estuary, North Sea, Essink et al. (1985) found decreasing densities of *N. diversicolor* toward a point of discharge of organic waste. They suggested two causes: mortality and migration. Such a migration was observed in the Oued Souss estuary in Morocco by Aït Alla et al. (2006b).

During the 3 years covered by the survey, eight cohorts were observed for the *N. diversicolor* population in the Authie versus six only in the Seine. This is in agreement with a longer reproductive period in the reference site than in the polluted site (February to May in the Authie and May in the Seine). During the same period (November 2002 to November 2003) in Poole Harbour, a marine site on the south coast of England, Le dit Durell et al. (in press) distinguished three peaks of recruitment each year. Further north, *N. diversicolor* populations generally have short spawning periods and only a single annual recruitment period, usually in spring or summer (Thames estuary, Dales 1950; Medway estuary, Kent, Wharfe 1977; east

TABLE 8.3
Characteristics of *Nereis diversicolor* Populations from Northern Europe to Morocco

Authors	Site	Density (ind.m^{-2})	Biomass (g.m^{-2})	Production (g.m^{-2})	P/B
Möller (1985) (juveniles)	Sweden	800–60000	0.3–2.8	1.6–7.0	3.5–9.0
Chambers and Milne (1975)	Ythan, U.K.	208–961	4.22–11.49	12.8	3.00
Nithart (1998)	Norfolk, U. K.	392	10.26	17.91	1.75
Essink et al. (1985)	Holland	3200	—	—	—
Heip and Herman (1979)	Belgium	5000–17000	13–39	61	2.5
Present work	Authie, France	672–3580	3.9–32.6	—	—
	Seine, France	80–920	0.6–3.4	—	—
Gillet (1990)	Loire, France	1700	16	34.7	2.2
Gillet and Torresani (2003)	Loire, France	300–2600	4.6–9.6	5.1–34.4	1.1–3. 6
Garcia-Arberas & Rallo (2002)	Spain	1060	8.57	16.87	1.97
Arias & Drake (1995)	Spain	1886	4.6	22.7	4.9
Fidalgo e Costa et al. (1998)	Portugal	433–15927	40.14–99.96	12.31–23.68	1.9–4. 0
Abrantes et al. (1999)	Portugal	43–718	15.9–74.2		4.4–7.9
Gillet (1993)	Bou Regreg, Morocco	620	15.35	66	4.3
Aït Alla et al. (2006b)	Souss, Morocco	507–4627	14.16–75.52	23.83–141.3	1.68–1.87

coast of England, Olive and Garwood 1981; western Sweden, Möller 1985). However, two spawning periods were reported in the Ythan estuary, Scotland (Chambers and Milne 1975) and in Norsminde Fjord, Denmark (Kristensen 1984). Further south, *N. diversicolor* populations have longer spawning periods and two recruitment periods in spring and autumn as for example in the Bou Regreg estuary (Gillet 1993), the Oued Souss (Aït Alla et al. 2006b), the Ria Formosa (Sprung 1994), the Canal de Mira (Abrantes et al. 1999), lagoonal estuarine systems of Portugal (Fidalgo e Costa et al. 1998), and the Loire estuary (Gillet 1990; Gillet and Torresani 2003). Changes in timing of spawning and life cycles in different geographical regions may be consequences of different local temperature regimes (Kristensen, 1984), since the inducing stimulus was reported to be a sharp temperature rise (Dales, 1950). However, as noted above, climatic factors were very similar between the sites of interest in the present study.

From the whole set of data obtained for *N. diversicolor* populations from the Seine and Authie estuaries, rational links have been identified between effects at different levels of biological integration. Reproductive success, for example, is related to energy allocation among mechanisms of defence, particularly detoxification processes involved in tolerance (Chapter 7), maintenance of basal metabolism, growth, and reproduction. The energy source for the organisms is food available in the field and is largely influenced by season (Chapters 12 through 14). *N. diversicolor* is considered an opportunistic species able to fulfill its energy needs with different diets (Olivier et al. 1995; Meziane and Rétière 2002). However, energy metabolism may be disturbed because feeding behaviour is impaired (Amiard-Triquet 2009). Effects of estuarine sediment contamination have been shown in laboratory and *in situ* assays of feeding in *N. diversicolor* (Moreira et al. 2006). Chemicals can also affect digestion by impairing digestive enzyme activity (De Coen and Janssen 1998).

Because *N. diversicolor* is a key species in assessing the structure and functioning of estuarine ecosystems in northwestern Europe, the cascading effects observed from infra-individual and population levels may be extended further to community level. A decrease in ragworm populations exerts consequences on food availability for their predators, but this effect has not yet been accurately assessed by comparison with other stressors linked to the reduction of natural areas in the Seine estuary (Chapter 1). In addition, the abundance of the population of *N. diversicolor* largely controls bioturbation efficiency. Decreases of sediment reworking, burrow wall surface, and pumping rates of *N. diversicolor* were observed from 2002 through 2004 in the Seine estuary (Gillet et al. 2008). Thus, in Chapter 15, the influence of bioturbation on geochemical processes in surficial sediments will be assessed.

REFERENCES

Abele, D. et al. 1998. Exposure to elevated temperatures and hydrogen peroxide elicits oxidative stress and antioxidant response in the Antarctic intertidal limpet *Nacella concinna*. *Comp. Biochem. Physiol.* 120B: 425–435.

Abrantes, A., F. Pinto, and M.H. Moreira. 1999. Ecology of the polychaete *Nereis diversicolor* in the Canal de Mira (Ria de Aveiro, Portugal): population dynamics, production and oogenic cycle. *Acta Oecol.* 20, 267–283.

Aït Alla, A. et al. 2006b. Response of *Nereis diversicolor* (Polychaeta, Nereidae) populations to reduced wastewater discharge in the polluted estuary of Oued Souss, Bay of Agadir, Morocco. *Estuar. Coast. Shelf Sci.* 70, 633–642.

Ait Alla, A. et al. 2006a. Tolerance and biomarkers as useful tools for assessing environmental quality in the Oued Souss estuary (Bay of Agadir, Morocco). *Comp. Biochem. Physiol.* 143C, 23–29.

Amiard, J.C. and C. Amiard-Triquet, Eds. 2008. *Les biomarqueurs dans l'évaluation de l'état écologique des milieux aquatiques.* Paris: Lavoisier.

Amiard-Triquet, C. 2009. Behavioral disturbances: the missing link between sub-organismal and supra-organismal responses to stress? Prospects based on aquatic research. *Human Ecol. Risk Ass.* (in press).

Andries, J.C. 2001. Endocrine and environmental control of reproduction in Polychaeta. *Can. J. Zool.* 79, 254–270.

Arias, A.M. and P. Drake. 1995. Distribution and production of the polychaete *Nereis diversicolor* in a shallow coastal lagoon in the Bay of Cadiz (SW Spain). *Cah. Biol. Mar.* 36, 201–210.

Bains, O.S. and C.J. Kennedy. 2004. Energetic costs of pyrene metabolism in isolated hepatocytes of rainbow trout, *Oncorhynchus mykiss. Aquat. Toxicol.* 67: 217–226.

Barnes, R.S.K. et al. 1988. *The Invertebrates. A New Synthesis*. Oxford: Blackwell Science.

Bartels-Hardege, H.D. and E. Zeeck. 1990. Reproductive behaviour of *Nereis diversicolor* (Annelida: Polychaeta). *Mar. Biol.* 106: 409–412.

Bianchini, A. et al. 2005. Mechanism of acute silver toxicity in marine invertebrates. *Aquat. Toxicol.* 72: 67–82.

Bryan, G.W., W.J. Langston, and L.G. Hummerstone. 1980. The use of biological indicators of heavy metal contamination in estuaries, with special reference to assessment of the biological availability of metals in estuarine sediments from the South-West Britain. *Occ. Publ. Mar. Biol. Ass. U.K.* 1, Plymouth, 73 pp.

Cairns, J. Jr. 1992. The threshold problem in ecotoxicology. *Ecotoxicol*, 1: 3–16.

Calow, P. 1991. Physiological costs of combating chemical toxicants: ecological implications. *Comp. Biochem. Physiol.* 100C: 3–6.

Cammen, L.M., 1987. Polychaetes. In *Animal Energetics*, Pandian, T.J. and Vernberg, F.J., Eds. New York: Academic Press, pp. 217–260.

Carr, R.S. 1981. Biochemical indices of stress in the sandworm *Neanthes virens* (Sars). Ph.D. Thesis, Texas A & M University.

Cassai, C. and P. Prevedelli. 1998. Reproduction effort, fecundity and energy metabolism in two species of the genus *Perinereis* (Polychaeta : Nereidae). *Invertebr. Reprod. Dev.* 34: 125–131.

Carr, R.S. and J.M. Neff. 1984. Field assessment of biochemical stress indices for the sandworm *Neanthes virens* (Sars). *Mar. Environ. Res.* 14: 267–279.

Chambers, R.M. and H. Milne. 1975. Life cycle and production of *Nereis diversicolor* O.F. Müller in the Ythan estuary, Scotland. *Estuar. Coast. Mar. Sci.* 3: 133–144.

Clark, R.B. 1961. The origin and formation of the heteronereis. *Biol. Rev. Camb. Philos. Soc.* 36: 199–236.

Clay, E. 1967. *Literature Survey of the Common Fauna of Estuaries*. I. *Nereis diversicolor* O.F. Müller, Marine Research Station, Brixham, Devon, UK, pp. 1–28.

Dales, R.P. 1950. The reproduction and larval development of *Nereis diversicolor* O.F. Müller. *J. Mar. Biol. Ass. U.K.* 29: 321–360.

De Coen, W.M. and C.R. Janssen. 1998. The use of biomarkers in *Daphnia magna* toxicity testing. I. The digestive physiology of daphnids exposed to toxic stress. *Hydrobiol.* 367: 199–209.

De Coen, W.M. and C.R. Janssen. 2003. The missing biomarker link: relationship between effects on the cellular energy allocation biomarker of toxicant-stressed *Daphnia magna* and corresponding population characteristics. *Environ. Toxicol. Chem.* 22: 1632–1641.

Desrosiers, G. et al. 1988. Comparaisons de critères utilisables pour l'étude de la structure des populations du polychète *Nereis virens* (Sars). *J. Can. Zool.* 66: 1454–1459.

Durou, C. and C. Mouneyrac. 2007. Linking steroid hormone levels to sexual maturity index and energy reserves in *Nereis diversicolor* from clean and polluted estuaries. *Gen. Comp. Endocrinol.* 150: 106–113.

Durou, C., C. Mouneyrac, and C. Amiard-Triquet. 2005. Tolerance to metals and assessment of energy reserves in the polychaete *Nereis diversicolor* in clean and contaminated estuaries. *Environ. Toxicol.* 20: 23–31.

Durou, C., C. Mouneyrac, and C. Amiard-Triquet. 2008. Environmental quality assessment in estuarine ecosystems: use of biometric measurements and fecundity of the ragworm *Nereis diversicolor* (Polychaeta, Nereididae). *Water Res.* 42: 2157–2165.

Durou, C. et al. 2007a. Biomonitoring in a clean and a multi-contaminated estuary based on biomarkers and chemical analyses in the endobenthic worm *Nereis diversicolor*. *Environ. Pollut.* 148: 445–458.

Durou, C. et al. 2007b. From biomarkers to population response in *Nereis diversicolor*: assessment of stress in estuarine ecosystems. *Ecotox. Environ. Saf.* 66: 402–411.

Essink, K. et al. 1985. Population dynamics of the ragworm *Nereis diversicolor* in the Dollard (Ems estuary) under changing conditions of stress by organic pollution. In *Marine Biology of Polar Regions and Effects of Stress on Marine Organisms*, Gray, J.S. and Christiansen, M.E., Eds. Chichester: John Wiley & Sons, pp. 585–600.

Fidalgo e Costa, P., R. Sarda, and L. Cancela da Fonsesca. 1998. Life cycle, growth and production of the polychaete *Nereis diversicolor* O.F. Müller in three lagoonal estuarine systems of the southwestern Portuguese coast (Odeceixe, Aljezur and Carrapateira). *Ecologie.* 29: 523–533.

Fischer, A. and U. Hoeger. 1993. Metabolic links between somatic sexual maturation and oogenesis in nereid annelids: a brief review. *Invertebr. Reprod. Dev.* 23: 131–138.

Fischer, A. and A. Dhainaut. 1985. The origin of the yolk in the oocytes of *Nereis virens* (Annelida, Polychaeta): electron microscopic and autoradiographic studies by use of unspecific and yolk-specific markers. *Cell Tissue Res.* 240: 131–138.

Fischer, A. and A. Dorresteijn. 2004. The polychaete *Platynereis dumerilii* (Annelida): a laboratory animal with spiralian cleavage, lifelong segment proliferation and mixed benthic/pelagic life cycle. *BioEssays* 26: 314–325.

Fletcher, C.R. 1974. Volume regulation in *Nereis diversicolor*-III: adaptation to a reduced salinity. *Comp. Biochem. Physiol.* 47A: 1221–1234.

Fritzsche, D. and J.A. von Oertzen, 1995. Metabolic responses to changing environmental conditions in the brackish water polychaetes *Marenzellaria viridis* and *Hediste diversicolor*. *Mar. Biol.* 121: 693–699.

Garcia-Alonso, J. and N. Rebscher. 2005. Estradiol signalling in *Nereis virens* reproduction. *Invertebr. Reprod. Dev.* 48: 95–100.

Garcia-Arberas, L. and A. Rallo. 2002. Life cycle, demography and secondary production of the polychaete *Hediste diversicolor* in a non-polluted estuary in the Bay of Biscay. *Mar. Ecol.* 23: 237–251.

Gillet, P. and S. Torresani. 2003. Structure of the population and secondary production of *Hediste diversicolor* (O.F. Müller, 1776), (Polychaeta, Nereidae) in the Loire estuary, Atlantic coast, France. *Estuar. Coast. Shelf Sci.* 56: 21–26.

Gillet, P. et al. 2008. Response of *Nereis diversicolor* population (Polychaeta, Nereididae) to the pollution impact: Authie and Seine estuaries (France). *Estuar. Coast. Shelf Sci.* 76: 201–210.

Gillet, P. 1990. Biomasse, production et dynamique des populations de *Nereis diversicolor* (annélide polychète) de l'estuaire de la Loire (France). *Oceanol. Acta* 13: 361–371.

Gillet, P. 1993. Impact de l'implantation d'un barrage sur la dynamique des populations de *Nereis diversicolor* (Annélide polychète) de l'estuaire du Bou Regreg, Maroc. *J. Rech. Océanogr.* 18: 15–18.

Golding, D.W. and E. Yuwono. 1994. Latent capacities for gametogenic cycling in the semelparous invertebrate *Nereis*. *Proc. Nat. Acad. Sci. USA* 91: 11777–11781.

Gray, J.S. 1976. The fauna of the polluted river Tees estuary. *Estuar. Coast. Mar. Sci.* 4: 653–676.

Grémare, A. and P.J.W. Olive. 1986. A preliminary study of fecundity and reproductive effort in two polychaetous annelids with contrasting reproductive strategies. *Intern. J. Invertebr. Reprod. Dev.* 9: 1–16.

Heip, C. and R. Herman. 1979. Production of *Nereis diversicolor* O.F. Muller (Polychaeta) in a shallow brackish water pond. *Estuar. Coast. Mar. Sci.* 8: 297–305.

Hoeger, U., N. Rebscher, and G. Geier. 1999. Metabolite supply in oocytes of *Nereis virens*: role of nucleosides. *Hydrobiology* 402: 163–174.

Hopkins, W.A., C.L. Rowe, and J.D. Congdon. 1999. Elevated trace element concentrations and standard metabolic rate in banded water snakes (*Nerodia fasciata*) exposed to coal combustion wastes. *Environ. Toxicol. Chem.* 18: 1258–1263.

Hummel, H. et al. 2000. The respiratory performance and survival of the bivalve *Macoma balthica* at the southern limit of its distribution area: a translocation experiment. *J. Exp. Mar. Biol. Ecol.* 251: 85–102.

Ivankovic, D. et al. 2005. Evaluation of the *Mytilus galloprovincialis* Lam. digestive gland metallothionein as a biomarker in a long-term field study: seasonal and spatial variability. *Mar. Pollut. Bull.* 50: 1303–1313.

Julian, D. et al. 2001. Influence of environmental factors on burrow irrigation and oxygen consumption in the mudflat invertebrate *Urechis caupo*. *Mar. Biol.* 139: 163–173.

Kirkegaard, J.B. 1970. Age determination of *Nephtys* (Polychaeta: Nephtyidae). *Ophelia* 7: 277–282.

Kristensen, E., 1983a. Ventilation and oxygen uptake by three species of *Nereis* (Annelida: Polychaeta). I. Effects of hypoxia. *Mar. Ecol. Prog. Ser.* 12: 289–297.

Kristensen, E., 1983b. Ventilation and oxygen uptake by three species of *Nereis* (Annelida: Polychaeta). II. Effects of temperature and salinity changes. *Mar. Ecol. Prog. Ser.* 12: 299–306.

Kristensen, E. 2004. Degradation of organic matter in irrigated burrows: what do we know? First Nereis Park Conference. Bioturbation: the ever changing seafloor. Carry-le-Rouet, France. 7–9 November 2004.

Kristensen, E. 1984. Life cycle and production in estuarine populations of the polychaetes *Nereis virens* and *Nereis diversicolor*. *Holarctic Ecol.* 7: 249–256.

Lafont, R. and M. Mathieu. 2007. Steroids in aquatic invertebrates. *Ecotoxicol.* 16: 109–130.

Lannig, G., J.F. Flores, and I.M. Sokolova. 2006. Temperature-dependent stress response in oysters, *Crassostrea virginica*: pollution reduces temperature tolerance in oysters. *Aquat. Toxicol.* 79: 278–287.

Le dit Durell, S.E.A. et al. 2008. Life history strategy and production in a heavily predated and disturbed population of the ragworm *Nereis diversicolor*. *Estuar. Coast. Shelf Sci.* (in press).

Lee, R.F., A. Walker, and D.J. Reish. 2005. Characterization of lipovitellin in eggs of the polychaete *Neanthes arenaceodentata*. *Comp. Biochem. Physiol.* 140B: 381–386.

Le Gal, Y. et al. 1997. Charge énergétique en adénylates (CEA) et autres biomarqueurs associés au métabolisme énergétique. In *Biomarqueurs en écotoxicologie: Aspects fondamentaux*, Lagadic, L. et al., Eds. Paris: Masson, pp. 241–285

Lucas, A. and P.G. Beninger. 1984. The use of physiological condition indices in marine bivalve aquaculture. *Aquaculture* 44: 187–200.

Maltby, L. et al. 2001. Linking individual-level responses and population-level consequences. In *Ecological Variability: Separating Natural from Anthropogenic Causes of Ecosystem Impairment*, Baird, D.J. and Burton, G.A.J., Eds. Society of Environmental Toxicology and Chemistry, pp. 27–82.

Marchand, J. et al. 2004. Physiological cost of tolerance to toxicants in the European flounder *Platichthys flesus*, along the French Atlantic Coast. *Aquat. Toxicol.* 70: 327–343.

Mayer, F.L. et al. 2002. Physiological and non specific biomarkers. In *Biomarkers: Biochemical, Physiological and Histological Markers of Anthropogenic Stress*, Huggett, R.J. et al., Eds. Boca Raton, FL: Lewis Publishers, pp. 5–85.

Mettam, C., V. Santhanam, and M.S. Havard. 1982. The oogenic cycle of *Nereis diversicolor* under natural conditions. *J. Mar. Biol. Ass. U.K.* 62: 637–645.

Meziane, T. and C. Retière. 2002. Croissance de juvéniles de *Nereis diversicolor* nourris avec des détritus d'halophytes. *Oceanol. Acta.* 25: 119–124.

Möller, P., 1985. Production and abundance of juvenile *Nereis diversicolor* and oogenic cycle of adults in shallow waters of Sweden. *J. Mar. Biol. Ass. U.K.* 65: 603–616.

Moreira, S.M. et al. 2006. Effects of estuarine sediment contamination on feeding and on key physiological functions of the polychaete *Hediste diversicolor*: laboratory and *in situ* assays. *Aquat. Toxicol.* 78: 186–201.

Mouneyrac, C. et al. 2006. *In situ* relationship between energy reserves and steroid hormone levels in *Nereis diversicolor* (O.F. Müller) from clean and contaminated sites. *Ecotox. Environ. Saf.* 65: 181–187.

Nielsen, A.M. et al. 1995. Feeding, growth and respiration in the polychaetes *Nereis diversicolor* (facultative filter-feeder) and *N. virens* (omnivorous): a comparative study. *Mar. Ecol. Prog. Ser.* 125: 149–158.

Nithart, M. 1998. Population dynamics and secondary production of *Nereis diversicolor* in a North Norfolk salt marsh (UK). *J. Mar. Biol. Ass. U.K.* 78: 131–143.

Nithart, M., E. Alliot, and C. Salen-Picard, 1999. Production, respiration and ammonia excretion of two polychaete species in a north Norfolk salt marsh. *J. Mar. Biol. Ass. U.K.* 79: 1029–1037.

Oglesby, L.C. 1969. Salinity-stress and dessication in intertidal worms. *Amer. Zool.* 9: 319–331.

Oglesby, L.C. 1978. Salt and water balance. In: *Physiology of Annelids*, Mill, P.J., Ed. London: Academic Press, pp. 555–658.

Olive, P.J.W. 1980. Growth lines in polychaete jaws. In *Skeletal Growth of Aquatic Organisms: Biological Records of Environmental Change,* Rhoads, D.C. and Lutz, R.A., Eds. New York: Plenum Press, pp. 561–592.

Olive, P.J.W. 1983. Annelida: Polychaeta: oogenesis, oviposition and oosorption. In *Reproductive Biology of Invertebrates*, Adivodi, K.G. and Adivodi, R.G. Eds. New York: John Wiley & Sons, pp. 357–422.

Olive, P.J.W. and R.B. Clark. 1978. Physiology of reproduction. In *Physiology of Annelids*, Mill, P.J., Ed. Academic Press: London, pp. 271–368.

Olive, P.J.W. and P.R. Garwood. 1981. Gametogenic cycle and population structure of *Nereis (Hediste) diversicolor* and *Nereis (Nereis) pelagica* from Northern-East England. *J. Mar. Biol. Ass. U.K.* 61: 193–213.

Olivier, M. et al. 1995. Réponses comportementales des polychètes *Nereis diversicolor* (O.F. Müller) et *Nereis virens* (Sars) aux stimuli d'ordre alimentaire: utilisation de la matière organique particulaire (algues et halophytes). *Can. J. Zool.* 73: 2307–2317.

Porcher, M. 1984. Biochemistry of oocyte differentiation in nereids. *Fortschr. Zool.* 29: 207–225.

Rowe, C.L. 1998. Elevated standard metabolic rate in a freshwater shrimp (*Palaemonetes paludosus*) exposed to trace element-rich coal combustion waste. *Comp. Biochem. Physiol.* 121A: 299–304.

Rowe, C.L. et al. 2001. Metabolic costs incurred by crayfish (*Procambarus acutus*) in a trace element-polluted habitat: further evidence of similar responses among diverse taxonomic groups. *Comp. Biochem. Physiol.* 129C: 275–283.

Saiz-Salinas, J.I. and G. Francés-Zubillaga. 1997. Enhanced growth in juvenile *Nereis diversicolor* after its exposure to anaerobic polluted sediments. *Mar. Pollut. Bull.* 34: 437–442.

Scaps, P. 2002. A review of the biology, ecology and potential use of the common ragworm *Hediste diversicolor* (O.F. Müller) (Annelida: Polychaeta). *Hydrobiology* 470: 203–218.

Schroeder, P.C., E. Hofmann, and K. Wallace, 1977. Posterior regeneration and endocrine control of postlarval development in *Nereis grubei* (Kinberg). In *Essays on Polychaetous Annelids*, Reish, D.J. and Fauchald, K., Eds. Los Angeles: University of Southern California Press, pp. 391–405.

Smith, R. 1976. Further observations on the reproduction of *Nereis diversicolor* (Polychaeta) near Tvärminne, Finland, and Kristineberg, Sweden. 13: 179–184.

Smith, R.I. 1977. Physiological and reproductive adaptations of *Nereis diversicolor* to life in the Baltic Sea and adjacent waters. In *Essays on Polychaetous Annelids*, Reish, D.J. and Fauchald, K., Eds. Los Angeles: University of Southern California Press, pp. 373–390.

Sprung, M. 1994. Macrobenthic secondary production in the intertidal zone of the Ria Formosa, a lagoon in Southern Portugal. *Estuar. Coast. Shelf. Sci.* 38: 539–558.

Tshischka, K., D. Abele, and H.O. Poertner, 2000. Mitochondrial oxyconformity and cold adaptation in the polychaete *Neris pelagica* and the bivalve *Arctica islandica* from the Baltic and White Seas. *J. Exp. Biol.* 203: 3355–3368.

Verslycke, T. et al. 2004. Cellular energy allocation and scope for growth in the estuarine mysid *Neomysis integer* (Crustacea: Mysidacea) following chlorpyrifos exposure: a method comparison. *J. Exp. Mar. Biol. Ecol.* 306: 1–16.

Voogt, P.A. 1974. Biosynthesis and composition of sterols in Annelida. I. Investigations on some polychaetes. *Neth. J. Zool.* 24: 22–31.

Wharfe, J.R. 1977. An ecological survey of the benthic invertebrate macrofauna of the lower Medway estuary, Kent. *J. Anim. Ecol.* 46: 93–113.

Widdows, J., C. Nasci, and V. Fossato. 1997. Effects of pollution on the scope for growth of mussels (*Mytilus galloprovincialis*) from the Venice Lagoon, Italy. *Mar. Environ. Res.* 43: 69–79.

Widdows, J. et al. 2002. Measurement of stress effects (scope for growth) and contaminant levels in mussels (*Mytilus edulis*) collected from the Irish Sea. *Mar. Environ. Res.* 53: 327–356.

Xie, L. and P.L. Klerks. 2004. Fitness cost of resistance to cadmium in the least killifish (*Heterandria formosa*). *Environ. Toxicol. Chem.* 23: 1499–1503.

Zeeck, E. et al. 1994. Sex pheromones in marine polychaetes: steroids from ripe *Nereis succinea*. *Steroids* 59: 341–344.

Zwarts, L. and P. Esselink. 1989. Versatility of male curlews *Numenius arquata* preying upon *Nereis diversicolor*: deploying contrasting capture modes dependent on prey availability. *Mar. Ecol. Prog. Ser.* 56: 255–269.

9 Historical Records of the *Nereis diversicolor* Population in the Seine Estuary

Christophe Bessineton

CONTENTS

9.1 INTRODUCTION

The Seine estuary is located at the mouth of a river basin that is highly populated and exposed to important pollutant inputs from industry and agriculture. These inputs are contaminating the food web. Le Havre harbor at the mouth of the estuary and Rouen harbor 80 km upstream generate important sea traffic that has required the fitting of dikes, the dredging of channels, and land reclamation to support ports and industry, changing drastically the morphology of the estuary. A decrease in intertidal area has accompanied these changes and mudflats have been progressively colonised by shore vegetation and brackish reed beds.

Several studies of intertidal macrozoobenthos populations have been performed for the past 25 years using similar sampling protocols. *Macoma balthica* and *Nereis diversicolor* communities have been identified in the northern mudflat of the Seine estuary. A community of *Abra alba* and *Pectinaria koreni* exists at the mouth of the estuary, on subtidal muddy sands (GEMEL 1984; GEMEL 1987; Duhamel and Mayot 2002; Dauvin 2005; Jourde and Dancie 2007). These two communities are of particular importance for the food webs of fish nurseries and crustaceans in the estuary. Intertidal areas are also important for shore birds and the Seine estuary is of importance for birds, designated as a Natura 2000 area and classed as a National Nature Reserve since 1998.

The aim of the chapter is to describe long-term population variations of the polychaete *Nereis diversicolor*, which is a key species of the mudflat ecosystem.

9.2   METHODOLOGY

Different authors have sampled the intertidal macrozoobenthos of the Seine estuary since 1980, using the same sampling protocols (GEMEL 1984; GEMEL 1987 to 1994; CSLHN 1999 and 2005; Duhamel and Mayot 2002; Mayot and Duhamel 2004; Dauvin 2005): three to five replicates taken with a PVC handle core of 15 cm diameter and 25 cm depth, sieving on a 1-mm mesh sieve, preservation of samples in 10% formaldehyde salted solution, identification of species, and numeration of individuals. Biomasses have been measured by some authors, but their methods are too different to be taken into account here. Biological sampling was accompanied by salinity measures and sediment sampling for granulometry. The grain size of sediment is expressed as a percentage of the total weight of the fraction <63 μm.

Station locations were transformed in the GPS WGS 84 projection system using Map Info 6.5 professional software. Locations of stations and seasons of sampling programmes were not identical among authors. Selection from the set of data has permitted the choice of sampling programmes carried out in September or October. Sample stations were combined into 15 sectors according to location upstream or downstream and their situation on the lower, middle, or upper slikke (See Figure 1.3 in Chapter 1 and Figure 9.1). Data selection retained 13 sampling programmes including the years 1987, 1990, 1992, 1994, 1996, 1997, and 2000 through 2006 (Table 9.1).

The Seine estuary is a macrotidal system; the maximum amplitude of the tide is 8.30 m. Altitudes are referred to the zero of marine charts at Le Havre (0 CMH). The altitude of the slikke is between 0 and 7.5 m CMH. The intertidal area above 7.5 m CMH is colonised by shore vegetation.

Important sedimentation occurs in the northern channel, and the altitudes of sectors vary over time, particularly sectors A, B, D, E, and F. Altitudes were measured in each sector for different programmes. In 1987, sampling stations were located on sounding charts prepared by the port of Rouen and older topographic data for altitudes above +5 m CMH, where hydrographic ships cannot navigate. In 1989, a topographic study using stereophotography was performed by the Direction Départementale de l'Equipement for the construction of the Normandy Bridge. This was calibrated in 1990 and 1995 with some field measures. From 1996 to 2004, the altitude of each sector was recorded by measuring the lengths of PVC sticks inserted into the sediment and regularly monitored with a differential GPS, initially during the MAST III INTRMUD program, and after 2000 by the staff of the Nature Reserve. For sites below +5 m CMH, the measures were compared with sounding data taken every year by the ports of Rouen or Le Havre. River flow data were obtained from the Service de la Navigation de la Seine (2008), determined at the Pose dam, 160 km upstream, where the tidal influence ends. After different tests, mean yearly river flow was chosen to examine correlations with benthos data.

9.3   RESULTS

Data showing densities of *Nereis diversicolor* (number of individuals per square meter), salinity, sediment granulometry (percent of <63-μm weight fraction), altitudes of sectors (referring to the zero of marine charts at Le Havre) of the Seine

FIGURE 9.1 Locations of different sectors combining stations studied by different authors and average densities (number of individuals per square meter) of *Nereis diversicolor* in each sector. For general context, see Figure 1.3.

TABLE 9.1
Sampling Stations Investigated by Several Laboratories Combined to Create Sectors Shown in Figure 9.1

			GEMEL 87[a]	GEMEL 90[b]	GEMEL 92[c]	GEMEL 94[d]	Intrmud 96–97[e]	RN 2000–2006[f]
Grande crique	A	Higher	1b			Disappeared		
	B	Middle	1, 2	1	1, 2	RB2, RB2		
	C	Lower		2, 2b	2A	RA0, RA1, RA2		
Amont PdN	D	Higher	3	3	3, 4	R1 4, R1 5	101, 104	901, 902
	E	Middle	4	4	5, 5A	R1 5A	107	903, 904
	F	Lower	5	4B	5B	R1 5B	112	905, 906
Aval PdN	G	Higher	6, 7	7	6		201, 204	801, 802
	H	Middle	8	8	7, 8		207	803, 804
	I	Lower		8B	8A		212	805
Dune	J	Higher	18	16	16, 17			501, 502
	K	Middle	19	17	17A			503, 504
	L	Lower		17B	17B			505, 506
D2	M	Higher	20				301, 304	302
	N	Middle	21				307	303, 304
	O	Lower	22				310	305

[a] GEMEL 1984, 1987; [b]GEMEL 1990; [c]GEMEL 1992; [d]GEMEL 1994; [e]CSLHN 1999; [f]Duhamel and Mayot 2002, Mayot and Duhamel 2003, 2004, and 2006, CSLHN 2006; Jourde and Dancie 2007. See reference list.

estuary for 1987, 1990, 1992, 1994, 1996, 1997, and 2000 to 2006, and river flow (cubic meters per second) for these periods are reassembled in Appendix 9.1.

The data were subjected to principal component analysis (PCA; Figure 9.2). Factor 1, which explains 40.36% of the variance, is negatively correlated with salinity and altitude whereas factor 2 (25.24% of the variance) is negatively correlated with the percentage of mud (<63 μm) and densities of *N. diversicolor* (Figure 9.2a). The river flow is poorly represented in the principal plan 1–2 (Figure 9.2a) and better represented on the third axis (Figure 9.2b).

To analyse spatial and temporal variations, average values were calculated by sector with all years and by year with all sectors. Average densities by sector (Figure 9.1) show clearly a gradient of *N. diversicolor* densities from east to west and from north to south where a large tidal channel with lower altimetry is located. Highest densities are in sectors B, C, D, E, G, and J. These sectors are located in the high and middle slikkes of the upper part of the studied area that have more mud (>80%) and lower salinity (10‰) (Table 9.2). Sectors K, L, M, and N exhibited intermediate values for both density and granulometry and the highest salinity measures. In agreement with the trends shown in the PCA (Figure 9.2), these results pointed at granulometry as a major factor controlling *N. diversicolor* populations. Sectors F, I, and L

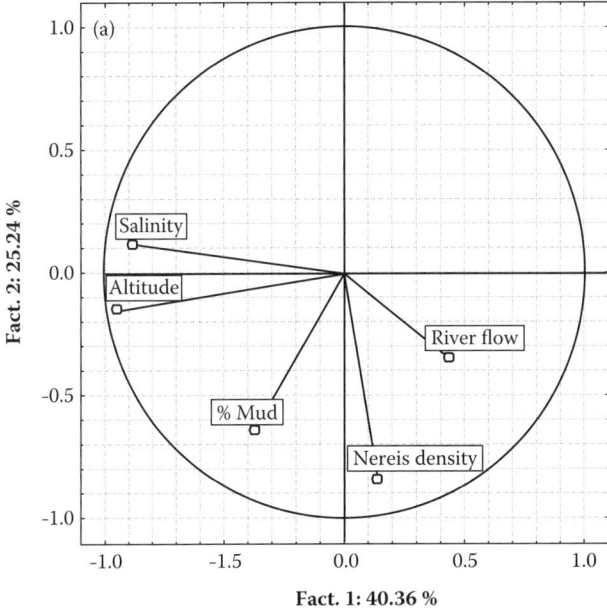

(a)

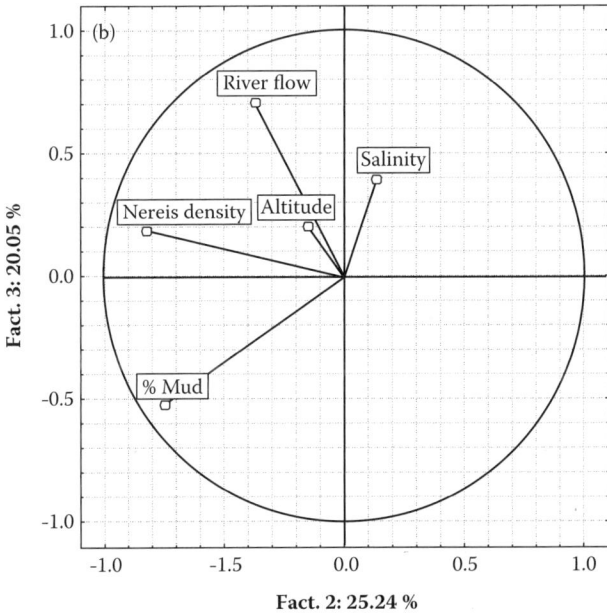

(b)

FIGURE 9.2   Principal component analysis (a: principal plan 1–2, b: principal plan 2–3) including biological (densities of *N. diversicolor*) and environmental parameters for different sectors of the Seine estuary from 1987 to 2006. See Appendix 9.1 for raw data.

TABLE 9.2
Average Values of Abiotic Parameters in Different Sectors

Sector	B	C	D	E	F	G	H	I	J	K	L	M	N
% mud	83.0	80.5	84.3	73.2	13.0	85.8	16.5	42.0	57.3	70.2	68.0	45.6	61.2
Altitude	7.2	4.5	8.1	6.6	3.2	8.0	5.7	3.0	7.1	5.1	2.3	5.4	3.3
Salinity	12.0	12.5	11.1	11.4	14.7	12.9	16.7	16.1	15.5	16.0	20.0	20.1	21.0

Notes: Mud content as percent of total weight of the grain size fraction <63 μm; altitude as height in meters above the zero of marine charts at Le Havre (CMH); salinity.

TABLE 9.3
Average Values of Mud Content, Salinity, Altitude, River Flow, and Density of *Nereis diversicolor,* 1987 to 2006

Year	1987	1990	1992	1994	1996	1997	2000	2001	2002	2003	2004	2005	2006
% mud	70.0	72.5	72.5	75.6	68.2	69.9	67.4	69.3	68.7	67.4	73.6		
S‰	20.2	15.6	13.9	11.4	14.8	15.6	16.0	15.5	16.0	15.5	13.5	15.9	15.3
Altitude (m CMH)	5.6	4.8	4.9	5.3	5.0	4.7	4.5	4.9	4.7	5.3	5.4	5.6	5.8
Flow river (m^3/s)	504.5	288.8	362.5	627.5	335.2	391.5	750.3	903.9	594.8	389.3	410.9	332.9	417.6
Ragworm density (ind./m^2)	1015	679	363	1161	429	476	536	743	328	235	322	190	227

have low density values; they are situated near the large and sandy tidal channel. Sector O is in this channel where the bottom is occupied by mobile sand not containing *N. diversicolor.* Sector A did not contain *N. diversicolor* in 1980 and 1987. It has been colonised by vegetation since the completion of the Normandy Bridge and was not sampled sufficiently to be taken into account in the statistical analysis. These two last sectors represent the upper and the lower limits of the *N. diversicolor* population in the northern channel of the Seine estuary.

Average values for each year (Table 9.3) do not show significant trends for mud content, salinity, and altitude. River flow shows important inter-annual variations. Average densities of *N. diversicolor* decreased generally during the period (Table 9.3). Figure 9.3 depicts an inverse relationship between these two parameters but no significant correlation exists. Nevertheless, a more detailed analysis shows that the data set is heterogeneous and can be separated into two periods: the first from 1987 to 1997 and the second from 2000 to 2006. Within each period, the same trend was recognized; during the latter period, a significant correlation (99% level) was shown between worm density and river flow.

The two linear regressions have the same origin, but data for the first period (1987 to 1997) reveal a slope twice as high as the second one (2000 to 2006) ($a_1 = 2.1441$ and $a_2 = 1.0082$, respectively). Further investigations have shown that despite the use

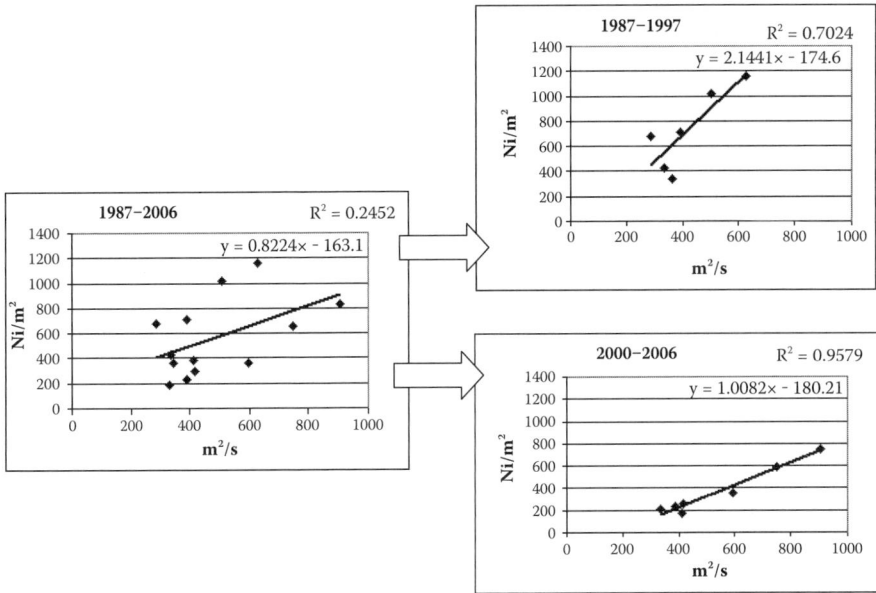

FIGURE 9.3 Relations between annual average densities of *Nereis diversicolor* and river flow from 1987 to 2006, 1987 to 1997, and 2000 to 2006.

of the same sampling protocol in the field by the different authors during the two periods, the analysis protocol changed at the end of 1997. It is assumed that this change in analysis method is an important factor for explaining the shift observed between the two slopes and must be considered in interpreting the long-term data series. In addition to the major importance of sediment granulometry in controlling population localisation of *N. diversicolor*, at least over recent years, river flow may be considered as influencing interannual variations of *N. diversicolor* densities.

As noted above, no *N. diversicolor* were found on the shore (sector A) or sand banks of the lower part (sector O). The limits of these habitats changed between 1987 and 2006. Table 9.4 shows the evolution of the surfaces of the different habitats in the northern part of the estuary measured with aerial pictures, bathymetric charts, and grain size analysis (SIRS 2007; GIP Seine Aval 2008; Cuvillez 2003; Port Autonome du Havre 2006; Cuvillez et al. 2008). Areas above +7.50 m CMH or lower than 5 m CMH did not contain *N. diversicolor*, as indicated above. The habitat of this species is between +5 and +7.5 m CMH. The most favourable habitat decreased from 886 ha in 1980 to 278 ha in 2005, corresponding to a loss of more than 70%.

9.4 DISCUSSION

To evaluate the food available for fish and birds preying on *Nereis diversicolor* in a first approach, the surface area of favourable habitat and the density of *N. diversicolor* were combined for each year. To avoid the heterogeneity of the data set already mentioned, densities were recalculated for 1987 through 1996 using the river flow

TABLE 9.4
Variation of Surface Areas (ha) of Habitats in Northern
Mudflat of the Seine Estuary, 1980 to 2005

Year	1980	1985	1990	1995	2000	2005
< 0 CMH	316	453	441	381	366	289
0–5 CMH	558	409	387	435	489	676
5–7.5 CMH	886	611	562	547	363	278
> 7.5 CMH	759	1046	1129	1156	1301	1276

Note: <0 CMH and 0–5 CMH: sand banks with mobile sand and strong currents; 5–7.50 CMH: mud flat; >7.50 CMH: shore vegetation. CMH: altitudes are referred to the zero of marine charts at Le Havre.

and linear regression for 2000 through 2006, which presents a better regression coefficient $y = 1.0082x-180.21$ ($R^2 = 0.9579$, Figure 9.3). Food availability indicators were calculated for each year by multiplying surface areas of the 5 to 7.50 m CMH layer by the recalculated densities, expressed as the percentage of the 1987 value (Table 9.5).

The pattern of change of this indicator shows a reduction of the food availability offered by *N. diversicolor* over the period studied. The food availability in 2006 is 26% of that in 1987. However, inter-annual variations are important. In 1994 and 2001, strong river flows compensated for the loss of surface area and the indicator is higher than in 1987.

9.5   CONCLUSIONS

A multi-year trend showed a decrease of population densities (number of individuals per square meter) of *Nereis diversicolor* with strong variations from one year to another. Several abiotic factors were considered in this chapter to explain the causes of these fluctuations. Regressions and PCA highlight the importance of sediment granulometry, and in the 2000 through 2006 period, a significant correlation was shown between worm density and river flow. Similarly, Mediterranean benthic communities located off the Rhone fluctuated in parallel with river flow (Harmelin-Vivien and Salen-Picard 2002). This is attributed to the varying quantities of suspended matter and associated nutrients brought to the sea.

Ecological studies on the north bank of the Seine estuary (Station PN, see Figure 1.3 in Chapter 1) in 2003 and 2004 revealed lower densities of diatoms (Chapter 12), foraminiferans (Chapter 13), and, at least in spring and summer, of nematodes (Chapter 14) compared to a reference estuary less impacted by anthropogenic activities.

Although these results were not obtained at the scale of the whole northern mudflat, it remains possible that *N. diversicolor* has experienced decreased food availability, even if this opportunistic species is considered capable of meeting its energy

TABLE 9.5
Food Availability Indicators

Year	1987	1990	1992	1994	1996	1997	2000	2001	2002	2003	2004	2005	2006
Densities recalculated	328	111	185	452	158	215	536	743	328	235	322	190	227
Mud surface area	657	592	549	505	462	440	375	353	332	310	288	266	245
Density × surface area	215,859	65,686	101,648	228,607	72,848	94,429	200,915	262,584	108,911	72,806	92,733	50,626	55,530
% of 1987 ragworm availability	100	30	47	106	34	44	93	122	50	34	43	23	26

Note: Surface area of layer 5 to 7.50 CMH multiplied by density of *Nereis diversicolor* recalculated relative to 1987 value. CMH: altitudes are referred to the zero of marine charts at Le Havre.

needs with different diets (Olivier et al. 1995; Meziane and Rétière 2002). In addition to reduced food bioavailability, feeding rate may be lowered in *N. diversicolor* living in contaminated sediments (Moreira et al. 2006). Activities of the amylase and cellulose digestive enzymes are lowered in *N. diversicolor* from contaminated sediments including specimens from the Seine estuary (J. Kalman personal communication). Finally, strategies allowing organisms to cope with chemical contaminants present an energy cost which can re-echo at the population level (Chapter 8).

Retrospective interpretation of data from different authors, even though they used the same sampling methodology in the same geographical area, may present heterogeneities. We found a difference of slopes in linear regressions of the river flow/density relationship over two periods, 1987 through 1997 and 2000 through 2006. The slope of the first period was twice as high as the slope of the second one. The second period was characterized by a good relationship between density of *N. diversicolor* and river flow and this seems to be an important controlling factor of interannual variations.

Harmelin-Vivien and Salen-Picard (2002) have shown a relationship between Rhone river flow (linked to climate) and sole fishery yields in the Mediterranean (Gulf of Lyons, France). This reaction was explained by the incorporation of terrestrial particulate organic matter into coastal food webs that induced a bottom-up transfer effect of trophic resources from one level up to another, such as polychaetes dominating benthic communities that constitute the main prey of the common sole located off the Rhone.

The habitat favourable for the species in the Seine estuary corresponds to an altitude between +5 and +7.50 m CMH and a mud content exceeding 30% (Appendix 9.1). The surface area of this optimal habitat reduced between 1987 and 2006 from 611 ha to 278 ha.

Nereis diversicolor is a key species of the mudflat ecosystem, situated at the bases of the food webs of many species of birds and fish. In a first approach, food availability derived from *Nereis diversicolor* was estimated as surface area multiplied by the density relative to 1987. In 2006, this resource represented 26% of the 1987 value. However, in some years, strong river flows may have compensated for the loss of surface area of habitat, as in 1994 and 2001.

REFERENCES

CSLHN (Cellule de Suivi du Littoral Haut Normand). 1999. Macrozoobenthos de l'estuaire de la Seine Prog. MAST III INTRMUD.Cellule de Suivi du Littoral Haut Normand, 2000 Le Havre.

CSLHN. 2005. Mesures compensatoire de Port 2000 : Création d'un nouveau méandre en amont de la Fosse Nord. Macrofaune benthique. Campagne de septembre 2004. Rapport d'étude pour le PAH.

CSLHN. 2006. Suivi du macrozoobenthos intertidal de la grande vasière nord de l'estuaire de la Seine. Campagne 2005. Rapport de la Cellule de Suivi du Littoral Haut Normand pour la Maison de l'Estuaire.

Cuvillez, A. 2003. Morphdynamique actuelle d'une vasière estuarienne intertidale, cas de la grande vasière de l'estuaire de la Seine. Couplage de la télé détection basse altitude et de l'altimétrie à haute résolution. Rap DEA Université du Havre..

Cuvillez, A. et al. 2008. Morphological responses of an estuarine intertidal mudflat to installations: annual and multi annual scale (Seine, France). *Geomorphologia* (submitted).

Dauvin, J.C. 2002. Gestion intégrée des zones côtières: outils et perspectives pour la préservation du patrimoine naturel. *Patrimoines naturels*, 57: 346.

Dauvin, J.C. 2005. The food web in the lower part of the Seine estuary: a synthesis of existing knowledge. *Hydrobiologia* 540: 13–27.

Duhamel, S. and S. Mayot. 2002. Suivi du macrozoobenthos intertidal de la grande vasière nord de l'estuaire de la Seine. Campagne 2000. Rapport de la Cellule de Suivi du Littoral Haut Normand pour la DIREN Haute Normandie.

GEMEL. 1984. Etude du Macrozoobenthos intertidal de l'estuaire de la Seine: relations avec le substrat. Actes du Museum de Rouen 1985–1993.

GEMEL. 1987. Actualisation de la cartographie biosédimentaire de la grande vasière de l'estuaire de la Seine. Rapport d'étude pour la Chambre de Commerce et d'Industrie du Havre, Le Havre, France.

GEMEL. 1990. Suivi biosédimentaire de la grande vasière nord de l'estuaire de la Seine au cours des travaux du Pont de Normandie- deuxième tranche d'études. Rapport d'étude pour la Chambre de Commerce et d'Industrie du Havre, Le Havre, France.

GEMEL. 1992. Suivi biosédimentaire de la Grande Vasière Nord de l'estuaire de la Seine au cours des travaux de construction du Pont de Normandie Dernière tranche d'études- Rapport d'étude pour la Chambre de Commerce et d'Industrie du Havre, Le Havre, France.

GEMEL. 1994. Suivi biosédimentaire de la grande vasière nord de l'estuaire de la Seine après la construction du Pont de Normandie. Rapport d'étude pour la Chambre de Commerce et d'Industrie du Havre, Le Havre, France.

GIP Seine Aval. 2008. Les vasières de l'estuaire de la Seine. Le point de vue du benthos. Bessineton, C. and Bacq, N. Eds. Fascicule Seine Aval (in press).

Goss-Custard, J.D. 2004. Modelling proposed mitigation areas for shorebirds: a case study on the Seine estuary, France. *Biol. Conserv.* 123 (205): 66–77.

Harmelin-Vivien, M. and C. Salen-Picard. 2002. Influence of climate and river run-off on sole fisheries in the Mediterranean. *French IGBP-WCRP News Letter* 14: 62–66.

Jourde, J. and S. Dancie. 2007. Suivi du macrozoobenthos intertidal de la grande vasière nord de l'estuaire de la Seine. Campagne 2006. Rapport de la Cellule de Suivi du Littoral Haut Normand pour la Maison de l'Estuaire.

Mayot, S. and S. Duhamel. 2003. Suivi du macrozoobenthos intertidal de la grande vasière nord de l'estuaire de la Seine. Campagne 2002. Rapport de la Cellule de Suivi du Littoral Haut Normand pour la Maison de l'Estuaire.

Mayot, S. and S. Duhamel. 2004. Suivi du macrozoobenthos intertidal de la grande vasière nord de l'estuaire de la Seine. Campagne 2003. Rapport de la Cellule de Suivi du Littoral Haut Normand pour la Maison de l'Estuaire.

Mayot, S. and S. Duhamel. 2006. Suivi du macrozoobenthos intertidal de la grande vasière nord de l'estuaire de la Seine. Campagne 2004. Rapport de la Cellule de Suivi du Littoral Haut Normand pour la Maison de l'Estuaire.

Meziane, T. and C. Retière. 2002. Croissance de juvéniles de *Nereis diversicolor* nourris avec des détritus d'halophytes. *Oceanol. Acta.* 25: 119–24.

Moreira, S.M. et al. 2006. Effects of estuarine sediment contamination on feeding and on key physiological functions of the polychaete *Hediste diversicolor*: laboratory and *in situ* assays. *Aquat. Toxicol.* 78: 186–201.

Olivier, M. et al. 1995. Réponses comportementales des polychètes *Nereis diversicolor* (O.F. Müller) et *Nereis virens* (Sars) aux stimuli d'ordre alimentaire: utilisation de la matière organique particulaire (algues et halophytes). *Can. J. Zool.* 73: 2307–2317.

Port Autonome du Havre. 2006. Levers aéro laser des années 2004 et 2006. Port Autonome du Havre, service des dragages et de l'Hydrographie. Le Havre, France.

Service de la Navigation de la Seine. 2008. Relevés journaliers des débits de la Seine au barrage de Poses de 1984 à 2006.

SIRS. 2007. Etude cartographique de l'occupation du sol entre 1966, 1973, 1985, 1994 et 1999. Système d'Information à Références Spatiales (SIRS), rapport d'étude pour la Maison de l'Estuaire, janvier 2007.

APPENDIX 9.1

Data records of densities of *Nereis diversicolor* (number of individuals per square meter), salinity, sediment granulometry (percent of weight fraction up to 63 μm), altitudes for different sectors (referring to zero of marine charts at Le Havre) of the Seine estuary for 1987, 1990, 1992, 1994, 1996, 1997, and 2000 to 2006, and river flow (cubic meters per second) for these periods.

Sector	Year	% of mud	Salinty (‰)	Flow River (m³/s)	Altitude (m CMH)	*Nereis* Density (Ni/m²)
B	1987	83.0	20.0	505	7	465
B	1990	83.0	10.0	289	7	1917
B	1922	83.0	10.0	363	7	513
B	1994	83.0	10.0	628	7	1613
B	2004	83.0	10.0	411	8	250
C	1990	80.5	14.8	289	4	625
C	1992	80.5	12.0	363	4	600
C	1994	80.5	12.0	628	4	944
C	2004	80.5	12.0	411	6	200
D	1987	84.3	20.0	505	8	2250
D	1990	84.3	15.8	289	8	150
D	1992	84.3	10.0	363	8	538
D	1994	84.3	10.0	628	8	1188
D	1996	84.3	10.0	335	8	877
D	1997	84.3	10.0	392	8	644
D	2000	84.3	10.0	750	8	1159
D	2001	84.3	10.0	904	8	1367
D	2002	84.3	10.0	595	8	442
D	2003	84.3	10.0	389	8	233
D	2004	84.3	10.0	411	8	325
D	2005	84.3	10.0	333	9	408
D	2006	84.3	8.0	418	9	617
E	1987	73.2	20.0	505	6	660
E	1990	73.2	17.2	289	5	700
E	1992	73.2	12.0	363	5	238
E	1994	73.2	12.0	628	5	900
E	1996	73.2	10.0	335	6	1056
E	1997	73.2	10.0	392	6	1559
E	2000	73.2	10.0	750	6	217
E	2001	73.2	10.0	904	6	733
E	2002	73.2	10.0	595	6	417
E	2003	73.2	10.0	389	8	209
E	2004	73.2	10.0	411	7	325
F	1987	56.9	27.5	505	3	980
F	1990	56.9	15.4	289	3	350

Sector	Year	% of mud	Salinty (‰)	Flow River (m³/s)	Altitude (m CMH)	*Nereis* Density (Ni/m²)
F	1992	56.9	13.0	363	3	425
F	1996	56.9	13.0	335	3	139
F	1997	56.9	13.0	392	3	109
F	2000	56.9	13.0	750	3	117
F	2001	56.9	13.0	904	3	142
F	2002	56.9	13.0	595	4	109
F	2003	56.9	13.0	389	3	100
F	2005	73.2	13.0	333	4	100
G	1987	85.8	20.8	505	8	1155
G	1990	85.8	13.9	289	8	325
G	1992	85.8	12.0	363	8	625
G	1996	85.8	12.0	335	8	644
G	1997	85.8	12.0	392	8	1142
G	2000	85.8	12.0	750	8	742
G	2001	85.8	12.0	904	8	1109
G	2002	85.8	12.0	595	8	384
G	2003	85.8	12.0	389	8	392
G	2004	85.8	12.0	411	8	209
G	2005	85.8	12.0	333	8	527
G	2006	85.8	12.0	418	8	492
H	1987	82.7	23.0	505	6	740
H	1990	82.7	16.1	289	5	1000
H	1992	82.7	16.1	363	5	350
H	1996	82.7	16.1	335	6	224
H	1997	82.7	16.1	392	6	580
H	2001	82.7	16.1	904	6	884
H	2002	82.7	16.1	595	6	209
H	2003	82.7	16.1	389	6	100
H	2004	82.7	16.1	411	7	84
H	2005	82.7	16.1	333	7	183
H	2006	82.7	16.1	418	6	58
I	1990	55.7	16.1	289	3	450
I	1992	55.7	16.1	363	3	175
I	1996	55.7	16.1	335	3	32
I	2000	55.7	16.1	750	3	50
I	2001	55.7	16.1	904	3	0
I	2002	55.7	16.1	595	3	9
I	2003	55.7	16.1	389	3	17
I	2005	30.0	16.1	333	4	0
I	2006	30.0	16.1	418	4	0
J	1987	57.3	12.5	505	7	80
J	1990	57.3	16.0	289	7	1083

Sector	Year	% of mud	Salinty (‰)	Flow River (m³/s)	Altitude (m CMH)	*Nereis* Density (Ni/m²)
J	1992	57.3	16.0	363	7	375
J	2000	57.3	16.0	750	7	992
J	2001	57.3	16.0	904	7	1059
J	2002	57.3	16.0	595	7	892
J	2003	57.3	16.0	389	7	175
J	2005	30.0	16.0	333	8	450
J	2006	30.0	16.0	418	8	642
K	1987	70.2	16.0	505	4	1590
K	1990	70.2	16.0	289	4	550
K	1992	70.2	16.0	363	4	75
K	2000	70.2	16.0	750	4	900
K	2002	70.2	16.0	595	4	683
K	2003	70.2	16.0	389	6	292
K	2004	70.2	16.0	411	6	175
K	2005	70.2	16.0	333	6	350
K	2006	70.2	16.0	418	6	125
L	1990	68.0	20/0	289	2	317
L	1992	68.0	20/0	363	2	75
L	2000	68.0	20/0	750	2	0
L	2001	68.0	20/0	904	2	608
L	2002	68.0	20/0	595	2	150
L	2003	68.0	20/0	389	2	117
L	2004	68.0	20/0	411	3	0
L	2005	68.0	20/0	333	3	47
L	2006	68.0	20/0	418	3	66
M	1987	45.6	21.5	505	5	240
M	1996	45.6	20.0	335	5	433
M	1997	45.6	20.0	392	5	254
M	2000	45.6	20.0	750	5	1125
M	2001	45.6	20.0	904	5	1500
M	2002	45.6	20.0	595	5	475
M	2003	45.6	20.0	389	6	884
M	2004	45.6	20.0	411	6	58
M	2005	45.6	20.0	333	6	17
M	2006	45.6	20.0	418	6	42
N	1987	61.2	21.0	505	3	1990
N	1996	61.2	21.0	335	3	27
N	2000	61.2	21.0	750	3	592
N	2002	61.2	21.0	595	3	175
N	2003	61.2	21.0	389	3	7
N	2004	61.2	21.0	411	4	59
N	2005	61.2	21.0	333	4	8

10 Ecological Status and Health of the Planktonic Copepod *Eurytemora affinis* in the Seine Estuary

Joëlle Forget-Leray, Sami Souissi, David Devreker,
Kevin Cailleaud, and Hélène Budzinski

CONTENTS

The concept of a biological indicator is based on the observation of the role of an organism in maintaining a function and/or balance of an ecosystem (Karr and Dudley 1981). The main use of a biological indicator is to characterize an actual state and to oversee and/or predict a significant change in the quality of a habitat (Gibson et al. 2000). For a biological indicator to be effective and functional, it must be sufficiently abundant during a great part of the year to permit easy sampling (Wilson 1994). However, these characteristics alone are not sufficient to qualify a species as an estuarine biological indicator.

In the case of monitoring the bioavailability and/or effect of chemical contamination, a biological indicator must be able to bioaccumulate a contaminant integrated across all bioavailable sources (also referred to as a biomonitor in this context) and/or show variation in selected biological parameters (biomarkers) at sublethal exposures (Purchase and Fergusson 1986; Wilson 1994; Mikac et al. 1996; MacFarlan et al. 2000). Moreover, in an estuary, life history traits of organisms also react to the natural fluctuations of environment parameters (salinity, hydrodynamics, temperature, etc.) in addition to the presence of anthropogenic factors (contaminants). It thus becomes difficult to dissociate the relative contributions of factors of natural origin and of anthropogenic origin.

Gibson et al. (2000) recognized an effective biological indicator as "a sign or a signal which translates a complex message, coming from several potential sources (i.e., natural and anthropogenic), in a simplified and useful way." The choice of a bioindicator species in an estuary will thus fall on a euryhaline species for which variations in salinity are not excessively stressful (Wilson 1994). Such an approach is possible only if the biology of the species and its interactions with its environment are well known, thus requiring a multidisciplinary approach (study of physical *in situ*, chemical and biological processes) that takes into account all sources of variability among potential stress factors and the targeted organism.

Recent research in the framework of the Seine Aval program has shown that the planktonic copepod *Eurytemora affinis* may be a candidate to serve as a bioindicator of water quality in the Seine estuary (Mouny 1998; Cailleaud 2006; Devreker 2007). Planktonic organisms are good indicators of estuarine conditions and particularly of salinity gradients because some of them are able to maintain their positions in specific water masses (Laprise and Dodson 1994). *E. affinis* occurs in great abundance over a large part of the year in the oligo-mesohaline part of the Seine estuary and its life history traits respond to high variations of salinity, temperature (Devreker 2007), and pollutant availabilities (Cailleaud 2006).

10.1 GEOGRAPHICAL DISTRIBUTION OF *EURYTEMORA AFFINIS*

Among estuarine planktonic copepod species, *Eurytemora affinis* is certainly the most represented and one of the most studied. It is present in many estuaries of the northern hemisphere and is often an essential basal organism in their food chains (Fockedey and Mees 1999; Winkler and Greve 2004). *E. affinis* can also inhabit freshwater environments (Lake Ohnuma, Ban 1994) and hypersaline environments (Lee 1999). In estuaries, this species is typical of the oligo-mesohaline zone, with salinity ranging from 0 to 18 and sometimes higher (von Vaupel-Klein and Weber 1975; Mouny et al. 1998; Lawrence et al. 2004).

The broad distribution of this species in the northern hemisphere shows its high level of adaptability to aquatic environments and particularly to brackish ones. However, differences exist at the genetic level (Lee 1999) between the multiple populations of *E. affinis* although they exhibit a certain morphological stasis (Lee and Frost 2002). By studying these genetic divergences, these authors have gathered these populations into different clades correlated with geographical distribution. The impossibility of interbreeding among some of these populations was also highlighted (Lee et al., unpublished data), even among geographically close populations, as were differences in tolerance to salinity (Lee and Peterson 2002, 2003). As a consequence, Lee (2000) suggested that *E. affinis* should not be considered a single species and be treated as a complex of cryptic species. The *E. affinis* population in the Seine estuary belongs to the European clade defined by Lee (1999, 2000) although the number of samples taken in Europe remains low.

10.2 LIFE CYCLE OF *EURYTEMORA AFFINIS*

10.2.1 INTRODUCTION

The life cycle of *E. affinis* is composed of six naupliar larval stages (N1 to N6), five juvenile copepodid stages (C1 to C5), and the reproductive adult stage (Figure 10.1).

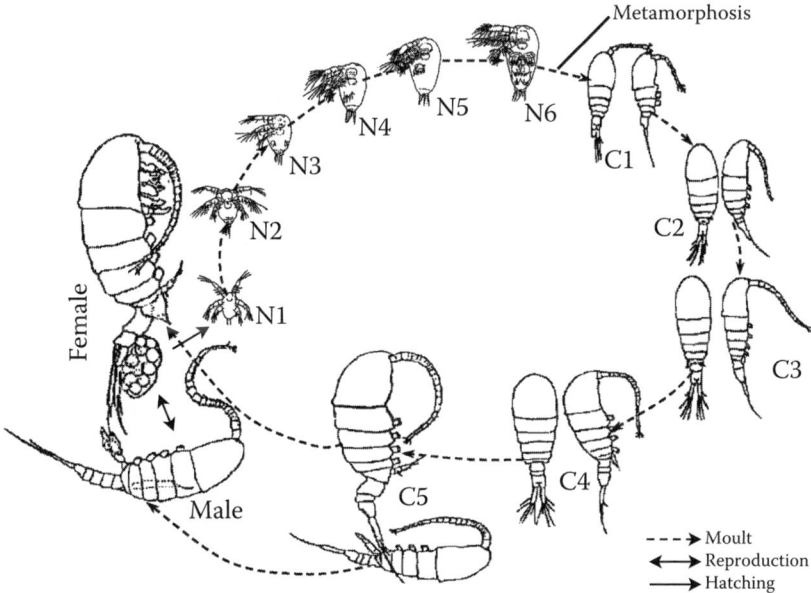

FIGURE 10.1   Details of developmental cycle of the copepod *Eurytemora affinis*. N1 to N6 correspond to naupliar stages. C1 to C5 correspond to copepodid stages. (Drawing of different stages according to S.K. Katona 1971 and modified by S. Souissi.)

1st reproductive cycle

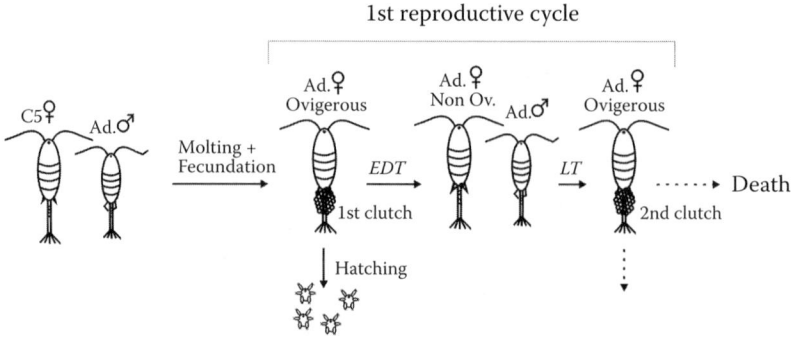

FIGURE 10.2 Reproductive cycle of *Eurytemora affinis*. Fertilization by male occurs after the last moult of a female (from C5 to adult) and leads to the production of an egg sac. After embryonic development (*EDT*), nauplii hatch from eggs and females are non-ovigerous. The time needed to produce a new egg sac (with or without new fertilization by a male) is called the latency time (*LT*). *EDT* + *LT* represents the inter-clutch time (*ICT*). This cycle may be repeated several times until a female's death. (Figure modified from Devreker et al., submitted.)

The naupliar stages are much differentiated morphologically compared to the copepodid stages that are more morphologically similar to adults. At each successive moult, the numbers of pairs of legs and segments increase from C1 to C4. Although sexual maturity is reached only during the moult between the pre-adult (C5) and adult stages, sexual morphological differentiation can be visible at C5, sometimes at C4 (Figure 10.1), at the level of the genital appendix.

A fertilized egg-bearing female will produce eggs that are carried in an oviferous sac in contrast to free-spawning calanoid copepods that release their eggs into the water. The male is attracted by the female via the detection of sexual pheromones released into the medium (Katona 1973; Lonsdale et al. 1998).

During the reproductive phase, the time separating the extrusion of two successive clutches, usually called the inter-clutch time (*ICT*), can be split into two parameters: embryonic development time (*EDT*) and latency time (*LT*). *EDT* is the time between the production of a clutch and the hatching of the nauplii (Figure 10.2). *LT* is the time between the hatching of nauplii and extrusion of a new clutch (Figure 10.2), and is a function of the maturation times of oocytes (Williamson and Butler 1987). The egg production rate (*EPR*) is calculated as a function of the number of eggs produced during cycles dependent on *EDT* and *LT*.

10.2.2 Post-Embryonic Development of *E. affinis* and Effects of Environmental Factors

Temperature is often considered the key external factor that affects the life history traits and population dynamics of copepods. Huntley and Lopez (1992) have shown that 90% of the growth variability in 33 copepod species may be explained by temperature alone. The variability in temperature characterizes seasonal succession and directly affects copepod reproduction and development (Ianora et al. 1992).

Temperature directly influences copepod metabolism and an increase of temperature increases the metabolic rate (Gillooly et al. 2001) and the amount of energy allocated for egg production.

In estuarine ecosystems, salinity is also an important factor affecting the distribution of copepod populations (Roddie et al. 1984; Devreker et al. 2004). Holste and Peck (2005) showed that salinities below 15 significantly reduced the hatching success of eggs produced by the free-spawning copepod *Acartia tonsa*. Kimmel and Bradley (2001) showed that salinity alters patterns of protein expression in *E. affinis*. Osmotic stress alters energy allocation to other biological processes such as reproduction. Bradley (1986) has shown that salinity and temperature also interact in *E. affinis*.

Devreker et al. (2007) have shown the clear effects of temperature and salinity on the post-embryonic development of *E. affinis* from the Seine estuary, a brackish environment. An increase in temperature from 10 to 15°C induced a decrease in mean development time from 9 to 6 days at salinities 5 and 15, respectively. A salinity increase from 15 to 25 engendered increases in both mortality rate and mean development time. Differences in development time and mortality between salinity 5 and 15 are very low for the Seine estuary population. These general effects of temperature and salinity on post-embryonic development are well known for *E. affinis* from different geographical populations.

Development times are generally calculated from the mean development times of copepods from N1 to adult for both sexes. Most populations of *E. affinis* inhabit brackish water (i.e., estuaries) or fresh water (two populations from lakes in Ohnuma and Michigan). In several studies (Heinle and Flemer 1975; Devreker et al. 2007), differences in post-embryonic development times between 10 and 15°C are relatively small. Below 10°C, the effect of low temperature is more pronounced. The difference in post-embryonic development time between 10 and 15°C is 9 days for males and 12 days for females; this difference reaches 23 and 25 days between 10 and 5°C (Vijverberg 1980) and 36 days between 14 and 4°C (V. Escaravage, personal communication). These results suggest adaptation of *E. affinis* to the temperate environment and show its capacity for development in colder environments.

The high mortality rate observed at 10°C for the more southern population from the Gironde estuary (Poli and Castel 1983) and at 20°C for the more northern population from Tjeukemeer, suggests local adaptations to warmer or colder climates as appropriate. However, it is difficult to draw a clear conclusion on the effect of temperature on post-embryonic development in *E. affinis* by comparing literature data. The food quality and quantity differ from one experiment to another, and our results showed high variability among different food sources that can be higher than environment-induced variability.

The development time is generally longer for copepodids than for nauplii. When copepods exhibited slow development rates, the probability of mortality during development increased, especially at high salinity. Copepods are considered stressed when they are not in their optimal salinity range (Gaudy et al. 2000; Hall and Burns 2002; Chinnery and Williams 2004). Physiological stress results in an increase in energy allocated to osmoregulation through the synthesis or degradation of non-essential protein to regulate intracellular organic solute concentration (Kimmel and Bradley 2001). Thus, the energy used for osmoregulation is not available for development, resulting in an increase in development time, and if the stress reaches a threshold, mortality may occur. Lee et al. (2003) showed that copepod populations derived from saline and freshwater habitats developed and survived better at, respectively, higher and lower salinities.

The longevity of *E. affinis* from the Seine estuary is more affected by salinity than by temperature. However, studies on other calanoid copepods have shown that longevity decreases strongly when temperature increases (Smyly 1980; Uye 1981; Ianora 1998). In both sexes, longevity in the adult stage was higher at 10°C than at 15°C, and lower than that at 10°C with lower food quality. Longevity of adult females determines the reproductive period of the life cycle and the number of clutches produced, so longevity is a very important parameter for population dynamics (Gehrs and Robertson 1975; Peterson 2001).

Experiments using individual-based protocols confirm high variability of development at the individual scale (Ban 1994; Devreker et al. 2007). Development time in the only non-feeding stage N1 was the shortest and had the lowest variability. Devreker et al. (2004) showed that stress induced by high salinities (30 and 35) affected development time in this stage. Higher individual variability appeared with the first feeding stages as observed for other calanoid copepods (Caramujo and Boavida 1999; Peterson and Painting 1990). Experiments with a mixed low-quality food diet showed an increase in development time in stages N6 and C1 (Devreker et al. 2007). Similarly, a decrease in development rate of N6 and C1 was observed in the presence of contaminants (Forget-Leray et al. 2005). Consequently, we suggest that this decrease appears when conditions are not optimal for the development of *E. affinis*.

Among the developmental stages, C4 and C5 show the slowest rates, especially for females. Katona (1970) showed morphological differences (i.e., body size) between males and females from stage C3 onward; moreover, a difference in development time between the sexes appeared in stage C4 and increased in C5. This delay in female development is well known and arises from larger morphological differentiation in C4 (Katona 1971) and maturation of oocytes in C5 (Smith and Lane 1985). A temperature decrease from 15 to 10°C increases the differences in development time between males and females, especially during the transition between groups C4–C5 and the adult stage (Ban 1994; Devreker et al. 2007). Increase in salinity from 15 to 25 increased the variability of the transition rate from C4–C5 to adult.

10.2.3 Reproduction and Effects of Environmental Factors

The reproduction of *E. affinis* from the Seine estuary is clearly affected by temperature and salinity. However, as for post-embryonic development, there are few differences between salinity levels 5 and 15 compared to level 25 (Devreker, submitted). The reproductive parameters studied in the laboratory at individual scale show high variability. Since the Seine estuary is highly variable in terms of temperature and salinity, this variability ensures that a portion of the population is likely to experience optimal conditions for reproduction. Only an individual-based protocol allows the observation of a large panel of reproductive parameters such as *EDT* and *LT* and effect of female age.

Temperature and salinity affected clutch size, *EDT*, and *LT* and, consequently, the *EPR* of *E. affinis*. *EDT* and *LT* increased when the temperature decreased as observed in other studies (Poli and Castel 1983; Ban and Minoda 1991; Matias-Peralta et al. 2005). Increases in *EDT* and *LT* are based on the temperature-dependent nature of metabolism, as was seen in post-embryonic development times (Gillooly et al. 2001). The effect of temperature on clutch size seems to depend on local environmental conditions. The majority of *E. affinis* populations show decreases in clutch size with increasing temperature. Clutch size is dependent on oocyte size and thus co-varies with the size of the female (Hirche et al. 1997); larger females are observed at lower temperatures.

Salinity stresses organisms when they are not at their optimal salinity (Hall and Burns 2002; Chinnery and Williams 2004). Energy reallocation may affect embryonic development, maturation of eggs, and egg production. The *EDT* and the *LT* are shortest and clutch size is lowest at salinity 25 compared to other salinities for the Seine estuary population (Devreker et al., submitted). As Souissi et al. (1997) have shown, the individual variability in the development time of copepods can increase when they are not at their optimal temperature. Jiménez-Melero et al. (2007) have also shown in another egg-bearing copepod, *Arctodiaptomus salinus*, that the variability in development time, *EDT* and *ICT* increased at high temperature when survival is reduced. We found higher individual variability in *EDT* and *LT* at salinity 25 compared to all other salinities at 15°C. A similar result was found for post-embryonic development (Devreker et al. 2007).

10.2.3.1   Variation at the Population Scale

The general effects of temperature and salinity observed on reproductive parameters are known for several populations of *E. affinis* from different geographical regions. In the Seine estuary, temperatures range from 2.5° to 25°C and salinities from 0 to 25. Reproductive parameters showed high variation among conditions and regions. For several populations of *E. affinis*, *ICT* and *EDT* showed longer duration at 10°C than at 15°C at which the duration was less than 2 days. Clutch size fluctuated from 8 to 72.1 eggs per clutch and the *EPR* from 0.31 to 34.2 eggs per female per day. Between 10°C and 15°C, two geographically distant populations from Chesapeake Bay (Lloyd 2006) and Tjeukemeer (Vijverberg 1980) showed a similar variation in mean values of *ICT* and *EDT*.

However, in several studies reported in the literature only the mean egg production rate and the mean clutch size were documented. Only the individual-based protocol used by Devreker et al. (submitted), Ban (1994), and Ishikawa et al. (1999) have permitted estimates of *EDT*, *LT*, longevity, and clutch sizes of adult females.

The variability of *E. affinis* reproductive parameters among different populations may have different origins as each population can be expected to be adapted to thrive in its native environmental conditions. The highest reproductive output of *E. affinis* from the Seine estuary occurred at low salinity and mid-range temperature (15°C). This pattern has been observed for post-embryonic development (Devreker et al. 2007) and survival under starvation of early naupliar stages (Devreker et al. 2004). These optimal conditions correspond to the distribution of *E. affinis* in the low salinity zone of the Seine estuary (Mouny and Dauvin 2002), with annual maximum density occurring when water temperature was approximately 15°C (Mouny 1998).

The range of clutch sizes observed *in situ* was also similar to the range observed in our experiments at the same temperatures (Devreker 2007). Differences in experimental protocols may serve as another source of variability among *E. affinis* populations. Particularly, estimation of the *EPR* can differ, depending on observation method. Gasparini et al. (1999) used the incubation method commonly used with free-spawning copepods that produce eggs continuously in contrast to egg-bearing copepods. Ask et al. (2006) improved this method by adjusting incubation duration as a function of the predicted *EDT* of *E. affinis* as a function of temperature, but they did not take into account the *LT*. Duration of acclimation to experimental conditions is also an important factor that can modify expression of biological parameters.

The importance of acclimation in embryonic development time experiments has already been shown by Andersen and Nieslen (1997). Thus, even if some parameters were similar to others measured in populations of the Seine estuary, Chesapeake Bay (Lloyd 2006), and Tjeukemeer (Vijverberg 1980), comparisons of data from the literature must be made with caution. We recommend common garden experiments (i.e., standardization of experimental protocols) to provide inter-comparative data and to test for adaptation.

10.2.3.2 Clutch Size Variability

Clutch size is not uniform and shows variability among females and among successive clutches of the same female (Devreker et al., submitted). At 15°C and low salinities, clutch size is dependent on female age. The lower number of eggs produced in the first clutch compared to other clutches under some conditions is probably linked to the energy loss during the last moult from C5 to the adult stage when oocytes mature (Fryd et al. 1991). However the mean clutch size decreased as females aged, as observed for *Cyclops kolensis* (Jamieson and Santer 2003), although observations of individuals showed that the clutch size decreased only when females reached a critical age (Devreker et al., submitted). Moreover, this critical age is reduced when conditions are more stressful, such as in the high salinity treatment. These processes have significant consequences at the population level, and should be included in individual based models of copepod population dynamics (Souissi et al. 2005).

10.2.3.3 Embryonic Development Time and Latency Time Variability

The absence of high variability in *EDT* for *E. affinis* (Devreker et al., submitted) has been observed by Jamieson and Santer (2003) for *Cyclops kolensis*. The *EDT* was highly temperature-dependent (Vijvenberg 1980; Ban 1994; Lloyd 2006) and salinity-dependent (Ishikawa et al. 1999) in *E. affinis*. In contrast, the *LT* showed high fluctuations, especially at 10°C, and clutches with low mean *LT* can be followed by clutches with high mean *LT* (Devreker et al., submitted). The mean time needed to produce the first clutch after moulting into the adult stage is shorter than the mean *LT* during a female's lifespan. This demonstrates that females can produce eggs early after moulting into adult stage (Ban 1994), which may be related to oocytes maturing in stage C5 as seen in the calanoid copepod *Diaptomus pallidus* (Williamson and Butler 1987).

Pre-adult C5 females can be observed with dark oocytes corresponding to their maturation stage (Williamson and Butler 1987). In some cases, females are able to extrude fertilized eggs shortly after the hatch of the previous clutch, suggesting that maturation of eggs in oocytes occurs simultaneously with embryonic development.

10.2.3.4   Egg Production Rate Variability

The variability in *EPR* has been correlated to male density, which can be limiting at times (Williamson and Butler 1987). Experiments have shown that reproductive success can be achieved in the presence of only one male (Devreker et al., submitted). Females do not need to mate every time to produce a viable clutch; storage of the spermatophore for fertilizing eggs is possible and has been observed in *E. affinis* (Katona 1975), and a species of egg-bearing copepod *Pseudodiaptomus* (Jacoby and Youngbluth 1983). The high number of males observed in the field (Devreker 2007) may be necessary to increase the probability of mating and serves to decrease the *LT*. Moreover, females are observed *in situ* and in cultures with many spermatophores, suggesting that sufficient numbers of males are present in the estuary.

10.3   ECOLOGY OF *EURYTEMORA AFFINIS* IN THE SEINE ESTUARY

In the Seine estuary, *E. affinis* can reach up to 90 to 99% of zooplankton species abundance in the low salinity zone (LSZ) between salinities 0.5 and 15 (Mouny and Dauvin 2002). Unlike in other European estuaries, *E. affinis* is the dominant species through most of the year and is rarely replaced by other estuarine copepods such as *Acartia* species in the LSZ (Mouny and Dauvin 2002; David et al. 2005; Devreker 2007).

Preliminary work linked the annual population dynamics of *E. affinis* to variations in temperature (Mouny 1998). In the Seine estuary, the population of *E. affinis* shows a high growth rate for temperatures between 10°C and 15°C in spring. Most frequently, maximum densities are observed at around 15°C in May and June as a result of higher reproduction rates and shorter development times (Devreker et al. 2007). However, during summer when temperatures exceed 18°C, the abundance of *E. affinis* decreases but the reasons remain uncertain. *In situ* observations are consistent with experimental work that shows optimal reproduction and a higher developmental rate at 15°C and at low (5 to 15) salinities (Devreker et al. 2007), with better naupliar survival in salinity 15 at 15°C than at 10°C (Devreker et al. 2004).

10.3.1   Population Dynamics at the Tidal Scale

The influence of the tidal cycle on the population structure of *E. affinis* in the Seine estuary was studied by sampling copepods at a fixed point near the Pont de Normandie (PN, see Chapter 1, Figure 1.3) over a 50-hour period.

The population structure of *E. affinis* in the Seine estuary is greatly influenced by the tidal cycle, with the copepods more abundant in bottom water. *E. affinis* was described as an epibenthic species by Sibert (1981) in the Nanaimo estuary with a capacity for active vertical migration (Kimmerer et al. 2002). This is supported by Morgan et al. (1997) who showed that the variation of abundance in adults and copepodids during the tidal cycle in the Columbia River estuary is the result of retention-adapted migration behaviour. The changes in the spatial (vertical) localization over the tidal scale have been suggested as adaptations of the copepods to avoid advection out of their salinity tolerance range (Kimmerer et al. 1998).

Roman et al. (2001) and Lloyd (2006) also suggested that the distribution of *E. affinis* adults and copepodids in the near bottom layer in Chesapeake Bay avoids advection in the surface current and maintains them in the maximum turbidity zone. Moreover, adults of *E. affinis* are known to show swimming behaviour that keeps them at a high-turbulence layer (Woodson 2005). Therefore turbulence and salinity variation play major roles in the distribution of this species during a tidal cycle.

Salinity varies inside an estuary as well as between estuaries. Generally the population of *E. affinis* is maintained mainly in the oligo-mesohaline zones of estuaries. Furthermore, the whole population of *E. affinis* is not concentrated in the same zone of the Seine estuary. Nauplii are concentrated in a salinity zone close to 5, whereas the adults are distributed across a broader and higher range of salinities around 15, from 5 to 20. This salinity range is optimal for the reproduction and development of the species.

As observed by Hough and Naylor (1992) for the Conwy estuary, *E. affinis* adults of the Seine estuary seem to have adapted an active behaviour of migration toward the bottom at the beginning of the flood right after the period of resuspension when current velocity decreases (Figure 10.3). This active behaviour enables them to avoid salinities above 20, a condition that seems more stressful for this population from results obtained in laboratory (Devreker et al. 2007). Little or no active migration of adults and copepodids occurs during the ebb. This migration behaviour is somewhat different from the synthetic diagram that includes active migration toward the surface in the period of flood as suggested by Morgan et al. (1997). This behavior concerns adult and copepodid stages, while naupliar stages are much more dependent on local hydrodynamics. In fact, naupliar stages show good survival in very low salinities and sufficient survival in strong salinities to survive transport toward the mouth of the Seine and return to the oligo-mesohaline zone of the estuary during a tidal cycle.

10.3.2 POPULATION DYNAMICS AT THE INTER-SEASONAL SCALE

The study of *E. affinis* population dynamics on an annual scale proceeded throughout 2005, which proved to be a peculiar year because it was characterized by a negative anomaly of the water temperature at the end of the winter. The winter period may be critical for the population growth of several copepod species (Molinero et al. 2005).

On an inter-seasonal scale, hydrodynamics provide two principal sources of variability that may affect the population dynamics of *E. affinis*. Flows that vary according to periods of precipitation or dryness that increase or decrease freshwater input define the periods of low water level. The lunar cycles that define the periods of neap and spring tides can also affect the intensity of mixing of water masses.

In addition to the negative temperature anomaly observed at the end of winter 2005, a flow anomaly characterized by periods of particularly low water level in winter, at the beginning of spring, and at the end of autumn compared to the previous years caused the copepod population to stay upstream in the estuary. In a similar situation of particularly low spring flow after winter precipitation, Kimmel et al.

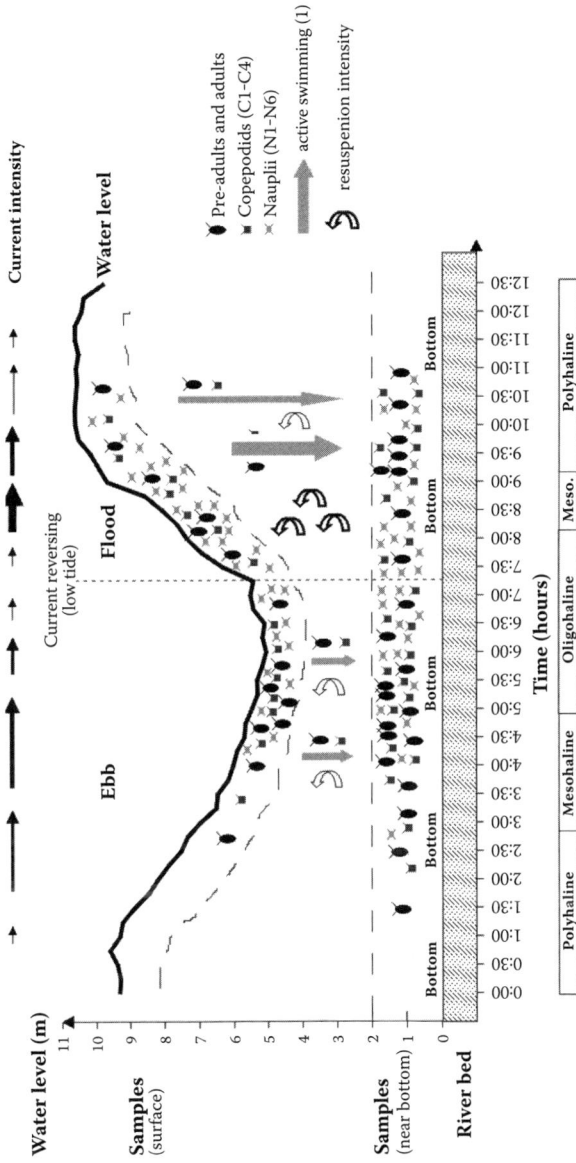

FIGURE 10.3 Dispersion of developmental stage of *E. affinis* in the middle part of the Seine estuary (Pont de Normandie, PN) as function of a mean tidal cycle. Widths of arrows on the top of the figure represent the magnitude of the water velocity during a length of time represented by their length. The different water masses have been identified at the bottom as functions of salinity range according to McLusky (1989): oligohaline zone [0.5 to 5], mesohaline [5 to 18], and polyhaline [18 to 25]. The population density increases during the ebb with low constant resuspension and hypothetical migration of adults (ovals) and copepodids (squares) that dominate the population from the polyhaline to the mesohaline zone in surface and bottom water. In the oligohaline zone around low slack, when current velocity is low, nauplii dominate the population. At the onset of the flood when current velocity is maximal, the population is resuspended; adult and copepodids start to migrate to the bottom water while current velocity decreases.

(2006) observed a transfer of the *E. affinis* habitat in the freshwater zones of the Chesapeake Bay during spring. They hypothesize that the position and the abundance of the *E. affinis* population, above or below average, can be predicted from winter precipitation levels.

Although hydrodynamics had an impact on the spatial distribution of the population, adults always showed capacities to maintain themselves in zones of salinity higher than the nauplii, increasing the size of their habitat. However the reduction in the salinity interval in which the *E. affinis* population develops and/or moves on an inter-seasonal scale also suggests a reduction in the size of the "optimal" habitat, confined in lower salinity zones, with the reduction of the flow. Moreover, although the *E. affinis* population is restricted to the oligo-mesohaline zone of the Seine estuary, it should not be forgotten that this species tends to invade fresh water (Lee 1999) using estuaries as transition interfaces. Many invasions have occurred over the past centuries and continue to take place. It may thus be possible to see a movement of the zone of distribution of this species in the long term, even seeing this phenomenon accelerated by anthropogenic modifications on the hydrodynamics of the Seine estuary such as the installation of bays and the climatic changes that tend, in some cases, to decrease river flow and thereby move the mixing zone upstream.

The second factor impacting the population dynamics of *E. affinis* inter-seasonally is temperature. In experiments focusing on reproductive rate (Devreker et al., submitted) and the rate of development (Devreker et al. 2007), *E. affinis* from the Seine estuary were more sensitive to temperature than to salinity. This interaction between life history traits and temperature had a direct effect on the *in situ* population dynamics. Initial work (1996 through 1998) on the population dynamics of *E. affinis* in the Seine estuary showed a high density of copepods during spring when the temperature was between 10°C and 15°C (Mouny 1998). The average winter temperature was close to 7°C. Therefore the increase in temperature between the end of winter and the beginning of spring seems to be associated with population growth. In the laboratory, the best hatching success, egg production rate, and survival during development are observed at 10°C and 15°C. Thereafter, summer temperatures at least equal to 20°C are associated with a reduction in fitness, specifically in the rate of recruitment into the population as characterized by a fall in the survival rate from egg to adult stage in the laboratory. Late winter temperatures below 5°C caused a change in primary production and probably in the recruitment of the *E. affinis* population. This phenomenon seems to cause an important and immediate deficit in the abundance of nauplii, the proportion of which strongly decreases in the population from the beginning of spring, perhaps due to a stronger production of diapause eggs or non-viable eggs.

In the Gironde estuary, *E. affinis* is replaced by *Acartia* species in summer as in many estuaries (Lawrence et al. 2004 and references therein); *Acartia* species are more competitive in warmer conditions (Bradley 1991). Moreover *A. tonsa* show high reproductive capacity in near-shore waters with high phytoplankton abundance (Paffenhöfer and Stearns 1988), as observed in spring in the Seine estuary. Kimmel et al. (2006) suggested that low rainfall in winter, resulting in low river discharge in spring, decreases the range of *E. affinis* habitat and increases that of *A. tonsa* during summer, as we observed in 2005. In 2005 only some individuals of *Acartia* species were found from March to June, the population of *Acartia* being most important in July and September (Beyrend-Dur and Devreker, unpublished data) during low mean river discharge. In July the habitats of *Acartia* species and *E. affinis* do not overlap; *Acartia* lives in higher salinity than *E. affinis*; their habitats do overlap in September.

10.3.3 ROLE OF *E. AFFINIS* IN THE SEINE ESTUARY FOOD WEB

The distribution of *E. affinis* in the estuary has a major impact on food web dynamics in the Seine estuary as the majority of predators of *E. affinis*, such as the mysid *Neomysis integer* or the decapod *Palaemon longirostris*, are concentrated in the bottom water (Mouny et al. 1998). The importance of predation pressure on the *E. affinis* population from the Seine estuary has already been investigated (Mouny 1998). However only the predation exerted on the adult copepods had been considered, while several studies have shown that some of the predators, mysids in particular, may prey on the smaller stages (Winkler and Greve 2004).

Since, as shown by Devreker et al. (2004), the nauplii can certainly survive in water of very low or very high salinity, selective predation may be one of the principal causes of their mortality as for copepodids whose abundance is much lower. This selective predation is very important to take into account in food web dynamic models because of its direct effect on the population structure on an inter-seasonal scale (Fulton 1982). Experiments undertaken in the laboratory on the effectiveness of mysid selectivity in turbid and turbulent environments and at different groups developmental stages and temperatures are essential for estimating stage-dependent predation pressures according to season. This process plays a fundamental role in the population dynamics of *E. affinis* in the Seine estuary.

The grazing pressure of adult copepods on phytoplankton, suspended detritus, and bacteria also plays an important role in the food web of the Seine estuary (Devreker 2007). Inter-seasonal variations in *E. affinis* population density do not follow the mean chlorophyll *a* (Chl *a*) concentrations in the Seine estuary, as observed by Lawrence et al. (2004) in different estuaries of Wayquoit Bay (Massachusetts, USA). Positive relationships have already been observed in the Westerschelde estuary (Escaravage and Soetaert 1995). In the Seine estuary, the Chl *a* concentration maximum value in 2005 was mainly characterized by blooms of large diatom species (Gómez et al., submitted). A high proportion of this chlorophyll is certainly not available because the cells may be too large for ingestion by copepods and especially by nauplii.

Another part of phytoplankton community consists of dinofagellates and chlorophytes (Gómez et al., submitted) that may be available for *E. affinis*. Their production in spring and summer, however, seemed to be insufficient for the growth of the *E. affinis* population in 2005, as there was no increase in population density. Lawrence et al. (2004) suggested that the lag of phytoplankton during a peak of copepod density may be attributed to strong grazing pressure of copepods on the phytoplankton. This can partially explain the negative relationship between copepod density and Chl *a* concentration in this study, and in preceding years in which copepod density decreased in summer (Mouny 1998) while mean monthly Chl *a* concentration increased in the same period.

Adults are not the only developmental stages that impact phytoplankton grazing. The nauplii, even though much smaller, play a significant part in the grazing pressure on the phytoplankton because of their very high density (Merrell and Stoecker 1998). Moreover, it is possible that their diet differs from the adult diet because they do not have the same capacity for selectivity (Allan et al. 1977; Richman et al. 1977). Moreover, naupliar stages do not have the possibility of feeding on large-sized particles such as large phytoplanktonic cells (diatoms) and bacteria associated with large detritus particles, as adults do. The grazing pressure may be exerted on different biological compartments with different intensities, with potential impact on the trophic assessment of the estuary of the Seine as suggested by Tackx et al. (2003) for the Schelde estuary.

E. affinis may thus exert a control on the phytoplankton populations and also on the populations of bacteria in the Seine estuary. However Boak and Goulder (1983) estimated that the grazing pressure of E. affinis on bacteria populations in the Humber estuary (England) would account for only 0.03% of the attached bacteria and 0.04% of the free bacteria, and thus produced very little impact on their populations. Supposing that the suspended matter contains an important source of potential food for E. affinis (bacteria associated with the particles, organic waste), the advantage of remaining at the bottom of the estuary would be associated with a feeding strategy. The major part of carbon exchange from the lower links of the food chain toward the higher links starts at the bottom of the estuary.

These complex processes of biotic and abiotic interaction reveal the difficulties of interpreting the population dynamics of an estuarine copepod. Relevant food chains are affected by hydrodynamic conditions and thus vary according to tidal regime (lunar cycles) and seasonal weather (temperature, rainfall, etc.). These interactions may impact carbon cycle assessments of the Seine estuary. However, the carbon assessments conducted on the Seine estuary (Garnier et al. 2001) have tended to underestimate the contributions of planktonic communities.

Taking into account the strong density of E. affinis in the Seine estuary (at least during the seasonal maximum abundance) and its role in the biotransformation of organic matter (through grazing), a total assessment of the contribution of the E. affinis population in the carbon budget of the estuary is necessary. This copepod biomass may represent a source of carbon available for both predators and also for the bacterial production that decomposes the carcasses of dead copepods (Tang et al. 2006). Such a carbon assessment based on the E. affinis population biovolume in the estuary must also consider hydrodynamic sources of variation.

10.4   SEASONAL VARIATION OF HYDROPHOBIC ORGANIC CONTAMINANT CONCENTRATIONS IN THE SEINE ESTUARY WATER COLUMN AND THEIR TRANSFER TO *EURYTEMORA AFFINIS*

E. affinis is fundamental in the ecological equilibrium of the local food web and may contribute to the transfer of contaminants to higher trophic levels. Among the most worrying contaminants, the hydrophobic organic contaminants (HOCs) including polycyclic aromatic hydrocarbons (PAHs), polychlorinated biphenyls (PCBs), alkyl-phenols (APs), and many pesticides have been detected at high levels in many North Atlantic estuaries (Ko and Baker 1995; Fernandez et al. 1997; Minier et al. 2006, Cailleaud et al. 2007a and b). High levels of PCBs and PAHs have been reported in the Seine River and in its estuary (Fernandez et al. 1997; Motelay-Massei et al. 2002; Minier et al. 2006; Cailleaud et al. 2007a). Some data are also available on the contamination by nonylphenol (NP) and nonylphenol polyethoxylate (NPnEO) (Cailleaud et al. 2007b).

These compounds are known to exhibit higher toxicities than their precursors and to mimic oestrogens in hormonal regulation. Thus, NPnEOs, nonylphenol poly-ethoxy carboxylates (NPnECs), and NPs are of toxicological interest because of their potential bioaccumulation in organisms that may be exposed via sediments or via the dissolved phase (Cross-Sorokin et al. 2003), and because of their acute and chronic toxicities (Brown et al. 1999; Forget-Leray et al. 2005).

The concentrations, distribution, and fate of these HOCs in estuarine waters, sed-iments, and biota are influenced by numerous factors (salinity, pH, temperature, sus-pended particulate material [SPM], river flow, tidal currents). Transfer mechanisms of these contaminants from the water column to organisms, especially to plank-tonic species that play a key role in the exchange of organic matter in the local food web, are not well defined. Although some studies have reported transfers of organic contaminants to zooplankton species (Harris et al. 1977; Cailleaud et al. 2007a,b; Magnusson et al. 2007), field data concerning bioaccumulation of HOCs by copep-ods are limited. Nevertheless, the understanding of these phenomena is essential for assessing toxicological risks to biota from contaminants. Extraction and quantifica-tion methods for PAHs and PCBs are detailed elsewhere (Cailleaud et al. 2007a,b).

10.4.1 Seasonal Variations of Total PCBs and PAHs in the Seine Estuary at Tancarville Station

Cailleaud et al. (2007a) measured total PCB (sum of 27 compounds studied) and PAH (20 compounds) contents in the dissolved phase in SPM and in *E. affinis* from November 2002 to February 2005, as shown in Table 10.1. Dissolved PCBs and PAHs range, respectively, from 2.0 to 21.2 ng.L^{-1} and from 2.9 to 23.9 ng.L^{-1} with similar profiles.

Maximum contamination levels reported during winter periods may be linked to remobilization of the sediment after intensive flood periods as reported by Kowalewska et al. (2003), by higher inputs of combustion-derived PAHs from indus-trial and domestic processes (Motelay-Massei et al. 2007), or by decreasing micro-bial and photodecomposition activities induced by lower temperatures (Witt 1995). The high PAH level in July 2003 may have been caused by atmospheric deposition (Motelay-Massei et al. 2002), surface runoff brought in by strong rainfall, or dredg-ing activities.

Concentrations of PAHs bound to particles vary between 499 and 5819 ng.g^{-1}. PCB concentrations in SPM are 10 times lower, ranging from 58 to 463 ng.g^{-1} (Table 10.1).

TABLE 10.1
PAH and PCB Concentrations in Dissolved Phase, SPM, *Eurytemora affinis*, and in Water Column of the Seine Estuary at Tancarville Station

| | Dissolved phase (ng.L−1) | | | | SPM (ng.g−1 dry weight) | | | | Eurytemora (ng.g−1 dry weight) | | | | Water column (ng.L−1) | | | |
| | n = 10 | | | | n = 12 | | | | n = 15 | | | | n = 10 | | | |
	Mean	Median	Min.	Max.	Mean	Median	Min.	Max.	Mean	Median	Min.	Max.	Mean	Median	Min.	Max.
CB 52	0.6	0.6	N.D	2.7	16	18	3	24	62	48	24	137	2.4	1.5	1.1	6.8
CB101	1.4	1.1	0.3	3.5	32	33	13	50	188	134	86	376	4.5	3.4	2.1	11.1
CB87	0.5	0.4	N.D.	1.5	10	10	5	23	27	19	5	75	1.5	1.1	0.7	4.4
CB118	1.3	1.1	0.5	3.0	25	25	5	42	94	86	31	247	3.4	2.8	6.8	1.9
CB153	0.7	0.5	0.1	2.5	39	33	7	129	203	203	83	373	3.6	3.1	1.5	6.9
CB138	0.8	0.6	N.D.	2.8	33	33	8	56	169	153	83	316	3.4	2.6	1.4	7.5
CB187	0.1	0.1	N.D.	0.2	9	8	1	17	46	43	20	96	0.7	0.7	0.4	1.5
CB180	0.2	0.1	N.D.	0.7	14	11	4	29	54	53	11	139	1.2	1	0.5	3
ΣPCB	7.0	5.6	2.0	21.2	227	223	58	463	976	788	383	1785	26.1	20.2	14.1	55.0

Phe	1.6	1.1	0.6	3.8	176	194	25	278	78	29	5	310	13	12	6	23
Fluo	1.4	0.8	0.3	3.1	377	354	59	614	70	37	14	406	26	25	13	45
Pyr	2.8	2.7	0.8	5.2	346	323	59	581	105	73	22	427	26	25	15	39
Triph+Chrys	0.9	0.6	0.4	1.6	277	263	44	430	82	53	16	300	20	19	10	34
Bbf+Bkf	0.7	0.4	N.D.	1.1	579	524	79	1124	120	59	15	559	36	32	18	70
B(e)P	0.5	0.3	N.D.	1.7	240	241	35	394	64	45	11	255	16	14	8	30
B(a)P	0.3	0.2	N.D.	1.4	278	262	42	513	54	22	9	283	18	17	9	34
IP	0.2	0.1	N.D.	1.0	261	246	39	505	53	34	N.D.	263	16	14	5	32
B(g,h,i)P	0.3	0.2	N.D.	1.4	243	242	33	406	54	34	9	258	16	14	4	30
ΣPAH	11.1	8.1	2.9	23.9	3361	3130	499	5819	879	478	165	3866	227	211	123	407

Note: Total PAHs (RPAHs) = sum of Phe, An, 1-, 2-, 3-, and 9-methylphenanthrene, 2-methylanthracene, Fluo, Pyr, BaA, Triph + Chrys, (B[b+j+k]F), BaP, BeP, Per, IP, DaA + DaC and BP.

Total PCBs (RPCBs) = congeners 8, 18, 29, 50 + 28, 52, 104, 44, 66, 101, 87, 154 + 77, 118, 188, 153, 105, 138, 126, 187, 128, 200, 180, 170, 195, 206, and 209.

N.D. = not detected. n = Number of sampling cruises.

From Cailleaud et al. 2007a.

Increasing levels have been measured for both PAHs and PCBs during winter, as a consequence of increased flow-inducing upstream erosion and resuspension of more contaminated particles. Total particulate PAHs and PCBs contribute to the contamination of the water column for 81 to 99% and for 47 to 89%, respectively. Cailleaud et al. (2007a) pointed out that concentrations of PAHs and PCBs in the Seine estuary are characteristic of an urban and industrialized area with levels 10 times higher than those measured in Chesapeake Bay (Gustafson and Dickhut 1997) or in Singapore coastal waters (Wurl and Obbard 2005), but comparable to levels reported in the Venice lagoon (Manodori et al. 2006).

Long et al. (1995) suggested two effect-based guidelines for both PAH and PCB concentrations in marine and estuarine sediments. Effect range low (ERL) and effect range median (ERM) are suitable methods for evaluating potential effects of these contaminants on Seine estuary organisms. According to Long, deleterious biological effects rarely occur if the total PAH and PCB concentrations in sediments are below their respective ERL values (4022 and 22.7 ng.g^{-1}); below the ERM guidelines (44792 and 180 ng.g^{-1}), adverse effects are likely to occur; and effects occasionally occur for concentrations equal to and above the ERL but below the ERM guidelines.

Cailleaud et al. (2007a) demonstrated that adsorbed PAHs present potential risks for organisms in winter periods, with concentrations above the ERL but below the ERM guidelines. Furthermore, adsorbed PCB concentrations are always above the ERL value and often above the ERM guideline. The risk associated with adsorbed PCBs is therefore important and higher than for PAHs. Reported values of PCB and PAH body burdens in the copepod E. affinis fall within a comparable range, 976 ng.g^{-1} and 879 ng.g^{-1}, respectively, both as means. For both contaminant groups, maximum levels were measured during winter periods.

In the Seine estuary, concentration factors (CFs) are in the range of 13,000 to 91,000 (for PCBs) and from 1,000 to 17,000 (for PAHs) (Cailleaud et al. 2007a), relatively high compared with usual reported bioconcentration factors for such organisms (Fisk et al. 2001). Furthermore, data about HOCs bioaccumulated by zooplankton in field studies are very limited, especially for estuarine species, and almost nonexistent for individual micro-zooplankton species. This void has been caused by the difficulties encountered in collecting sufficient supplies of micro-sized organisms to ensure accurate and precise quantification of contaminant burdens.

Harding et al. (1997) and Peters et al. (2001) reported low levels of PCBs (<100 ng.g^{-1}) for marine species from the southern Gulf of the St Lawrence and the southern North Sea. In the case of PAH bioaccumulation by zooplankton, field data are even scarcer. Nevertheless, Nour El-Din and Abdel-Moati (2001) reported PAH levels below 40 ng.g^{-1} in mixed marine copepods from the Arabian Gulf. In contrast, contaminant body burdens in the study of Cailleaud et al. (2007a) are distinctly higher, probably caused by higher exposure levels in estuarine ecosystems.

In estuarine copepods, metabolism reduced to a minimum to maintain physiological activity at low temperatures may imply low biotransformation of contaminants during winter periods (Gyllenberg and Lundqvist 1979; Pagano and Gaudy 1986; Gaudy et al. 2000). Furthermore, E. affinis are exposed for long periods and can bioaccumulate larger amounts of PCBs and PAHs.

In summary, total PAH and PCB concentrations in the Seine estuary exhibit seasonal variations, especially in particles and in *E. affinis*. Furthermore, *E. affinis* may bioaccumulate large quantities of PAHs and PCBs (1 g.g⁻¹ range for maximum). These very high levels have been measured for both contaminants during winter seasons, in SPM (when flow increased) and in *E. affinis*. For other seasons, contaminant concentrations were lower and quite regular. PAHs in SPM were mainly of pyrolytic origin, dominated by 4-ring compounds, while PAH sources in the dissolved phase were more diffuse and dominated by 3- and 4-ring compounds. The most bioavailable PAH compounds in the water column were also the most bioaccumulated ones in *E. affinis*. Besides, high proportions of carcinogenic PAHs have been measured in the copepods. PCB patterns in the water column principally contained penta-CB congeners, while the most abundant congeners measured in *E. affinis* were penta- and hexa-CB compounds.

When the results were combined in a principal component analysis (for details see Cailleaud et al. 2007a), the uptake of PCBs and PAHs by *E. affinis* was a function of the exposure time since the highest concentrations of contaminants in *E. affinis* were measured for samples with the highest percentages of adults.

10.4.2 Seasonal Variation of Total Alkylphenols (APs) in the Seine Estuary at Tancarville Station

Cailleaud et al. (2007b) measured alkylphenol (AP) (nonylphenol polyethoxy carboxylate [NPnEC], nonylphenol polyethoxylate [NPnEO], and nonylphenol [NP]) contents in the dissolved phase in SPMs and in *E. affinis* from November 2002 to January 2004 as shown in Table 10.2.

Dissolved AP concentrations ranged between 399 and 2214 ng. L⁻¹. The highest concentrations were measured during winter and summer periods and the lowest

TABLE 10.2
Average Alkylphenol Concentrations in Dissolved Phase, in SPM and Totaled in Water Column in the Seine Estuary at Tancarville Station, November 2002 to January 2004

	Dissolved Phase (ng.L⁻¹)				SPM (ng.g⁻¹ dry weight)				Water Column (ng.L⁻¹)			
	n = 10				n = 10				n = 10			
	Mean	Median	Min.	Max.	Mean	Median	Min.	Max.	Mean	Median	Min.	Max.
NP1EC	647	536	259	1455	204	183	18	515	661	544	278	1480
NP2EC	24	0	N.D.	147	N.D.	N.D.	N.D.	N.D.	24	0	N.D.	147
NP1EO	69	57	41	149	231	110	40	692	86	82	45	183
NP2EO	78	56	33	224	167	50	21	757	89	59	37	260
4-NP	110	79	15	386	1672	850	289		215	168	78	467
ΣAPEs	929	843	399	2214	2434	1173	405	9636	1079	1133	545	2390

Note: Total alkylphenols (ΣAPs); sum of 4-NP, NP1EO, NP2EO, NP1EC and NP2EC; N.D. : not detected; n = number of sampling cruises.

ones during spring and autumn periods. One possible explanation of the decrease of total AP concentrations in the water may be higher bacterial activity when water temperature increases and lower biodegradation rates during winter seasons (Jonkers et al. 2005).

Maximum biological production in the Seine Estuary is temperature-dependent and is observed during spring and autumn. The high dissolved concentration measured in July 2003 may have been caused by recent local AP discharges. This local emission probably originates from the Tancarville sewage treatment plant (STP) and from the petrochemical site of Port Jérôme less than 1 km upstream of the sampling site. The Tancarville STP is not a new technology plant, suggesting inadequate treatment of wastewater. In addition, dissolved APs show seasonal variation, similar to those observed for dissolved PAHs and PCBs (Cailleaud et al. 2007a).

Total concentrations of APs adsorbed on particles vary from 405 to 9636 ng.g^{-1} (dry weight) (Table 10.2). Significant increases of total AP concentrations were measured during winter periods, similar to those observed for PAHs and PCBs adsorbed on particles for the same period (Cailleaud et al. 2007a). These concentrations are in agreement with those reported for the Scheldt estuary (Jonkers et al. 2003). Reviews on AP distributions have reported NP and NPEO concentrations in sediments ranging from <0.003 to 72 µg.g^{-1} and 0.004 to 38 µg. g^{-1}, respectively (Thiele et al. 1997; Ying et al. 2002), with higher NP proportions adsorbed on SPM than dissolved in water, because of the hydrophobic nature of these compounds. These observations confirm that the Seine estuary sediments were highly polluted by AP compounds for the period studied (Cailleaud et al. 2007b).

SPM accounted for 3 to 32% of total AP contamination in the Seine estuary water column (16 ± 8% as a mean) (Cailleaud et al. 2007b). On the one hand, SPM contribute to 52 ± 20% on average of total NP contamination of the water column of the Seine estuary. On the other hand, NP1EO, NP2EO, and NP1EC are principally found in the dissolved phase (83 ± 9%, 90 ± 4%, and 97 ± 2%, respectively). 4-NP has a pronounced lipophilic character and is therefore more likely to be associated with organic matter on particles, while NPEOs and more particularly NPECs are more hydrophilic. Thus, the partition between particles and dissolved water for different AP compounds is close to the one described by Jonkers et al. (2005) for the Dutch coastal zone.

Total AP body concentrations in the copepod *E. affinis* vary from 3423 to 6406 ng.g^{-1} (Cailleaud et al. 2007b). NP2EO and 4-NP were detected in all copepod samples while NPECs were never detectable. NP2EO is the most important compound bioaccumulated by *E. affinis*. NPEO and NP bioconcentration factors in biological matrices are classically calculated using field (caging) or laboratory experiments (McLeese et al. 1981; Ekelund et al. 1990; Smith and Hill 2004). However, the occurrence of NPEOs, NPECs, and NP in natural biological matrices has been poorly reported. Verslycke et al. (2005) reported high NP2EO (220 to 981 ng.g^{-1} lipid weight) and NP (206 to 435 ng.g^{-1} lipid weight) concentrations in mysid crustaceans.

NP has been detected in the Norway lobster *Nephrops norvegicus* and the spot tail mantis shrimp *Squilla mantis* from the Adriatic Sea at levels ranging, respectively, from 274 to 399 and 118 to 254 ng.g^{-1} lipid weight (Ferrara et al. 2005). Nevertheless, considering high AP levels measured in the water column and in the copepods collected in the Seine estuary (Cailleaud et al. 2007b), physiological disorders may occur.

A PNEC value may be calculated by dividing the database LC_{50} of one compound (when no other aquatic toxicity data are available) for the most sensitive species in the ecosystem by an extrapolation factor of 1000, according to a 2003 European Community Technical Guidance Document, in order to protect the largest proportion of the aquatic community from toxic risks. NP1EC, NP1EO, and NP2EO PNEC values are, respectively, 2.0, 0.11, and 0.11 $\mu g.L^{-1}$ (Fenner et al. 2002).

The European Union (EU 2005) proposed 0.33 $\mu g.L^{-1}$ as the PNEC value for NP in both fresh and salt water. In the Seine estuary, total APs may account for a risk (total AP concentration/sum of each compound PNEC) always greater than 1. Considering the proposed PNEC sediment value of 180 $\mu g.kg^{-1}$ dry weight (according to EU 2005), NP concentrations in SPM at Tancarville station indicate a risk from this compound that always exceeds 2 (2 to 39). The SPM level of the Seine estuary is also potentially highly harmful to estuarine organisms. Recent studies have shown that 4-NP, at levels similar to those measured by Cailleaud et al. (2007b), affected the development time of the copepod *Tigriopus japonicus* (Marcial et al. 2003). Seo et al. (2006) also observed that sublethal 4-NP concentrations were toxic to copepods at the gene level. Ghekiere et al. (2006) reported vitellin induction in the mysid *Neomysis integer* after exposure to low NP levels. Endocrine disruptions were reported in amphipods *Corophium volutator* exposed to high levels NP (Brown et al. 1999). Furthermore, high AP levels bioaccumulated by *E. affinis* may be harmful to their predators, thus endangering the local food web equilibrium, as this species plays a key role in the Seine estuary.

In summary, APs were present at high levels both in the dissolved phase and in SPM of the Seine estuary. The dissolved phase and SPM contributed 84 and 16%, respectively, of total AP contamination in the water column. Total AP maximum levels in the dissolved phase and in SPM were observed during winter periods while significant decreases were observed during spring and autumn. These declines may be ascribed to maximum biological activities during these seasons. Particularly high NP and NP2EO concentrations were measured in all *E. affinis* samples, providing insight into the high capacity of this species to bioaccumulate APs. Besides, levels of APs measured in the water column were always at levels that could present risks for aquatic species. Thus, the endocrine disruptions (inter-sex fish) observed in the Seine river and bay may be related to high AP concentrations (Minier et al. 2000).

REFERENCES

Allan, J.D. et al. 1977. Grazing in juvenile stages of some estuarine calanoid copepods. *Mar. Biol.* 43: 317–331.

Andersen, C.M. and T.G. Nielsen. 1997. Hatching rate of the egg-carrying estuarine copepod *Eurytemora affinis*. *Mar. Ecol. Prog. Ser.* 160: 283–289.

Ask, J., M. Reinikainen, and U. Båmstedt. 2006. Variation in hatching success and egg production of *Eurytemora affinis* (Calanoida, Copepoda) from the Gulf of Bothnia, Baltic Sea, in relation to abundance and clonal differences of diatoms. *J. Plankton Res.* 28: 683–694.

Ban, S. 1992. Effects of photoperiod, temperature, and population density on induction of diapause egg production in *Eurytemora affinis* (Copepoda: Calanoida) in lake Ohnuma, Hokkaido, Japan. *J. Crustacean Biol.* 12: 361–367.

Ban, S. 1994. Effect of temperature and food concentration on post-embryonic development, egg production and adult body size of calanoid copepod *Eurytemora affinis*. *J. Plankton Res.* 16: 721–735.

Ban, S. and T. Minoda. 1991. The effect of temperature on the development and hatching success of diapause and subitaneous eggs in *Eurytemora affinis* (Copepoda: Calanoida) in lake Ohnuma, Hokkaido, Japan. Proceedings of Fourth International Conference on Copepoda. *Bull. Plankton Soc. Jpn.* Special volume, 299–308.

Ban, S. and T. Minoda. 1992. Hatching of diapause eggs of *Eurytemora affinis* (Copepoda: Calanoida) collected from lake-bottom sediments. *J. Crustacean Biol.* 12: 51–56.

Boak, A.C. and R. Goulder. 1983. Bacterioplankton in the diet of the calanoid copepod *Eurytemora* sp. in the Humber estuary. *Mar. Biol.* 73: 139–149.

Bradley, B.P. 1986. Genetic expression of temperature tolerance in the copepod *Eurytemora affinis* in different salinity and temperature environments. *Mar. Biol.* 91: 561–565.

Bradley, B.P. 1991. Distribution of copepods in coastal zone waters: seasonal succession in Chesapeake Bay. *Bull. Plankton Soc. Jpn.* Special volume, 129–131.

Brown, R.J., M. Conradi, and M.H. Depledge. 1999. Long-term exposure to 4-nonylphenol affects sexual differentiation and growth of the amphipod *Corophium volutator* (Pallas, 1766). *Sci. Total Environ.* 233: 77–88.

Cailleaud, K. 2006. La qualité de l'eau en estuaire de Seine: Analyse pluridisciplinaire de la contamination organique via l'utilisation du modèle animal *Eurytemora affinis*: Etude *in situ* et expérimentale. Ph.D. Thesis, University of Bordeaux 1.

Cailleaud, K. et al. 2007a. Seasonal variations of hydrophobic organic contaminant concentrations in the water column of the Seine estuary and their transfer to a planktonic species *Eurytemora affinis* (Calanoïda, copepoda). 1. PCBs and PAHs. *Chemosphere* 70: 270–280.

Cailleaud, K. et al. 2007b. Seasonal variation of hydrophobic organic contaminant concentrations in the water column of the Seine estuary and their transfer to a planktonic species *Eurytemora affinis* (Calanoïda, copepod). 2. Alkylphenol polyethoxylates. *Chemosphere* 70: 281–287.

Caramujo, M.J. and M.J. Boavida. 1999. Characteristics of the reproductive cycles and development times of *Copidodiaptomus numidicus* (Copepoda: Calanoida) and *Acanthocyclops robustus* (Copepoda: Cyclopoida). *J. Plankton Res.* 21: 1765–1778.

Chinnery, F.E. and J.A. Williams. 2004. The influence of temperature and salinity on *Acartia* (Copepoda: Calanoida) nauplii survival. *Mar. Biol.* 145: 733–738.

Cross-Sorokin, M.Y. et al. 2003. Uptake and depuration of 4-nonylphenol by the benthic invertebrate *Gammarus pulex*: How important is feeding rate? *Environ. Sci. Technol.* 37: 2236–2241.

Dauvin, J.C. et al. 2000. Main characteristics of the boundary layer macrofauna in the English Channel. *Hydrobiologia* 426: 139–156.

David, V. et al. 2005. Long-term changes of the zooplankton variability in a turbid environment: the Gironde estuary (France). *Estuar. Coast. Shelf Sci.* 64: 171–184.

Devreker, D. 2007. Dynamique de population du copépode *Eurytemora affinis* dans l'estuaire de la Seine: approche combinée in situ multi-échelle et expérimentale. Ph.D. Thesis University of Le Havre.

Devreker, D., S. Souissi, and L. Seuront. 2004. Development and mortality of the first naupliar stages of *Eurytemora affinis* (Copepoda, Calanoida) under different conditions of salinity and temperature. *J. Exp. Mar. Bio. Ecol.* 303: 31–46.

Devreker, D. et al. 2007. Effects of salinity and temperature on the post-embryonic development of *Eurytemora affinis* (Copepoda; Calanoida) from the Seine estuary: a laboratory study. *J. Plankton Res.* 29: 117–133.

Devreker, D. et al. Effects of salinity, temperature and individual variability on the reproduction of *Eurytemora affinis* (Copepoda; Calanoida) from the Seine estuary: a laboratory study. *J. Exp. Mar. Biol. Ecol.* (submitted).

EC. 2003. Technical Guidance Document on Risk Assessment. European Commission Joint Research Center, EUR 20418 EN/2.

Ekelund, R. et al. 1990. Bioaccumulation of 4-nonylphenol in marine animals—A re-evaluation. *Environ. Pollut.* 64: 107–120.

Escaravage, V. and K. Soetaert. 1995. Secondary production of the brackish copepod communities and their contribution to the carbon fluxes in the Westerschelde estuary (The Netherlands). *Hydrobiologia* 311: 103–114.

European Union, 2005. Common implementation strategy for the water framework directive: 4-nonylphenol (branched) and nonylphenol (final version).

Fenner, K. et al. 2002. Including transformation products into the risk assessment for chemicals: the case of nonylphenol ethoxylate usage in Switzerland. *Environ. Sci. Technol.* 36: 1147–1154.

Fernandez, M.B. et al. 1997. Polyaromatic hydrocarbon (PAH) distributions in the Seine River and its estuary. *Mar. Pollut. Bull.* 34: 857–867.

Ferrara, F. et al. 2005. Alkylphenols and alkylphenols ethoxylates: contamination of crustaceans and fishes from the Adriatic Sea (Italy). *Chemosphere* 59: 1145–1150.

Fisk, A.T. et al. 2001. Persistent organic pollutants (POPs) in a small, herbivorous, Arctic marine zooplankton (*Calanus hyperboreus*): trends from April to July and the influence of lipids and trophic transfer. *Mar. Pollut. Bull.* 43: 93–101.

Fockedey, N. and J. Mees. 1999. Feeding of the hyperbenthic mysid *Neomysis integer* in the maximum turbidity zone of the Elbe, Westerschelde and Gironde estuaries. *J. Mar. Syst.* 22: 207–228.

Forget-Leray, J. et al. 2005. Impact of endocrine toxicants on survival. development, and reproduction of the estuarine copepod *Eurytemora affinis* (Poppe). *Ecotoxicol. Environ. Saf.* 60: 288–294.

Freire, C.A. et al. 2003. Adaptive patterns of osmotic and ionic regulation, and the invasion of fresh water by the palaemonid shrimps. *Comp. Biochem. Physiol.* A136: 771–778.

Fryd, M., O.H. Haslund, and O. Wohlgemuth. 1991. Development, growth and egg production of the two copepod species *Centropages hamatus* and *Centropages typicus* in the laboratory. *J. Plankton Res.* 13: 683–689.

Fulton, R.S. III. 1982. Predatory feeding of two marine mysids. *Mar. Biol.* 72: 183–191.

Garnier, J. et al. 2001. Lower Seine river and estuary (France): carbon and oxygen budgets during low flow. *Estuaries* 24: 964–976.

Gasparini, S., J. Castel, and X. Irigoien. 1999. Impact of suspended particulate matter on egg production of the estuarine copepod, *Eurytemora affinis*. *J. Mar. Syst.* 22: 195–205.

Gaudy, R., G. Cervetto, and M. Pagano. 2000. Comparison of the metabolism of *Acartia clausi* and *A. tonsa*: influence of temperature and salinity. *J. Exp. Mar. Biol. Ecol.* 247: 51–65.

Gehrs, C.W. and A. Robertson. 1975. Use of life tables in analysing the dynamics of copepod populations. *Ecology* 56: 665–672.

Ghekiere, A., T. Verslycke, and C. Jansen. 2006. Effects of methoprene, nonylphenol, and estrone on the vitellogenesis of the mysid *Neomysis integer*. *Gen. Comp. Endocr.* 147: 190–195.

Gibson, G.R. et al. 2000. Estuarine and Coastal Marine Waters: Bioassessment and Biocriteria Technical Guidance. EPA 822-B-00-024. Washington, D.C.

Gillooly, J.F. et al. 2001. Effects of size and temperature on metabolic rate. *Science* 293: 2248–2251.

Gómez, F., D. Devreker, and S. Souissi. Seasonal and tidally driven distribution of microphytoplankton in the macrotidal Seine Estuary (France). *Estuar. Coasts* (submitted).

Gustafson, K.E. and R.M. Dickhut. 1997. Distribution of polycyclic aromatic hydrocarbons in southern Chesapeake Bay water: evaluation of three methods for determining freely dissolved water concentration. *Environ. Toxicol. Chem.* 16: 452–461.

Gyllenberg, G. and G. Lundqvist. 1979. The effects of temperature and salinity on the oxygen consumption of *Eurytemora hirundoides* (Crustacea, Copepoda). *Ann. Zool. Fennici* 16: 205–208.

Hall, C.J. and C.W. Burns. 2002. Effects of temperature and salinity on the survival and egg production of *Gladioferens pectinatus* Brady (Copepoda: Calanoida). *Estuar. Coast. Shelf Sci.* 55: 557–564.

Harding, G.C. et al. 1997. Bioaccumulation of polychlorinated biphenyls (PCBs) in the marine pelagic food web, based on seasonal study in the southern Gulf of St. Lawrence, 1976–1977. *Mar. Chem.* 56: 145–179.

Harris, R.P. et al. 1977. Accumulation of ^{14}C-1-naphtalene by an oceanic and an estuarine copepod during long-term exposure to low-level concentrations. *Mar. Biol.* 42: 187–195.

Heinle, D.R. and D.A. Flemer. 1975. Carbon requirements of a population of the estuarine copepod *Eurytemora affinis*. *Mar. Biol.* 31: 235–247.

Hirche, H.J., U. Meyer, and B. Niehoff. 1997. Egg production of *Calanus finmarchicus*: effect of temperature, food and season. *Mar. Biol.* 127: 609–620.

Holste, L. and M.A. Peck. 2005. The effects of temperature and salinity on egg production and hatching success of Baltic *Acartia tonsa* (Copepoda: Calanoida): a laboratory investigation. *Mar. Biol.* 148: 1061–1070.

Hough, A.R. and E. Naylor. 1992. Endogenous rhythms of circatidal swimming activity in the estuarine copepod *Eurytemora affinis* (Poppe). *J. Exp. Mar. Bio. Ecol.* 161: 27–32.

Huntley, M.E. and M.D.G. Lopez. 1992. Temperature dependent production of marine copepods: a global synthesis. *Am. Nat.* 140: 201–242.

Ianora, A. 1998. Copepod life history traits in subtemperate regions. *J. Mar. Syst.* 15: 337–349.

Ianora, A., M.G. Mazzocchi, and R. Grottoli. 1992. Seasonal fluctuations in fecundity and hatching success in the planktonic copepod *Centropages typicus*. *J. Plankton Res.* 14: 1483–1494.

Ishikawa, A., S. Ban, and N. Shiga. 1999. Effects of salinity on survival. and embryonic and postembryonic development of *Eurytemora affinis* from a freshwater lake. *Plankton Biol. Ecol.* 46: 113–119.

Jacoby, C.A. and M.J. Youngbluth. 1983. Mating behavior in three species of *Pseudodiaptomus* (Copepoda: Calanoida). *Mar. Biol.* 76: 77–86.

Jamieson, C.D. and B. Santer. 2003. Maternal aging in the univoltine freshwater copepod *Cyclops kolensis*: variation in egg sizes, egg development times, and naupliar development times. *Hydrobiologia* 510: 75–81.

Jiménez-Melero, R. et al. 2007. Post-embryonic developmental plasticity of *Arctodiaptomus salinus* (Copepoda: Calanoida) at different temperatures. *J. Plankton. Res.* 29: 553–567.

Jonkers, N., R.W.P.M. Laane, and P. De Voogt. 2003. Fate of nonylphenol ethoxylates and their metabolites in two Dutch Estuaries: evidence of biodegradation in the field. *Environ. Sci. Technol.* 37: 321–327.

Jonkers, N., R.W.P.M. Laane, and P. De Voogt. 2005. Sources and fate of nonylphenol ethoxylates and their metabolites in the Dutch coastal zone of the North Sea. *Mar. Chem.* 96: 115–135.

Karr, J.R. and D.R. Dudley. 1981. Ecological perspectives on water quality goals. *Environ. Manage.* 5: 55–68.

Katona, S.K. 1970. Growth characteristics of the copepods *Eurytemora affinis* and *E. herdmanii* in laboratory cultures. *Helgoland Wiss. Meer.* 20: 373–384.

Katona, S.K. 1971. The developmental stages of *Eurytemora affinis* (Poppe, 1880) (Copepoda, Calanoida) raised in laboratory cultures, including a comparison with the larvae of *Eurytemora americana* Williams, 1906, and *Eurytemora herdmani* Thompson & Scott, 1897. *Crustaceana* 21: 5–20.

Katona, S.K. 1975. Copulation in the copepod *Eurytemora affinis* (Poppe, 1880). *Crustaceana* 28: 89–95.

Kimmel, D.G. and B.P. Bradley. 2001. Specific protein responses in the calanoid copepod *Eurytemora affinis* (Poppe, 1880) to salinity and temperature variation. *J. Exp. Mar. Biol. Ecol.* 266: 135–149.

Kimmel, D.G., W.D. Miller, and M.R. Roman. 2006. Regional scale climate forcing on zooplankton dynamics in Chesapeake Bay. *Estuar. Coasts* 29: 375–387.

Kimmerer, W.J., J.R. Burau, and W.A. Bennett. 1998. Tidally oriented vertical migration and position maintenance of zooplankton in a temperate estuary. *Limnol. Oceanogr.* 43: 1697–1709.

Kimmerer, W.J., J.R. Burau, and W.A. Bennett. 2002. Persistence of tidally oriented vertical migration by zooplankton in a temperate estuary. *Estuaries* 25: 359–371.

Ko, F.C. and J.E. Baker. 1995. Partitioning of hydrophobic organic contaminants to resuspended sediments and plankton in the mesohaline Chesapeake Bay. *Mar. Chem.* 49: 171–188.

Kowalewska, G. et al. 2003. Transfer of organic contaminants to the Baltic in the Odra Estuary. *Mar. Pollut. Bull.* 46: 703–718.

Laprise, R. and J.J. Dodson. 1994. Environmental variability as a factor controlling spatial patterns in distribution and species diversity of zooplankton in the St. Lawrence estuary. *Mar Ecol. Prog. Ser.* 107: 67–81.

Lawrence, D., I. Valiela, and G. Tomasky. 2004. Estuarine calanoid copepod abundance in relation to season, salinity, and land-derived nitrogen loading, Waquoit Bay, MA. *Estuar. Coast. Shelf Sci.* 61: 547–557.

Lee, C.E. 1999. Rapid and repeated invasions of fresh water by the copepod *Eurytemora affinis*. *Evolution* 53: 1423–1434.

Lee, C.E. 2000. Global phylogeography of a cryptic copepod species complex and reproductive isolation between genetically proximate "populations." *Evolution* 54: 2014–2027.

Lee, C.E. and B.W. Frost. 2002. Morphological stasis in the *Eurytemora affinis* species complex (Copepoda: Temoridae). *Hydrobiologia* 480: 111–128.

Lee, C.E. and C.H. Peterson. 2002. Genotype-by-environment interaction for salinity tolerance in the freshwater-invading copepod *Eurytemora affinis*. *Physiol. Biochem. Zool.* 75: 335–344.

Lee, C.E. and C.H. Peterson. 2003. Effects of developmental acclimation on adult salinity tolerance in the freshwater-invading copepod *Eurytemora affinis*. *Physiol. Biochem. Zool.* 76: 296–301.

Lee, C.E., J.L. Remfert, and G.W. Gelembuik. 2003. Evolution of physiological tolerance and performance during freshwater invasions. *Integr. Comp. Biol.* 43: 439–449.

Lloyd, S. 2006. Zooplankton ecology in the Chesapeake bay estuarine turbidity maximum, with emphasis on the calanoid copepod *Eurytemora affinis*. University of Maryland, College Park.

Long, E.R. et al. 1995. Incidence of adverse effects within ranges of chemical concentrations in marine and estuarine sediments. *Environ. Manage.* 19: 81–97.

Lonsdale, D.J., M.A. Frey, and T.W. Snell. 1998. The role of chemical signals in copepod reproduction. *J. Mar. Sys.* 15: 1–12.

MacFarlane, G.R., D.J. Booth, and K.R. Brown. 2000. The semaphore crab, *Heloecius cordiformis*: bio-indication potential for heavy metals in estuarine systems. *Aquat. Toxicol.* 50: 153–166.

Magnusson, K. et al. 2007. Bioaccumulation of ^{14}C-PCB 101 and ^{14}C-PBDE 99 in the marine planktonic copepod *Calanus finmarchicus* under different food regimes. *Mar. Environ.* 63: 67–81.

Manodori, L. et al. 2006. PCBs and PAHs in sea-surface microlayer and subsurface water samples of the Venice Lagoon (Italy). *Mar. Pollut. Bull.* 52: 184–192.

Marcial. S.H., A. Hagiwara, and T.W. Snell. 2003. Estrogenic compounds affect development of harpacticoid copepod *Tigriopus japonicus. Environ. Toxicol. Chem.* 22: 3025–3030.

Matias-Peralta, H. et al. 2005. Effects of some environmental parameters on the reproduction and development of a tropical marine harpacticoid copepod *Nitocra affinis* f. *californica* Lang. *Mar. Poll. Bull.* 51: 722–728.

McLeese, D.W. et al. 1981. Lethality and accumulation of alkylphenols in aquatic fauna. *Chemosphere* 10: 723–730.

McLusky, D.S. 1989. *The Estuarine Ecosystem*, 2nd ed. London: Chapman & Hall.

Merrell, J.R. and D.K. Stoecker. 1998. Differential grazing on protozoan microplankton by developmental stages of the calanoid copepod *Eurytemora affinis* Poppe. *J. Plankton Res.* 43: 289–304.

Mikac, N. et al. 1996. Uptake of mercury species by transplanted mussels *Mytilus galloprovincialis* under estuarine conditions (Krka river estuary). *Sci. Total Environ.*184: 173–182.

Minier, C. et al. 2006. A pollution monitoring pilot study involving contaminant and biomarker measurements in the Seine estuary, France, using zebra mussels (*Dreissena polymorpha*). *Environ. Toxicol. Chem.* 25: 112–119.

Minier, C. et al. 2000. Flounder health status in the Seine Bay: a multibiomarker study. *Mar. Environ. Res.* 50: 1–5.

Molinero, J.C. et al. 2005. Phenological changes in the Northwestern Mediterranean copepods *Centropages typicus* and *Temora stylifera* linked to climate forcing. *Oecologia* 145: 640–649.

Morgan, C.A., J.R. Cordell, and C.A. Simenstad. 1997. Sink or swim? Copepod population maintenance in the Columbia River estuarine turbidity maxima region. *Mar. Biol.* 129: 309–317.

Motelay-Massei, A. et al. 2002. Atmospheric deposition of toxics into the Seine estuary, France: example of polycyclic aromatic hydrocarbons. *Atmos. Chem. Phys. Discuss.* 2: 1351–1369.

Motelay-Massei, A. et al. 2004. Distribution and spatial trends of PAHs and PCBs in soils in the Seine River basin, France. *Chemosphere.* 55: 555–565.

Motelay-Massei, A. et al. 2007. PAHs in the bulk atmospheric deposition of the Seine river basin: source identification and apportionment by ratios, multivariate statistical techniques and scanning electron microscopy. *Chemosphere* 67: 312–321.

Mouny, P. 1998. Structure spatio-temporelle du zooplancton et du suprabenthos de l'estuaire de la Seine. Dynamique et rôle des principales espèces dans la chaîne trophique pélagique. Thèse de doctorat, Université Pierre et Marie Curie, Paris.

Mouny, P. and J.C. Dauvin. 2002. Environmental control of mesozooplankton community in the Seine estuary (English Channel). *Oceanol. Acta* 25: 13–22.

Mouny, P. et al. 1998. Biological components from the Seine estuary: first results. *Hydrobiologia* 373–374: 333–347.

Nour El-Din, N.M. and M.A.R. Abdel-Moati. 2001. Accumulation of trace metals, petroleum hydrocarbons, and polycyclic aromatic hydrocarbons in marine copepods from the Arabian Gulf. *B. Environ. Contam. Tox.* 66: 110–117.

Paffenhöfer, G.A. and D.E. Stearns. 1988. Why is *Acartia tonsa* (Copepoda: Calanoida) restricted to near-shore environments? *Mar. Ecol. Prog. Ser.* 42: 33–38.

Pagano, M. and R. Gaudy. 1986. Biologie d'un copépode des mares temporaires du littoral méditerranéen français, *Eurytemora velox*. II. Respiration et excrétion. *Mar. Biol.* 90: 551–564.

Peters, L.D., C. Porte, and D.R. Livingstone. 2001. Variation of antioxidant enzyme activities of Sprat (*Sprattus sprattus*) larvae and organic contaminant levels in mixed zooplankton from the southern North Sea. *Mar. Pollut. Bull.* 42: 1087–1095.

Peterson, W.T. 2001. Patterns in stage duration and development among marine and freshwater calanoid and cyclopoid copepods: a review of rules, physiological constraints, and evolutionary significance. *Hydrobiologia* 453–454: 91–105.

Peterson, W.T. and S.J. Painting. 1990. Developmental rates of the copepods *Calanus australis* and *Calanoides carinatus* in the laboratory, with discussion of methods used for calculation of development time. *J. Plankton Res.* 12: 283–293.

Pigliucci, M. 2005. Evolution of phenotypic plasticity: where are we going now? *Trends Ecol. Evol.* 20: 481–486.

Poli, J.M. and J. Castel. 1983. Cycle biologique en laboratoire d'un copépode planctonique de l'estuaire de la Gironde: *Eurytemora hirundoides* (Nordquist, 1888). *Vie et Milieu* 33: 79–86.

Purchase, N.G. and J.E. Fergusson. 1986. *Chione (austrovenus) stutchburyi*, a New Zealand cockle, as a bio-indicator for lead pollution. *Env. Poll. Ser. B, Chem. Phys.* 11: 137–151.

Richman, S., D.R. Heinle, and R. Huff. 1977. Grazing by adult estuarine calanoid copepods of the Chesapeake Bay. *Mar. Biol.* 42: 69–84.

Roddie, B.D., R.J.G. Leakey, and A.J. Berry. 1984. Salinity, temperature tolerance and osmoregulation in *Eurytemora affinis* (Poppe) (Copepoda: Calanoida) in relation to its distribution in the zooplankton of the upper reaches of the Forth estuary. *J. Exp. Mar. Biol. Ecol.* 79: 191–211.

Roman, M.R., D.V. Holliday, and L.P. Sanford. 2001. Temporal and spatial patterns of zooplankton in the Chesapeake Bay turbidity maximum. *Mar. Ecol. Prog. Ser.* 213: 215–227.

Seo, J.S. et al. 2006. Small heat shock protein 20 gene (Hsp20) of the intertidal copepod *Tigriopus japonicus* as a possible biomarker for exposure to endocrine disruptors. *B. Environ. Contam. Tox.* 76: 566–572.

Sibert, J.R. 1981. Intertidal hyperbenthic populations in the Nanaimo estuary. *Mar. Biol.* 64: 259–265.

Smith, M.D. and E.M. Hill. 2004. Uptake and metabolism of technical nonylphenol and its brominated analogues in the roach (*Rutilus rutilus*). *Aquat. Toxicol.* 69: 359–369.

Smith, S.L. and P.V.Z. Lane. 1985. Laboratory studies of the marine copepod *Centropages typicus*: egg production and development rates. *Mar. Biol.* 85: 153–162.

Smyly, W.J.P. 1980. Effect of constant and alternating temperatures on adult longevity of the freshwater cyclopoid copepod, *Acanthocyclops viridis* (Jurine). *Arch. Hydrobiol.* 89: 353–362.

Souissi, S., F. Carlotti, and P. Nival. 1997. Food- and temperature-dependent function of moulting rate in copepods: an example of parameterization for population dynamics models. *J. Plankton Res.* 19: 1331–1346.

Souissi, S. et al. 2005. Describing space–time patterns in aquatic ecology using IBMs and scaling and multi-scaling approaches. *Nonlin. Anal. Real World Appl.* 6: 705–730.

Tackx, M.L. et al. 2003. Selective feeding of *Eurytemora affinis* (Copepoda, Calanoida) in temperate estuaries: model and field observations. *Estuar. Coast. Shelf Sci.* 56: 305–311.

Tang, K.W., C.S. Freund, and C.L. Schweitzer. 2006. Occurrence of copepod carcasses in the lower Chesapeake Bay and their decomposition by ambient microbes. *Estuar. Coast. Shelf Sci.* 68: 499–508.

Thiele, B., K. Günther, and M.J. Schwuger. 1997. Alkylphenol ethoxylates: trace analysis and environmental behavior. *Chem. Rev.* 97: 3247–3272.

Uye, S.I. 1981. Fecundity studies of neritic calanoid copepods *Acartia clausi* Giesbrecht and *A. steueri* Smirnov: a simple empirical model of daily egg production. *J. Exp. Mar. Biol. Ecol.* 50: 255–271.

Verslycke, T.A. et al. 2005. Flame retardants, surfactants and organotins in sediment and mysid shrimp of the Scheldt estuary (The Netherlands). *Environ. Pollut.* 136: 19–31.

Vijverberg, J. 1980. Effect of temperature in laboratory studies on development and growth of Cladocera and Copepoda from Tjeukemeer, The Netherlands. *Freshwater Biol.* 10: 317–340.

von Vaupel-Klein, J.C. and R.E. Weber. 1975. Distribution of *Eurytemora affinis* (copepoda: calanoida) in relation to salinity: field and laboratory observations. *Neth. J. Sea Res.* 9: 297–310.

Williamson, C.E. and N.M. Butler. 1987. Temperature, food and limitation of copepod reproductive rates: separating the effects of multiple hypotheses. *J. Plankton Res.* 9: 821–836.

Wilson, J.G. 1994. The role of bioindicators in estuarine management. *Estuaries.* 17: 94–101.

Winkler, G. and W. Greve. 2004. Trophodynamics of two interacting species of estuarine mysids, *Praunus flexuosus* and *Neomysis integer*, and their predation on the calanoid copepod *Eurytemora affinis*. *J. Exp. Mar. Biol. Ecol.* 308: 127–146.

Witt, G. 1995. Polycyclic aromatic hydrocarbons in water and sediment of the Baltic Sea. *Mar. Pollut. Bull.* 31: 237–248.

Woodson, C.B. 2005. Thin layers: physical and chemical cues contributing to observed copepod aggregations. Ph.D. Georgia Institute of Technology, Atlanta.

Wurl, O. and J.P. Obbard. 2006. Distribution of organochlorine compounds in the sea surface microlayer, water column, and sediment of Singapore's coastal environment. *Chemosphere* 62: 1105–1115.

Ying, G.G., B. Williams, and R. Kookana. 2002. Environmental fate of nonylphenols and APE: a review. *Environ. Int.* 28: 215–226.

11 From Pollution to Altered Physiological Performance
The Case of Flatfish in the Seine Estuary

Christophe Minier and Rachid Amara

CONTENTS

11.1 CONTAMINATION OF THE AQUATIC ENVIRONMENT

The massive increase of the human population produced an exponential increase in technological and industrial development around the world. This outburst of advancements has greatly increased awareness and concern for the quality of the environment over recent decades. No single region on earth can now be considered free from any pollutant. Many thousands of persistent organic pollutants (POPs) have been produced and in part released into the environment. They can now be found even in remote regions where they have never been used nor produced (Muir et al. 1988; Simonich and Hites 1995; Vilanova et al. 2001; Derocher et al. 2003).

The common understanding that pollution can be found everywhere is accompanied by the fear that POPs may affect human health and development. A range of environmental contaminants have been shown to affect the health of wildlife and human populations. The consequences include loss of biological biodiversity, increased occurrence of diseases, alteration of habitats, and thus impaired sustainability of ecosystems and their recreational or economic value.

Aquatic environments are particularly affected in this scenario. Most humans live near coasts, directly impacting the coastal environments. Coastal habitats are also considered the ultimate sinks for many chemicals released from anthropogenic activities, due to direct (e.g., industrial and domestic effluents) or diffuse (e.g., agriculture and urban runoff) discharges or atmospheric processes (Stegeman and Hahn 1994). Many thousands of POPs released during human activities ultimately contaminate estuaries and other coastal areas. Their low water solubilities allow them to bind strongly to particulate matter in the water column, and as a result can contaminate sediments for decades, thus exposing plants and animals living in contact with these sediments (Jones and de Voogt 1999; Larsson et al. 2000).

11.2   FISH AS BIOINDICATORS

The presence of a xenobiotic compound in an ecosystem does not alone imply deleterious effects. Pollutants interact with environmental and biological systems based on their intrinsic physicochemical properties and reactivity, yielding a characteristic pattern of environmental and internal exposure concentrations for each pollutant. Final exposure and effect assessment according to this concept will always be subject to uncertainty as a result of inherent variability and complexity of both environmental and biological systems (Figure 11.1).

Environmental chemists and toxicologists have extensively studied the exposure, fate, and effects of chemical contaminants in aquatic ecosystems, but the complexity of the systems seldom led to clear connections between pollutants and their ecological effects. If an environment is highly complex, the complexity is often multiplied in an organism (and the environment will contain hundreds of different types) due to a wide diversity of cells, physiological conditions, and abilities to respond to a compound. The abilities of various pollutants (and their derivatives) to mutually impose their toxic actions upon several cellular or physiological processes and the long periods required before deleterious effects on populations tend to manifest themselves only complicate risk assessments relative to the presence of xenobiotics.

For several reasons, fish species have attracted considerable interest in studies assessing biochemical and physiological responses to environmental contaminants (Powers 1989; Van der Oost et al. 2003). Fish can be found virtually everywhere in the aquatic environment, and they play a major ecological role in aquatic food webs because they function as carriers of energy from lower to higher trophic levels (Beyer 1996). The understanding of toxicant uptake, behaviour, and responses in fish therefore presents high ecological relevance. Fish represent increasing food sources for humans, thus contributing to the contamination of human tissues and disease (FAO 2007). Studies of fish biology may also benefit from knowledge of other vertebrate species such as mammals. Fish are usually more representative of the toxicological mechanisms that may affect human health than, for example, invertebrate species (Ketata et al. 2008).

Fish have often been used to study their body burdens and the transfer of pollutants in the food web (see Chapter 6). They can be good indicators of the bioaccumulation resulting from the contamination of the environment, except in the case of some compounds such as polycyclic aromatic hydrocarbons (PAHs) that fish can metabolise (Van der Oost et al. 2003). Flatfish have been extensively used in environmental studies. They live in close contact with sediments and thus may be readily exposed to the associated POPs.

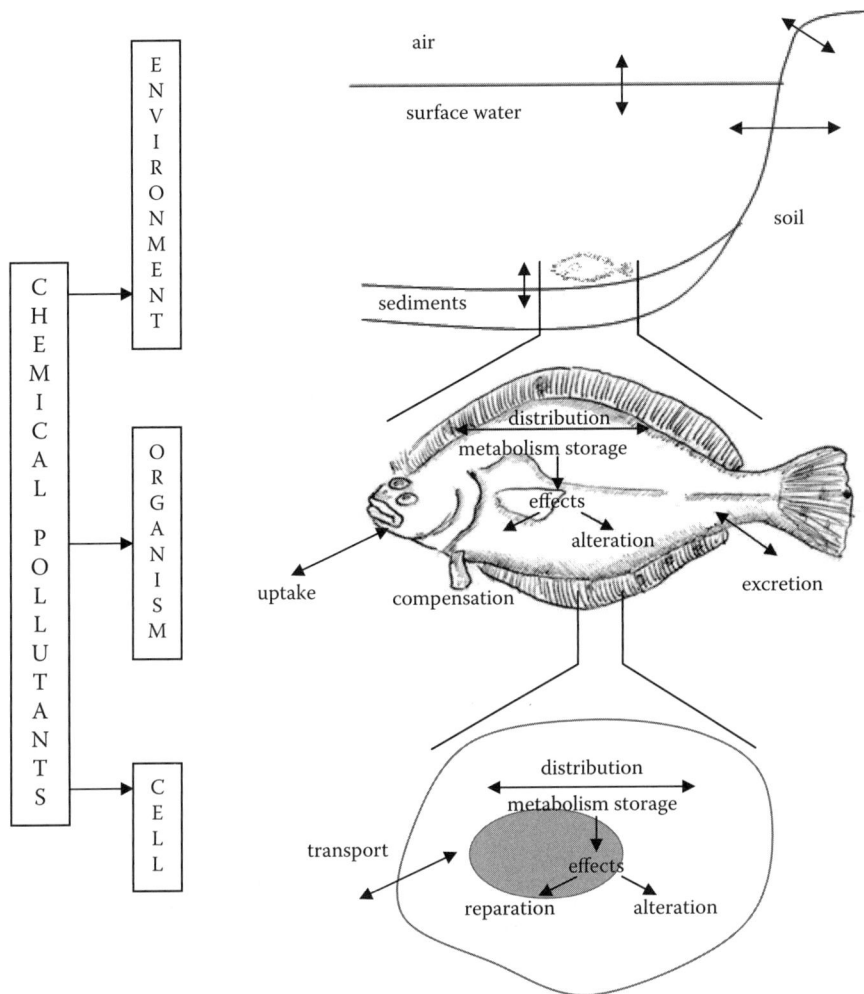

FIGURE 11.1 Pollutant input, distribution, and fate in an environmental and biological system. Illustration of key features and commonalities between exposure and effect assessment. (Adapted from Schwarzenbach, R.P. et al. 2006. *Science* 313: 1072–1077. With permission.)

Both organic compounds and metals have been measured in tissues from flounder (*Platichthys flesus*) and sole (*Solea solea*) along the French coast (IFREMER 2007). When considering the results, it is obvious that the Seine estuary is particularly notable. Fish body burdens are high for organic contaminants, especially polychlorinated biphenyls (PCBs) and dichloro-diphenyl-tetrachlorohexane (DDT) and its metabolites. Although the Gironde estuary may be more impacted by certain metals such as cadmium, very high levels are also found in the Seine (Table 11.1). Mean body concentrations of PCBs (sums of seven congeners) varied from 850 (Jaouen-Madoulet 2000) to 1975 ng.g^{-1} ww (IFREMER 2007) in flounder. Liver concentrations were

TABLE 11.1

Chemical Pollutant Body Concentrations in Adult Flounder and Sole from the Seine Bay

Pollutant		Σ7PCB	Σ20PAH	ΣDDT	Lindane	HgCH₃	Cd
Concentration	Flounder	1977	45	136	0.7	420	2
(ng.g⁻¹ ww)	Sole	535	—	12	0.3	270	3

Sources: IFREMER 2007 and Jaouen-Madoulet 2000.

usually three times those reported in the whole body (Jaouen-Madoulet 2000). A clear pollution gradient can be seen from the bay to upstream of the Seine estuary, indicating that most pollution comes from the river (IFREMER 2007).

11.3 EFFECTS OF POLLUTION ON PHYSIOLOGICAL PERFORMANCE

Habitat quality for fish cannot be measured directly because it depends on a range of interdependent factors that determine the survival and, ultimately the future reproductive success of the juveniles (Gibson 1994). However, habitat-specific growth and conditions can serve as useful proxies for habitat quality because they integrate the effects of multiple environmental factors (Vinagre et al. 2005). The recent use of enclosures and individually marked juvenile flounder has been successfully employed to estimate local growth, from which the qualities of various habitats have been inferred (Tarpgaard et al. 2005).

Several indices effectively estimate fish condition, for example, morphometric techniques such as Fulton's K index and biochemical indices such as RNA/DNA ratios and a lipid storage index based on the ratio of triacylglycerol (TAG) quantity to sterol (ST) quantity (reserve lipids to structural lipids). Fish growth rate was estimated from otolith microstructure analysis (daily growth increments).

All these indices appeared altered when studying juvenile flatfish sampled in the Seine estuary compared to those sampled in less polluted estuaries such as the Authie and Canche (Amara et al. 2007). For example, the growth rate and condition indices of 0-group juvenile flounder caught in the Seine estuary were significantly lower than those caught in the Authie estuary. Growth rate, Fulton's K condition index, and the RNA/DNA ratio were reduced by 26, 17, and 27%, respectively (Figure 11.2). On a local scale, indicators of growth and condition were also used successfully to compare habitat quality along an estuarine gradient (Figure 11.3). Juvenile fish exhibited the slowest growth and lowest condition indices in the more polluted site (Morin et al. 2005). Our studies indicate that both growth and condition indices measured on juvenile flatfish were well correlated with anthropogenic disturbance (e.g., sediment chemical contaminants) and provided useful tools to compare and assess habitat quality.

Adult flounder living in the Seine estuary were also studied and the condition indices and growth rates were compared to results from other major French estuaries

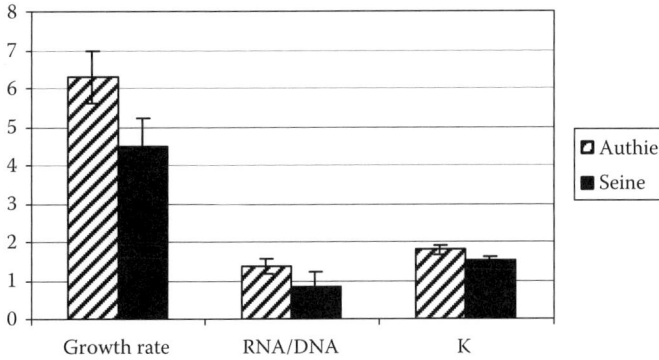

FIGURE 11.2   Growth rate (micrometres per day growth increment in otolith), Fulton's condition index (K in milligrams per cubic millimetre), and metabolic activity (RNA/DNA) of juvenile flounder (mean age 45 days) from the Seine and Authie estuaries. All physiological parameters are significantly different in the two estuaries ($P < 0.05$, $n = 30$). Bars indicate standard errors.

and a reference site that was clearly less contaminated (Marchand et al. 2004). Fish from the Seine and Gironde estuaries showed more pronounced reductions in growth rate, with 40 to 46% lower values when compared to the reference site. Condition index was the lowest in the Seine and represented only 79% of the value recorded for the reference site.

Juvenile sole (*Solea solea*) were also used to assess the habitat quality. This species of economic importance develops in close contact with bottom sediment and is restricted to very shallow coastal waters during the first year of life (Amara 2003; Eastwood et al. 2003). Following settlement on the nursery, 0-group sole, like many flatfish species, show a strong decline in swimming activity levels during summer, with relatively low activity continuing until November (Durieux et al. 2007). Because of the strong habitat association characteristic of this species, growth rates and condition indices of 0-group sole analysed were assumed to reflect the quality of the nursery grounds where they were caught. Results of the study showed decreases in both condition index and growth rate of sole from the Seine estuary when compared to the Authie estuary (Figure 11.4), to other estuaries from the northern French coast (Amara et al. 2007), and more generally to other French estuaries (Gilliers et al. 2006). In these studies, growth rate was generally the more sensitive variable, and sole sampled in the Seine estuary showed 3 to 19% reductions of this parameter when compared to the less contaminated sites.

The hypothesis of the role of contaminants as major factors in the general reduction of growth and condition in juvenile flounder and sole must be considered with caution. Many variables linked to external factors and not related to pollution (e.g., ambient temperature, salinity, nutrient inputs, drainage of basin area) may interfere with responses to chemical stress. However, temperature and salinity were not greatly different among the nursery sites used to study juvenile sole and flounder. Furthermore, no effects of hydrological conditions or differing food availability could be discerned.

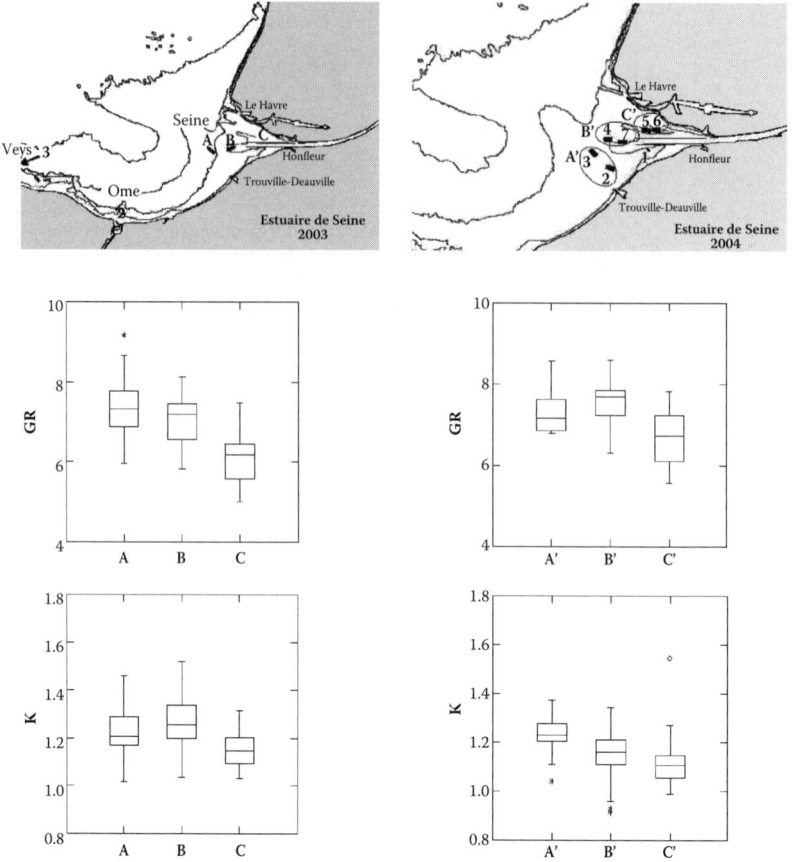

FIGURE 11.3 Growth rate (micrometres per day growth increment in otolith) and Fulton's condition index (K in milligrams per cubic millimetre) of juvenile sole collected along Seine estuary gradient. Bars indicate standard errors. (*Source:* Morin, J. et al. 2005. Rapport Seine Aval.)

Seabed substrata (sand and muddy sand) in the studied areas supported abundant and uniform distributions of benthic fauna that are likely to serve as prey for juvenile flatfish and should not be limiting for the growth of 0-group fish (Van der Veer et al. 1991; Amara et al. 2001). Measurements of benthic invertebrate biomass even showed that the mouth of the Seine estuary hosted the highest biomass among the studied zones (Amara et al. 2007). As previously stated, the nursery located in the Seine estuary is characterised by high concentrations of a variety of contaminants and thus may explain the altered physiological performances of the fish.

In the work of Amara and collaborators (2007), most of the variability in fish growth and condition was explained by concentrations of sediment-bound contaminants. Many laboratory-based studies have underlined the impacts of xenobiotics on reduction in fish growth (Rowe 2003; Claireaux et al. 2004; Alquezar et al. 2006). Finally, the general trend observed along the transect from the mouth of the Seine

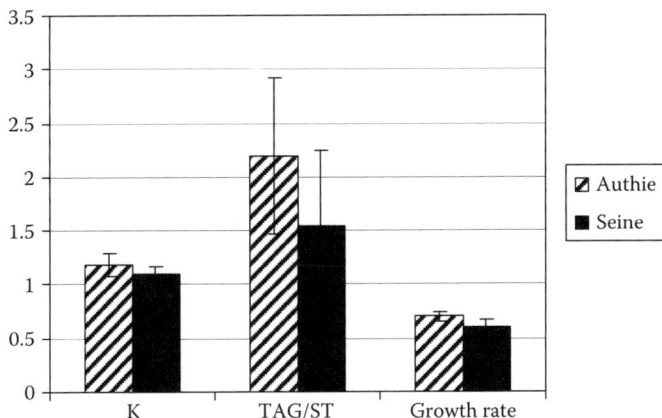

FIGURE 11.4 Growth rate (millimetres per day growth increment in otolith), Fulton's condition index (K in milligrams per cubic millimetre), and nutritional status (TAG/ST = ratio of quantity of reserve lipids to quantity of structural lipids) of juvenile sole (mean age 142 ± 16 days) from the Seine and Authie estuaries. Bars indicate standard errors (n = 143). (*Source:* Amara, R. et al. 2007. *Mar. Ecol. Prog. Ser.* 351 201–208. With permission.)

estuary to the Seine Bay corroborates other studies performed in relation to pollution gradients (Burke et al. 1993).

11.4 BIOCHEMICAL EFFECTS OF POLLUTION ON FLATFISH

Exposure to contaminants may induce protection and detoxification mechanisms in the exposed animals that avoid more profound toxic effects. Based on this assumption, flounder from the Seine estuary are characterised by high expression of the multixenobiotic resistance (MXR) system (Minier et al. 2000). This defence system corresponds to the expression of membrane transport proteins that export from cells a wide variety of structurally unrelated compounds and maintain intracellular concentrations below toxic levels (Minier et al. 1999). It thus corresponds to a first line of defence against many contaminants present in the environment and, as observed in flounder, is expressed in high amounts in other organisms living in the Seine estuary (Minier et al. 2006).

The system may nevertheless be overwhelmed by or unable to cope with certain pollutants, and toxic compounds would then interact with different cellular compartments and molecules. Cellular responses to contaminants such as PAHs and PCBs generally involve induction of biotransformation enzymes such as cytochrome P450. These enzymes are able to metabolise lipophilic compounds into more polar products that are more easily excreted (Stegeman 2000).

Assessing EROD activity in fish liver constitutes a reliable way to evaluate the extent to which fish are in need of this type of metabolising enzyme when exposed to contaminants (Van der Oost et al. 2003). EROD activity has been assessed in flounder living in many European estuaries. Results showed that fish from the Seine estuary are characterised by high levels of biotransformation activity related to the expression of cytochrome P450 (Minier et al. 2000), a result corroborated by measurements on other fish species (Burgeot et al. 1994).

Expression of defence mechanisms implies energetic costs that may reduce the energy available for other functions including growth and reproduction. The concomitant high expression and activity of defence and metabolising systems, together with the general reduction in condition index and growth rates observed in flatfish living in the Seine estuary, may illustrate this trade-off of available energy. Allocation of limited resources to detoxification and other processes (compensation, repair systems) may be linked to the physiological trend assessed in contaminated populations and reported to affect survival of organisms (Forbes 1999; Handy et al. 1999; Van Straalen and Hoffman 2000). Despite the abundance of food sources, the nutritional status of flounder and sole was affected (Figures 11.3 and 11.4). This may indicate that biological and physiological processes (predation, digestion, regulation of metabolic activities) may be affected or adapted to local environmental conditions.

Indeed, tolerance to chemical stress may result in phenotypic and genetic adaptation (Gillespie and Guttman 1999; Tanguy et al. 2002). In successive studies, Marchand and collaborators (2003 and 2004) identified a particular pattern of expression of allozymes in a population of flounder from the Seine estuary and hypothesised that the pattern may correspond to a "resistant character" associated with the expression of particular alleles in contaminated systems. Interestingly, the studied loci from which the specific pattern was derived included enzymes such as phospho-glucomutase, amino aspartate transferase, and glucose phosphate isomerase that are linked to the central metabolic pathway and energy production.

11.5 REPRODUCTIVE EFFECTS AND ENDOCRINE DISRUPTION

In association with reduced fish growth and condition indices, a general decrease of the gonado-somatic index is generally observed in flounder populations from the Seine estuary when compared to other estuaries. At maturity (February), female gonads only represent 6 to 14% of the total body weight in the Seine estuary flounder. These values are only half those generally observed in other French estuaries (Marchand et al. 2004) and in the North Sea (Gallien-Landriau 2003). Furthermore, assessing the number of mature oocytes that passed primary vitellogenesis (diameter >375 nm; Janssen et al. 1995; Le Duff 1997) indicated that fewer oocytes were present in mature gonads of fish sampled in the Seine estuary (Marchand et al. 2004).

Density of oocytes that reach maturity and may effectively contribute to reproduction reduced dramatically only half the values assessed in other estuaries (Figure 11.5). Although relative fecundity may be linked to age, it did not account for the results from the study because no significant age differences were recorded among the sampled flounder (mean age 2.8 ± 0.2 years). Thus, investigations of the physiological responses of flounder populations to contamination indicated a general decrease of fecundity in the Seine estuary with respect to reference sites and other French estuaries. A combination of reduced gonadal growth and altered production of mature oocytes would greatly impair overall fecundity.

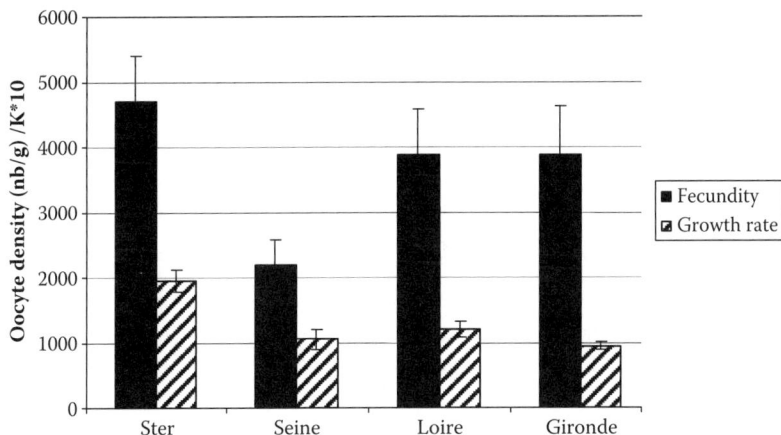

FIGURE 11.5 Relative fecundity (expressed in number of oocytes g^{-1} of ovary) and growth rate (in mg mm^{-3} * 10) of mature female European flounder from French estuaries (mean ± 95% confidence interval). (After Marchand et al. 2004.)

Reduced fecundity may further reflect the cost of tolerance to stress that leads to a reduction of the energy available for reproduction. It is also possible that altered or inadequate endocrine signalling may contribute to the observed effects. Indeed, as in other vertebrates, reproduction in flounder depends heavily on and is regulated by hormones that ensure timely, organised, and coordinated development.

Environmental endocrine disruptors (EDCs) are present in the Seine estuary (see Chapters 3, 6, and 10). These include xeno-estrogens such as steroids (natural and synthetic), alkylphenols, and bisphenol A (Gallien-Landriau 2003). Other EDCs may disrupt other key mechanisms; the endocrine system is highly complex and includes many molecules and organs. For example, PCBs have been shown to interact with the biosynthesis pathways of steroids that are implicated in the regulation of metabolic activity and growth, i.e., corticosteroids and reproduction (sexual) sexual steroids (Benachour et al. 2007).

Evidence for endocrine disruption can be assessed by measuring the induction of vitellogenin (vtg) synthesis in male fish. This phospho-lipoprotein is under estradiol control and is stored in oocytes to allow early development of the offspring. Although this protein has no physiological function in males and may lead to reduced fertility (Nash et al. 2004), 67% of the flounder sampled in the Seine estuary from 1998 to 2003 carried more than 1 $\mu g.mL^{-1}$ vtg in their plasma. This concentration in flounder is considered abnormal and indicative of significant effect of EDCs (Allen et al. 1999). Concentrations up to 1472 $\mu g.mL^{-1}$ were measured in the samples; the mean concentration was 125 $\mu g.mL^{-1}$ (Gallien-Landriau 2003).

Comparison of these results with other reported values shows that vtg plasma concentrations in flounder from the Seine estuary are among the three highest recorded in flounder across Europe. Values are below those measured in the Mersey and Tyne (Allen et al. 1999; Kleinkauf et al. 2004) but higher than those in the Thames and the Scheldt (Vethaak et al. 2005). Endocrine disruption is further exemplified by the occurrence of intersex flounder. These fish develop ovotestes, i.e., testes that differentiate both male and female cells. Four percent of the male flounder from the Seine estuary were found to be intersex fish after histological examination of their testes. Perhaps not surprisingly, intersex flounder have been found only in the three estuaries in which male flounder exhibited the highest mean plasma vtg concentrations (although it should be noted that intersex fish did not exhibit the highest vtg concentrations).

11.6 ECOLOGICAL RELEVANCE

The heavy pollution of the Seine estuary with compounds such as PCBs, PAHs, DDT, metals, and other compounds that constitute a complex and potentially toxic mixture is not under debate. The question is whether wild populations exposed to these pollutants are at risk. Studies of juvenile flatfish are thus ecologically relevant because coastal marine and estuarine shallow areas fulfil an important function in fish stock renewal, as nursery grounds for commercially important marine fish species (Beck et al. 2001).

Poor growth and nutritional status of juveniles may weaken overall condition, potentially making these individuals more vulnerable to predation, physiological stress, and disease (Adams 2002). The observed low growth rates and condition of juvenile flatfish may dramatically lower over-winter survival of juveniles living in polluted nurseries, thereby reducing recruitment to commercially exploitable stocks.

Chemical measurements showed that adult flatfish accumulate pollutants to a high extent in the Seine estuary and may represent a risk for human consumption. Concomitantly, flounder showed important biochemical, histological, and physiological alterations. The sum of these analyses clearly indicates that fish are suffering from pollution. Reduction of condition indices, growth rates, and fertility suggest that abundant food sources alone cannot guarantee high biological and physiological performance in the Seine estuary. Fish populations may have to be maintained by the selection of particular alleles or genotypes (Marchand et al. 2004) or by a constant influx of fish from other areas.

The usual question that follows biological indices that indicate that a given wild population is at risk is whether a "real" effect on the population was assessed. Is the studied species disappearing from the polluted area? Although it may appear illogical to wait for a population to collapse before taking action, environmental managers may in fact consider that pollution is one factor that leads to an equilibrium that does not alter the functioning of an ecosystem. Unlike results from acute exposure assessments, very few studies have provided evidence that chronic exposure to low doses of pollutants has led to the disappearance of a given species. However, this statement does not imply that this has not been the case even if it has not been clearly demonstrated.

Several fish species that lived in the Seine estuary and even supported fishing activities have disappeared (Guezenec et al. 1999). The recent pollution of a lake with EDCs led to the collapse of the resident fish population (Kidd et al. 2007). But disappearance of a population may not be the only outcome that can be expected. Reduced abundance may be observed, as in the case of sole in the Seine estuary (Gilliers et al. 2006). In this study, the reduction of fish growth performance and density correlated to levels of contamination of the studied sites. Low density in the lower part of the estuary has been reported for several fish species (Mouny et al. 1998) and may affect fishing activities.

It is well known from toxicological studies that only the more resistant animals survive high doses when treated with several doses of toxic compounds. Treatment can select resistant individuals. Marchand et al. (2004) identified a "resistant phenotype" in a population of flounder from the Seine estuary that fits with this perspective. Flounder may not disappear from the Seine estuary but only the tolerant individuals may survive. Flounder and sole are the more abundant fish species of the estuary (Mouny et al. 1998). They may also be among the more resistant species.

ACKNOWLEDGMENTS

The authors would like to thank all our colleagues, especially J. Morin who participated in this work and helped generate the data. This study is part of the Seine Aval and European Interreg IIIa program titled Risk Analysis associated with Endocrine Disruption in the Manche Regions (RAED).

REFERENCES

Adams, S.M. 2002. *Biological Indicators of Aquatic Ecosystem Stress*. Bethesda, MD: American Fisheries Society.

Allen, Y. et al. 1999. Survey of estrogenic activity in United Kingdom estuarine and coastal waters and its effects on gonadal development of the flounder *Platichthys flesus. Environ. Toxicol. Chem.* 18: 1790–1800.

Alquezar, R.S., J. Markich, and D.J. Booth. 2006. Effects of metals on condition and reproductive output of the smooth toadfish in Sydney estuaries, south-eastern Australia. *Environ Pollut.* 142: 116–122.

Amara, R. et al. 2001. Feeding ecology and growth of 0-group flatfishes (sole, dab and plaice) on a nursery ground (Southern Bight of the North Sea). *J. Fish Biol.* 58: 788–803.

Amara, R, Y. Désaunay, and F. Lagardère. 1994. Seasonal variation in growth of larval sole *Solea solea* L. and consequences on the success of larval immigration. *Neth. J. Sea Res.* 32: 287–298.

Amara, R. 2003. Ichthyodiversity and population dynamics of 0-group flatfish on a nursery ground in the southern bight of the North Sea (France). *Environ Biol Fishes.* 67: 191–201.

Amara, R. et al. 2007. Growth and condition indices in juvenile sole *Solea solea* measured to assess the quality of essential fish habitat. *Mar. Ecol. Prog. Ser.* 351: 201–208.

Beck, M.W. et al. 2001. The identification, conservation, and management of estuarine and marine nurseries for fish and invertebrates, *Bioscience* 51: 633–641.

Benachour, N. et al. 2007. Cytotoxic effects and aromatase inhibition by xenobiotic endocrine disrupters alone and in combination *Toxicol. Applied Pharmacol.* 222: 129–140.

Beyer, J. 1996. Fish biomarkers in marine pollution monitoring: evaluation and validation in laboratory and field studies. Academic thesis, University of Bergen, Norway.

Burgeot, T. et al. 1994. Monitoring biological effects of contamination in marine fish along French coasts by measurement of ethoxyresorufin-O-deethylase activity. *Ecotox. Environ. Saf.* 29: 131–147.

Burke, J.S., D.S. Peters, and P.J. Hanson. 1993. Morphological indices and otolith microstructure of Atlantic croaker, *Micropogonias undulatus*, as indicators of habitat quality along an estuarine pollution gradient. *Environ Biol Fishes*. 36: 25–33.

Claireaux, G.Y. et al. 2004. Influence of oil exposure on the physiology and ecology of the common sole *Solea solea*: experimental and field approaches. *Aquat Living Resour*. 17: 335–351.

Derocher, A.E. et al. 2003. Contaminants in Svalbard polar bear samples archived since 1967 and possible population level effects. *Sci. Total Environ*. 301: 163–174.

Durieux, E.D. et al. 2007. Temporal changes in lipid condition and parasitic infection by digenean metacercariae of young-of-year common sole *Solea solea* (L.) in an Atlantic nursery ground (Bay of Biscay, France). *J. Sea Res*. 57: 162–170.

Eastwood, P.D. et al. 2003. Estimating limits to the spatial extents and suitability of sole (*Solea solea*) nursery grounds in the Dover strait. *J. Sea Res*. 50: 151–165.

FAO (Food and Agriculture Organization). 2007. Future prospects for fish and fishery products. FAO Fisheries Circular 972/4, Part 1, FIEP/C972/4, Rome.

Forbes, V.E. 1999. *Current Topics in Ecotoxicology and Environmental Chemistry: Genetics and Ecotoxicology*. London: Taylor & Francis.

Gallien-Landriau, I. 2003. Etude de l'altération fonctionnelle du système reproducteur par les perturbateurs endocriniens. Caractérisation des effets, identification des xéno-estrogènes impliqués et conséquences sur les populations de poissons en estuaire et Baie de Seine. Ph.D. thesis, University of Le Havre. Le Havre, France.

Gibson, R.N. 1994. Impact of habitat quality and quantity on the recruitment of juvenile flatfishes. *Neth. J. Sea Res*. 32: 191–206.

Gilliers, C. et al. 2006. Are growth and density quantitative indicators of essential fish habitat quality? An application to the common sole *Solea solea* nursery grounds. *Estuar. Coast. Shelf Sci*. 69: 96–106.

Gillespie, R.B. and S.I. Guttman. 1999. Chemical-induced changes in the genetic structure of populations: effects on allozymes. In *Genetics and Ecotoxicology*, Forbes, V., Ed. London: Taylor & Francis, pp. 55–77.

Guézennec, L. 1999. Seine-Aval: un estuaire et ses problèmes. Programme scientifique Seine Aval. Editions IFREMER. Plouzané, France.

Handy, R.D. et al. 1999. Metabolic trade-off between locomotion and detoxification for maintenance of blood chemistry and growth parameters by rainbow trout (*Oncorhynchus mykiss*) during chronic dietary exposure to copper. *Aquat. Toxicol*. 47: 23–41.

IFREMER (Institut Français pour l'Exploitation de la Mer). 2007. Niveaux de concentration dans les produits de la pêche côtière française atlantique. Data available at http://www. ifremer.fr/envlit/.

Janssen, P.A.H., J.G.D. Lambert, and H.J.T. Goos. 1995. The annual ovarian cycle and the influence of pollution on vitellogenesis in the flounder, *Pleuronectes flesus*. *J. Fish Biol*. 47: 509–523.

Jaouen-Madoulet, A. 2000. Distribution et effets biologiques des PCB et HAP dans les organismes de l'estuaire de Seine. Ph.D. thesis, University of Le Havre. Le Havre, France.

Jones, K.C. and P. de Voogt. 1999. Persistent organic pollutants (POPs): state of the science. *Environ. Pollut*. 100: 209–221.

Ketata, I. et al. 2008. Endocrine-related reproductive effects in molluscs. *Comp. Biochem. Physiol*. 147: 261–270.

Kidd, K.A. et al. 2007. Collapse of a fish population after exposure to a synthetic estrogen. *Proc. Nat. Acad. Sci. USA* 104: 8897–8901.

Kleinkauf, A. et al. 2004. Abnormally elevated VTG concentrations in flounder (*Platichthys flesus*) from the Mersey estuary (UK): a continuing problem. *Ecotox. Environ. Saf*. 58: 356–364.

Larsson, P. et al. 2000. Persistent organic pollutants (POPs) in the pelagic systems. *Ambio* 29: 202–209.

Le Duff, M. 1997. Cinétique de l'ovogenèse et stratégies de ponte chez les poissons téléostéens en milieu tempéré. Ph.D. thesis. Université de Bretagne Occidentale.

Marchand, J. et al. 2004. Physiological cost of tolerance to toxicants in the European flounder *Platichthys flesus*, along the French Atlantic coast. *Aquat. Toxicol.* 70: 327–343.

Marchand, J. et al. 2003. Responses of European flounder *Platichthys flesus* populations to contamination in different estuaries along the Atlantic coast of France. *Mar. Ecol. Prog. Ser.* 260: 273–284.

Minier, C., N. Eufemia, and D. Epel. 1999. The multixenobiotic resistance phenotype as a tool to biomonitor the environment. *Biomarkers.* 4: 442–454.

Minier, C. et al. 2006. Multixenobiotic resistance protein expression in *Mytilus edulis*, *Mytilus galloprovincialis* and *Crassostrea gigas* from the French coasts. *Mar. Ecol. Prog. Ser.* 322: 155–168.

Minier, C. et al. 2000. Flounder health status in the Seine Bay: a multibiomarker study. *Mar. Environ. Res.* 50: 373–377.

Morin, J. et al. 2005. Identification des habitats essentiels de nourriceries de poissons à partir d'indicateurs faunistiques - Qualité de ces habitats pour les juvéniles de soles en estuaire de Seine. Rapport Seine Aval.

Mouny, P. et al. 1998. Biological components from the Seine estuary: first results. *Hydrobiologia.* 373–374: 333–347.

Muir, D.C.G., R.J. Norstrom, and M. Simon, 1988. Organochlorine contaminants in artic marine food chains: accumulation of specific polychlorinated biphenyls and chlordane-related compounds. *Environ. Sci. Technol.* 22: 1071–1079.

Nash, J.P. et al. 2004. Long-term exposure to environmental concentrations of the pharmaceutical ethynyl estradiol causes reproductive failure in fish. *Environ. Health Perspect.* 112: 1725–1733.

Powers, D.A. 1989. Fish as model systems. *Science* 246: 352–358.

Rowe, C.L. 2003. Growth responses of an estuarine fish exposed to mixed trace elements in sediments over a full life cycle. *Ecotox. Environ. Saf.* 54: 229–239.

Schwarzenbach, R.P. et al. 2006. The challenge of micropollutants in aquatic systems. *Science* 313: 1072–1077.

Simonich, S.L. and R.A. Hites. 1995. Global distribution of persistent organochlorine compounds. *Science* 269: 1851–1854.

Stegeman, J.J. 2000. Cytochrome P450 gene diversity and function in marine animals: past, present, and future *Mar. Environ. Res.* 50: 61–62.

Stegeman, J.J. and M.E. Hahn. 1994. Biochemistry and molecular biology of monooxygenases: current perspectives on forms, functions, and regulation of cytochrome P450 in aquatic species. In *Aquatic Toxicology: Molecular, Biochemical and Cellular Perspectives*, Malins, D.C. and Ostrander, G.K. Boca Raton, FL: Lewis Publishers, pp. 87-204.

Tanguy, A. et al. 2002. Polymorphism of metallothionein genes in the Pacific oyster *Crassostrea gigas* as biomarker of response to metal exposure. *Biomarkers* 7: 439–450.

Tarpgaard, E. et al. 2005. Using short-term growth of enclosed 0-group European flounder, *Platichthys flesus*, to assess habitat quality in a Danish bay. *J. Appl. Ichthyol.* 21: 53–63.

Van der Oost, R., J. Beyer, and N.P.E. Vermeulen. 2003. Fish bioaccumulation and biomarkers in environmental risk assessment: a review. *Environ. Toxicol. Pharmacol.* 13: 57–149.

Van der Veer, H.W., R. Dapper, and J.I. Witte. 2001. The nursery function of the intertidal areas in the western Wadden Sea for 0-group sole *Solea solea* L. *J. Sea Res.* 45: 271–279.

Van Straalen, N.M. and A.A. Hoffman. 2000. Review of experimental evidence for physiological costs of tolerance to toxicants. In *Demography in Ecotoxicology*, Kammenga, J.E. and Laskowski, R., Eds. New York: John Wiley & Sons, pp. 147–161.

Vethaak, A.D. et al. 2005. Integrated assessment of estrogenic contamination and biological effects in the aquatic environment of The Netherlands. *Chemosphere* 59: 511–524.

Vilanova, R. et al. 2001. Organochlorine pollutants in remote mountain lake waters. *J. Environ. Qual.* 30: 1286–1295.

Vinagre, C. et al. 2008. Habitat specific growth rates and condition indices for the sympatric soles *Solea solea* (Linnaeus, 1758) and *Solea senegalensis* Kaup 1858, in the Tagus estuary, Portugal, based on otolith daily increments and RNA/DNA ratio. *J. Appl. Ichthyol.* 24: 163–169.

12 Diatoms
Modern Diatom Distribution in the Seine and Authie Estuaries

Florence Sylvestre

CONTENTS

12.1 INTRODUCTION

Diatoms (order Bacillariophyceae) are eukaryotic, single-celled, photosynthetic algae that are present in almost every habitat on Earth (Figure 12.1). Diatoms are particularly known for the intricate geometries and spectacular patterns of their silica-based cell walls. These patterns are species-specific and are precisely reproduced during each cell division cycle, demonstrating genetic control of this biomineralization process (Kröger et al. 2000).

Diatoms are also well known for their ecological significance. In temperate oceans, diatoms dominate phytoplankton assemblages and blooms and are responsible for a considerable proportion of the world's primary net production. In estuarine and shallow coastal environments, diatoms are considered the most important components of the benthic microalgal assemblages (McLusky 1989). Exposed mudflats and sandflats are often described as "unvegetated" simply because vascular plants such as seagrasses are absent. However, the use of this adjective is inaccurate because a diverse assemblage of cyanobacteria and eukaryotic microalgae thrive in such sediments.

This microphytobenthos has been recently reviewed and referred to as a "secret garden" (MacIntyre et al. 1996; Miller et al. 1996). In Western Europe, where intertidal mudflats are generally devoid of macrophytic vegetation (McLusky 1989), microphytobenthos often becomes the main primary producer (Admiraal 1984; Colijn and de Jonge 1984; Underwood and Kromkamp 1999). It thus represents a

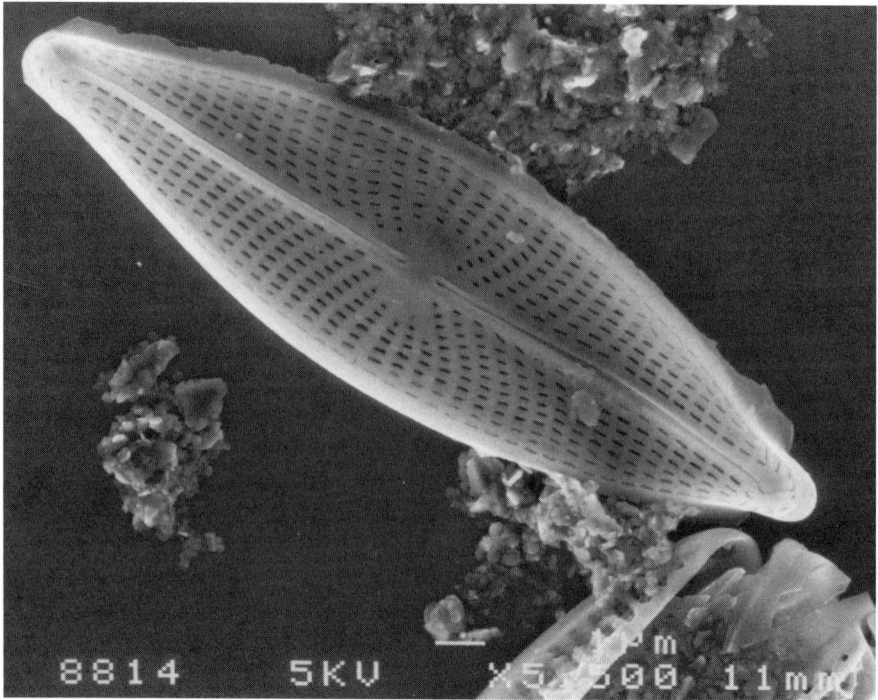

FIGURE 12.1   Example of a diatom: *Navicula phyllepta*.

major energy source that directly supplies the benthic food web (Herman et al. 2000) and also the pelagic food web (Riera and Richard 1996) through resuspension into the water column during high tide (Delgado et al. 1991; Lucas et al. 2000). Several studies have shown highly diverse and productive diatom assemblages in coastal and shelf sediments (Admiraal et al. 1982; Riaux 1983; Riaux-Gobin et al. 1998); for instance, the chlorophyll *a* content per square meter of shelf bottom may exceed that of the entire overlying water column (Cahoon et al. 1990).

Ecological studies of estuarine and shallow coastal water diatoms are fewer than studies of freshwater and open ocean habitats. Some work has been carried out to elucidate the interactions among diatom populations on mudflats and various environmental factors that explain their spatial and temporal distributions (Admiraal 1977a, b, and c, 1984; Admiraal and Peletier 1979 and 1980; Admiraal et al. 1982 and 1984), photosynthetic capacities (Blanchard et al. 1997), and productivity (Blanchard and Guarini 1998). Other studies have focused on the stabilization of estuarine sediments by benthic diatom assemblages and their ability to increase resistance to erosion, mostly caused by tidal currents, wind-induced waves, and bioturbation (Holland et al. 1974; Paterson 1995).

This chapter aims to describe a seasonal distribution based on diatom species composition (total assemblages, e.g., benthic plus planktonic forms after deposition) over a 2-year period in two estuaries, the Authie and the Seine. Species composition will be discussed with respect to (1) fresh, estuarine, and marine water influences, and (2) impacts produced by natural and human activities as revealed by these bioindicators.

12.2 MATERIALS

This study was conducted on a total of 27 samples, collected seasonally from two sites each in the Authie and Seine estuaries: Authie Nord [AN], Authie Port [AP], and Honfleur [HON], and Pont de Normandie [PN] (see Figure 1.3 in Chapter 1). The four stations were sampled on May 19, August 26, and November 27, 2003. In February 2003, only two samples were collected at Authie Nord [AN 2/03] and Pont de Normandie [PN 2/03]. Stations AN, AP, and PN were sampled on February 17, May 17, August 31, and November 24, 2004. In February 2004, two samples were collected at AN, distinguishing slikke [ANSL 2/04] and shore [ANSH 2/04] samples, respectively.

All samples were collected at low tide in the intertidal area. At each station, in order to limit the bias due to small-scale patchiness (centimetre to decametre), the uppermost layer (0 to 1 cm) of sediment was scraped off randomly at several points, over a surface of about 1 m². This procedure, categorized as *pseudoreplication* by sampling strategists (Hurlbert 1984), also minimizes bias resulting from the perturbation caused by repeated sampling at the same stations. The subsamples were mixed together, homogenized, and stored in formol to preserve living cells.

12.3 METHODS

Cell diatom concentrations were determined for each sample of a known volume. The samples (dry sediment plus added water) were allowed to settle in a sedimentation chamber and diatoms were counted using a Zeiss-Axiovert 25 inverted microscope (Lund et al. 1958). Dead and living diatom cells were quantified. An aliquot of Lugol's iodine solution was added to the samples to dye the cells (Riaux 1983).

Diatom taxonomic and counting analyses were conducted on sediment subsamples weighing ~500 to 1000 mg. Samples were processed with 10% HCl and 30% H_2O_2 to remove carbonates and organic matter, respectively. Small quantities of samples (0.2 ml) were evaporated onto coverslips that were subsequently mounted onto glass slides with Naphrax as a mounting medium. Diatoms were identified under a Nachet NS 400 light microscope [differential interference contrast (DIC), Normanski optics, 1000× magnification, N.A. = 1.32). A minimum of 200 diatom valves were counted for each sample.

Diatom identifications were based on a variety of taxonomic monographs including Lebour (1930), Cupp (1943), Hendey (1964), Krammer and Lange-Bertalot (1986, 1988, and 1991), Simonsen (1987), and Ehrlich (1995). Specimens were identified to their lowest taxonomic level (variety) following the species concept of Krammer and Lange-Bertalot. Data on the ecology of the diatom taxa were compiled from these references in addition to Admiraal (1977a, b, and c and 1984), Navarro (1982), John (1983), Poulin et al. (1984, 1986), Bérard-Therriault et al. (1987), Snoeijs (1993), Snoeijs and Vilbaste (1994), Witkowski (1994), Snoeijs and Potapova (1995), Snoeijs and Kasperovičiene (1996), Snoeijs and Balashova (1998), and Witkowski et al. (2000).

The relative abundances of the taxa were calculated as percentages of the total count. The specific diversity for each sample analysed was explored using two indices: (1) the Shannon–Wiener diversity index (H'; Shannon and Wiever 1963) of the assemblages was calculated from the relative abundance data (H' = $-\Sigma p_i log_2 p_i$, where p_i is the proportion of species i in the sample), and (2) the equitability or regularity index of Pielou (E = H'/log_2S, where S is the number of species).

Due to the size of the data matrix generated by the number of diatom taxa identified, we used a cluster analysis (instead of principal component analysis [PCA]; see Guiot and Roux 1993) to examine trends in the dataset. This method makes it possible to group samples (classes) based on the relative abundance of diatom taxa. It is a hierarchical, agglomerative process based on the variance of the intragroup distances. Instead of a direct processing of the species percentage table, we computed the usual Euclidean distances on samples from factor analysis coordinates (FACs), stemming from an earlier FAC analysis. After condensing the samples into classes, we determined the diatom species responsible for these classes. We used the VARCAR program (M. Roux, personal communication) to select the dominant, subdominant, and associated species for each class. The selections were based on the mean intra-class value, the mean inter-class value, and correlation rate corresponding to the inter-class variation compared with total variation.

12.4   RESULTS AND DISCUSSION

Based on the examination of 27 samples examined, the diatom flora is mainly composed of living cells (Figure 12.2). The number of diatoms was higher in May 2003 and 2004 in response to the availability of nutrients during the spring. Nevertheless, the highest number of cells was reached in August 2003, particularly for stations AN and AP. I consider this month as an exceptional situation, reflecting one of the warmest summers of past decades (Trenberth et al. 2007). This high productivity in August 2003 was probably in response to the high temperature recorded. The temperature of mud during emersion is a driver of the photosynthetic capacity of the system (Blanchard and Guarini 1998). From a comparison of stations, we observed higher numbers of diatoms at AN and AP than at HN and PN.

Of the 27 samples examined, 208 taxa belonging to 83 genera were identified, representing a highly diversified flora based on the number of samples (27). Assemblage diversity indices (Shannon H', equitability E) ranged from 1.52 to 4.36 and from 0.38 to 0.83, respectively. No significant differences between stations were noted and no seasonal trend was evident for species diversity (Figure 12.3).

The diatom flora was dominated by benthic brackish species commonly observed in estuarine conditions. Marine species were also present in the assemblage, but few pelagic marine forms characteristic of deep water ocean were noted. A few benthic oligosaline to freshwater forms were also identified. A cluster analysis was applied to the 27 samples and yielded 8 groups characterized by dominant, subdominant, and associated species (Table 12.1). The cluster analysis split the samples collected at station PN into 4 groups; samples AN 2/03 and ANSH 2/04 constituted group 5; and samples collected at stations AP and AN comprised group 6. The three samples from station HN were split into classes 7 and 8 and one was grouped in class 6.

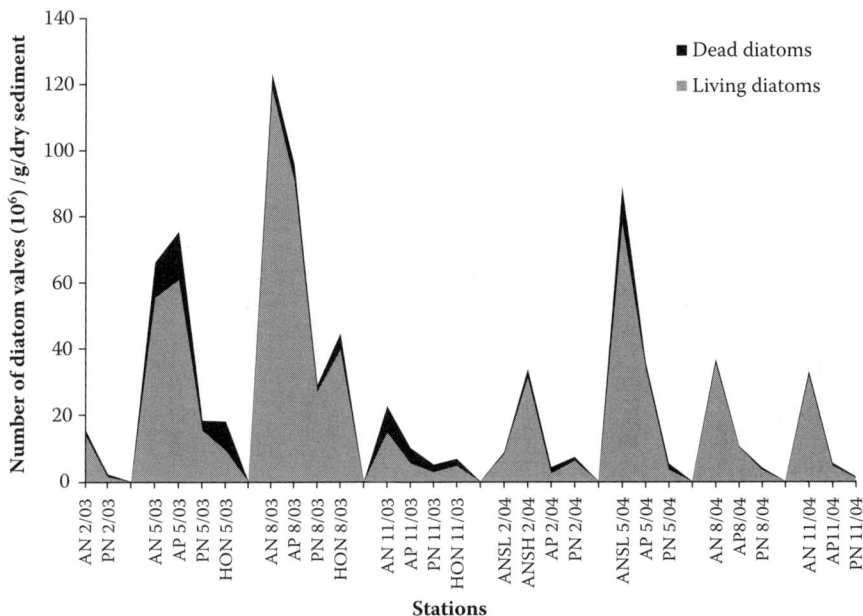

FIGURE 12.2   Number of diatom valves (10^6) per gram of dry sediments for 27 samples collected between February 2003 and November 2004 at Authie Nord [AN], Authie Port [AP], Honfleur [HON], and Pont de Normandie [PN] stations. Samples usually collected on the slikke and occasionally on the schorre [SH].

Group 1 consisted of two samples dominated by *Opephora mutabilis* and the subdominant *Achnanthes delicatula* species. *O. mutabilis* is widespread in marine to brackish water, abundant in the Baltic Sea and the North Sea (Witkowski et al. 2000). *A. delicatula* is mainly a mesohaline form, usually reported from brackish littoral waters but it can also occur in fresh water (Ehrlich 1995). These species are associated with two marine forms, a benthic one, *Delphineis minutissima*, and a planktonic one, *Thallassionema nitzschoides*.

Group 2 consists of sample PN 8/03, dominated by *Navicula incerta* and *Amphora coffeaeformis*. *N. incertata* is a cosmopolitan species, widespread in coastal environments (Hendey 1964). *A. coffeaeformis* is a widespread holo-euryhaline species regarded as one of the most common taxa in brackish to highly saline environments, observed in both shallow coastal and marine environments and inland saline waters (Noël 1982; Gasse et al. 1989; Ehrlich 1995; Sylvestre et al. 2001). These species are associated with two marine species from the genus *Opephora: O. mutabilis*, and *O. pacifica*; *O. pacifica* is widespread on European coasts and in the Pacific Ocean (Witkowski et al. 2000).

Group 3 included four samples (PN 5/03, PN 11/03, PN 2/04, PN 9/04) dominated by *Cyclotella meneghiniana*, *Paralia sulcata*, and the subdominant *Cyclotella striata*. These species are associated with tychoplanktonic species from the genus *Fragilaria*, a marine benthic form, *Delphineis minutissima*, and an epiphytic species, *Cocconeis placentula euglypta*, that can tolerate a broad range of salinity (Noël

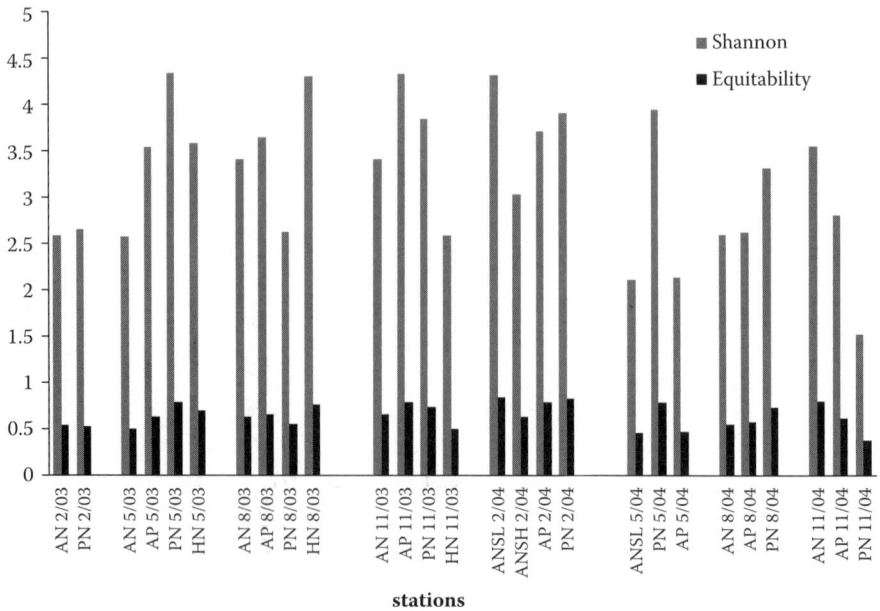

FIGURE 12.3 Shannon and equitability indices for 27 samples collected between February 2003 and November 2004 at Authie Nord [AN], Authie Port [AP], Honfleur [HN], and Pont de Normandie [PN] stations. Samples usually collected on the slikke and occasionally on the schorre [SH].

1982; Gasse et al. 1989; Sylvestre et al. 2001). *C. meneghiniana* is a euryhaline tychoplanktonic species that can thrive in pelagic habitats and very shallow waters (Hendey 1964). It has been found in a broad range of fresh and saline waters and is frequently observed on macrophytes in littoral deposits (Germain 1981). *P. sulcata* is a common species from shallow marine coastal environments (Hendey 1964; Navarro 1982); *C. striata* is common in marine and brackish waters, often abundant in estuaries in spring plankton (Hendey 1964).

Sample PN 5/04 is individualized by the cluster analysis because of its dominance by *Navicula incertata* associated with three others species: *N. phyllepta, N. perminuta*, and *N. eidrigiana*. These species are commonly observed in shallow coastal environments, particularly in European intertidal mudflats (Germain 1981; Riaux 1983; Admiraal 1984).

Group 5 is composed of two samples collected in 2004 from the schorre part of station AN. Both samples were dominated by *Navicula microdigitoradiata*, associated with *Paralia sulcata*, and *Nitzschia clausii*. *N. microdigitoradiata* is well known in European intertidal mudflats, scattered but locally abundant (Witkowski et al. 2000). *Nitzschia clausii* is an epipelic brackish water species usually observed in intertidal mudflats (Sylvestre et al. 2004).

Group 6 includes 15 samples, essentially those collected at stations AP and AN and a single sample collected in November 2003 at HN (HN 11/03). This group is dominated by small-size common species of estuarine intertidal

TABLE 12.1
Results of Cluster Analysis Applied to 27 Samples Studied

Cluster Analysis Group	Samples	Diatom Assemblage		
		Dominant	Subdominant	Associated
1	PN 2/03, PN 11/04	Opephora mutabilis	Achnanthes delicatula	Delphineis minutissima Thalassionema nitzschoides
2	PN 8/03	Navicula incertata Amphora coffeaeformis	Opephora mutabilis	Opephora pacifica
3	PN 2/04, PN 5/03, PN 11/03, PN 9/04	Cyclotella meneghiniana Paralia sulcata	Cyclotella striata	Cocconeis placentula euglypta Delphineis minutissima Fragilaria construens Fragilaria species
4	PN 5/04	Opephora species Navicula incertata	Navicula phyllepta Navicula perminuta Navicula eidrigiana	Coscinodiscus sp. Nitzschia epithemoides epithemoides Nitzschia pusilla Opephora krumbeirii
5	ANO 2/04, AN 2/03	Navicula microdigitoradiata	Paralia sulcata	Nitzschia clausii

(continued on next page)

TABLE 12.1 (continued)
Results of Cluster Analysis Applied to 27 Samples Studied

A	B	C		
		Diatom Assemblage		
Cluster Analysis Group	Samples	Dominant	Subdominant	Associated
6	AN 11/03, AP 5/04, HN 11/03, AN 5/03, AP 8/03, AN 5/04, AN 8/04, AP 8/04, AP 11/04, AN 11/04, AP 5/03, AP 11/03, ANI 2/04, AP 2/04, AN 8/03	*Navicula phyllepta*	*Paralia sulcata* *Navicula gregaria*	*Craticula halophila* *Gyrosigma peisonis* *Nitzschia hungarica* *Odontella aurita* *Rhaphoneis amphiceros* *Thallassionema nitzschoides*
7	HN 5/03	*Stephanodiscus hantzschii*		*Cyclotella meneghiniana* *Navicula arenaria*
8	HN 8/03	*Navicula citrus*	*Cymatosira belgica* *Gyrosigma fasciola*	*Amphora sp.1* *Delphineis minutissima* *Odontella aurita* *Pleurosigma aestuarii* *Skeletonema costatum*

Note: A: Number of sample classes. B: Samples included in each class. C: Dominant, subdominant, and associated species defined by the VARCAR program.

Navicula phyllepta; the subdominant species of this group is the marine tychoplank-tonic *Paralia sulcata* and the benthic *Navicula gregaria*. Marine (*Odontella aurita, Rhaphoneis amphiceros, Thallassionema nitzschoides*) and brackish water benthic (*Craticula halophila*) and epipelic species (*Gyrosigma peisonis, Nitzschia hunga-rica*) are associated with this assemblage.

Groups 7 and 8 consist of single samples (HN 5/03 and HN 8/03, respectively). HN 5/03 is dominated by *Stephanodiscus hantzschii*, an allochthonous freshwater species, frequent in areas adjoining wastewater inflow (Witkowski et al. 2000). This species is associated with the oligosaline–fresh water *Cyclotella meneghiniana* and *Navicula arenaria*, which is reported as an epipelic, cosmopolitan species fre-quently observed in intertidal flats of estuaries (Witkowski 1994). HN 8/03 is domi-nated by *Navicula citrus*, associated with *Cymatosira belgica, Gyrosigma fasciola*, and marine (*Delphineis minutissima, Odontella aurita, Skeletonema costatum*) and epipelic species *Pleurosigma aestuarii*.

Seasonal fluctuations of the dominant diatom species at stations AN and AP are demonstrated by the rapid growth of *Navicula phyllepta* in May (Figure 12.4). This species stays at the same level of growth or decreases slightly during the summer (August) before finally decreasing or disappearing during the winter (November, February). At station AN, *Navicula phyllepta* is replaced by *Navicula microdigi-toradiata* in February. Otherwise, during the winter (February and November), the diatom assemblages are dominated by marine species, as particularly observed at AP. At AN, marine species were more abundant in February 2004 in the sample col-lected on the slikke and in November 2004.

For stations HN and PN, seasonal fluctuations in the diatom assemblage distribu-tion were less evident because sampling covered only 3 months in 2003 for HN; data for February and May in 2004 were missing for PN. Nevertheless, these two stations were dominated by marine species rather than oligosaline–freshwater species, com-pared to stations AN and AP. For instance, station PN was dominated by *Opephora mutabilis* or *Cyclotella meneghiniana*. An anomaly is sample HN 5/03 exhibiting dominance of *Stephanodiscus hantzschii* recognized as a freshwater species and observed in wastewater. This diatom assemblage may represent a situation indicating the presence of waste effluent; but it is difficult to say whether its presence is excep-tional based on a short-term event or recurrent.

In summary, this study has shown in the Authie estuary a seasonal distribution of diatom assemblages with a dominance of small species belonging to the genus *Navicula* as commonly observed in European intertidal mudflats (Riaux 1983; Admiraal et al. 1984; Oppenheim 1988, 1991; Underwood 1994; Haubois et al. 2005). The colonisation ability of these small fast-growing diatoms may likely rep-resent an advantage (over larger slow-growing cells) in intertidal mudflats that can be characterized by physical disturbances. Small cells are biologically more active due to larger surface-to-volume ratios allowing higher division rates (Williams 1964; Admiraal 1977a), higher photosynthetic rates, and higher nutrient absorption (Banse 1976; Taguchi 1976; Sournia 1981; Hudon and Legendre 1987).

In the Seine estuary, the diatom flora differs from the Authie estuary; the assem-blages are mostly dominated by marine rather than oligosaline species, reflecting the influence of tidal movements. It is also composed of centric species that live on

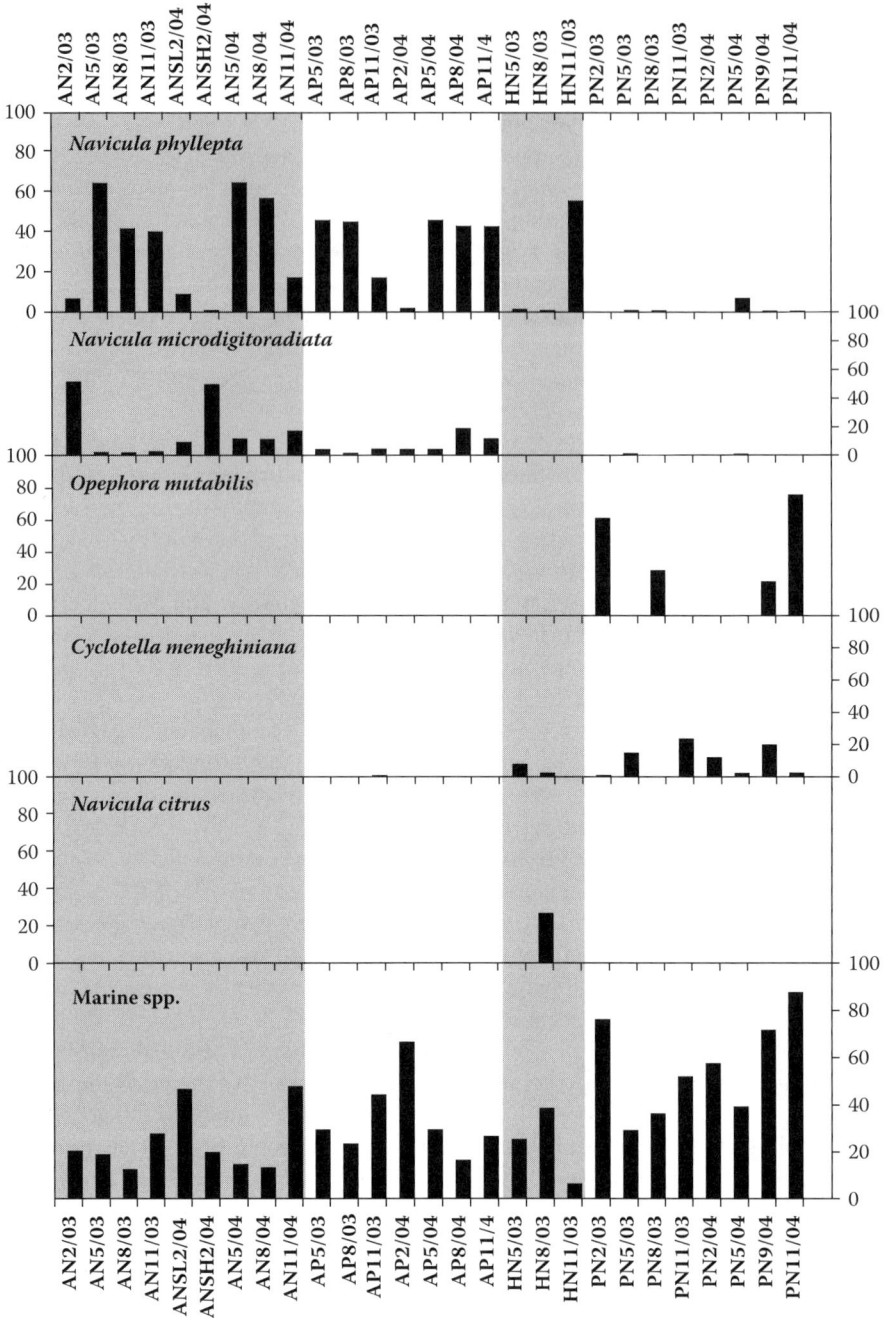

FIGURE 12.4   Seasonal variations of dominant diatom species grouped by stations as a function of time at Authie Nord [AN], Authie Port [AP], Honfleur [HN], and Pont de Normandie [PN] stations. Samples usually collected on the slikke and occasionally on the schorre [SH].

intertidal and subtidal sediments. These diatoms are readily transported upward into the water column where they survive to return to the sediments under calm conditions (Admiraal 1984). Moreover, even though seasonal sampling was too incomplete in the Seine estuary to show structural successions in the diatom distribution, some patchiness may be related to a disturbance by wastewater effluents in the ecosystem (e.g., dominance of *Stephanodiscus hantzschii* in sample HN 5/03).

Several parameters (e.g., salinity, temperature, irradiance, oxygen, nutrient availability, inundation, and emersion cycles) play an important role in explaining the structural succession of diatom flora in estuaries. In this comparative study of an estuary devoid of important human activity (the Authie) and the Seine estuary that concentrates almost 40% of the national economic activity and 50% of the river traffic, the description of the diatom distribution suggests control of the structural development of diatom assemblages by hydrodynamic conditions. The lack of structural succession and the dominance of marine species in the diatom assemblages of the Seine estuary may be down due to the higher energy environment (reflecting the morphology of the estuary) and the economic activity along the estuary. The morphology of the estuary, its mudflats, their hydrodynamism, and hydraulic movements sustained by the natural tidal movements and by the riverine transportation constrain the growth dynamics of the local diatoms.

REFERENCES

Admiraal, W. 1977a. Experiments with mixed populations of benthic estuarine diatoms in laboratory microecosystems. *Bot. Mar.* 20: 479–485.

Admiraal, W. 1977b. Salinity tolerance of benthic estuarine diatoms as tested with a rapid polarographic measurement of phytosynthesis. *Mar. Biol.* 39: 11–18.

Admiraal, W. 1977c. Tolerance of estuarine benthic diatoms to high concentrations of ammonia, nitrite ion, nitrate ion and orthophosphate. *Mar. Biol.* 43: 307–315.

Admiraal, W. 1984. The ecology of estuarine sediment-inhabiting diatoms. *Prog. Phycol. Res.* 3: 269–322.

Admiraal, W. and H. Peletier. 1979. Sulphide tolerance of benthic diatoms in relation to their distribution in an estuary. *British Phycol. J.* 14: 185–196.

Admiraal, W. and H. Peletier. 1980. Distribution of diatom species on an estuarine mud-flat and experimental analysis of the selective effect of stress. *J. Exp. Mar. Biol. Ecol.* 46: 157–175.

Admiraal, W., H. Peletier, and H. Zomer. 1982. Observations and experiments on the population dynamics of epipelic diatoms from an estuarine mudflat. *Estuar. Coast. Shelf Sci.* 14: 471–487.

Admiraal, W., H. Peletier, and T. Brouwer. 1984. The seasonal succession patterns of diatom species on an intertidal mudflat: an experimental analysis. *Oikos.* 42: 30–40.

Banse, K. 1976. Rates of growth, respiration and photosynthesis of unicellular algae as related to cell size. *J. Phycol.* 12: 135–140.

Bérard-Therriault, L., A. Cardinal, and M. Poulin. 1987. Les diatomées (Bacillariophyta) benthiques de substrats durs des eaux marines et saumâtres du Québec. 8. Centrales. *Nat. Can.* 114: 81–103.

Blanchard, G.F. et al. 1997. Seasonal effect on the relationship between the photosynthetic capacity of intertidal microphytobenthos and temperature. *J. Phycol.* 33: 723–728.

Blanchard, G.F. and J.M. Guarini. 1998. Temperature effects on microphytobenthos productivity in temperate intertidal mudflat. *Vie et Milieu* 48: 271–284.

Cahoon, L.B., R.S. Redman, and C.R. Tronzo. 1990. Benthic microalgal biomass in sediments of Onslow Bay, North Carolina. *Estuar. Coast. Shelf Sci.* 31: 805–816.

Colijn, F. and V.N. de Jonge. 1984. Primary production of microphytobenthos in the Ems–Dollard Estuary. *Mar. Ecol. Prog. Ser.* 14: 185–196.

Cupp, E.E. 1943. *Marine Plankton Diatoms of the West Coast of North America.* Berkeley: University of California Press.

Delgado, M., V.N. de Jonge, and H. Peletier. 1991. Experiments on resuspension of natural microphytobenthos populations. *Mar. Biol.* 108: 321–328.

Ehrlich, A. 1995. *Atlas of Inland Water Diatom Flora of Israël.* Jerusalem: Geological Survey of Israel.

Gasse, F. et al. 1989. Biological remains, geochemistry and stable isotopes for the reconstruction of environmental and hydrological changes in the Holocene lakes from North Sahara. *Palaeogeogr. Palaeoclimatol. Palaeoecol.* 60: 1–46.

Germain, H. 1981. *Flore des diatomées.* Coll.Faunes et Flores actuelles. Paris: Soc. Nouv. Ed Boubée.

Guiot, J. and M. Roux. 1993. Reconstitution statistique des environnements passés à partir de données paléoécologiques. In *Biométrie et Environnement,* Lebreton, J.D. and Asselain, B., Eds. Paris: Masson, pp. 123–149.

Haubois, A.G. et al. 2005. Spatio-temporal structure of the epipelic diatom assemblage from an intertidal mudflat in Marennes–Oléron Bay, France. *Estuar. Coast. Shelf Sci* 64: 385–394.

Hendey, N.I. 1964. Bacillariophyceae (diatoms). In *An Introductory Account of the Smaller Algae of British Coastal Waters: Fishery Investigations.* London: HMSO.

Herman, P.M. et al. 2000. Stable isotopes as trophic tracers: combining field sampling and manipulative labelling of food resources for macrobenthos. *Mar. Ecol. Prog. Ser.* 204: 79–92.

Holland, A.F., R.G. Zingmark, and J.M. Dean, 1974. Quantitative evidence concerning the stability of sediments by marine benthic diatoms. *Mar. Biol.* 27: 191–196.

Hurlbert, S.J. 1984. Pseudoreplication and the design of ecological experiments. *Ecol. Monogr.* 54: 187–211.

Hudon, C. and P. Legendre. 1987. The ecological implications of growth forms in epibenthic diatoms. *J. Phycol.* 23: 434–441.

John, J. 1983. The Diatom Flora of the Swan River Estuary, Western Australia. *Biblio. Phycol.* Band 64. Vaduz: J. Cramer.

Krammer, K. and H. Lange-Bertalot. 1986. *Süsswasserflora von Mitteleuropa. Bacillariophyceae. Teil 1: Naviculaceae.* Jena: Gustav Fischer Verlag,.

Krammer, K. and H. Lange-Bertalot. 1988. *Süsswasserflora von Mitteleuropa. Bacillariophyceae. Teil 2: Bacillariaceae, Epithemiaceae, Surirellaceae.* Jena: Gustav Fischer Verlag.

Krammer, K. and H. Lange-Bertalot. 1991. *Süsswasserflora von Mitteleuropa. Bacillariophyceae. Teil 3: Centrales, Fragilariaceae, Eunotiaceae.* Jena: Gustav Fischer Verlag.

Kröger, N. et al. 2000. Species-specific polyamines from diatoms control silica morphology. *PNAS* 97: 14133–14138.

Lebour, M.V. 1930. *The Plankton Diatoms of Northern Seas.* London: Ray Society.

Lucas, C.H. et al. 2000. Benthic-pelagic exchange of microalgae at a tidal flat. 1. Pigment analysis. *Mar. Ecol. Prog. Ser.* 196: 59–73.

Lund, J.W.G., C. Kipling, and E.D. Le Cren. 1958. The inverted microscope method of estimating algal numbers and the statistical basis of estimations by counting. *Hydrobiologia* 11: 143–170.

MacIntyre, H.L., R.J. Geider, and D.C. Miller. 1996. Microphytobenthos: the ecological role of the "secret garden" of unvegetated, shallow-water marine habitats. I. Distribution, abundance and primary production. *Estuaries* 19: 186–201.

McLusky, D.S. 1989. *The Estuarine Ecosystem.* London: Chapman & Hall.

Miller, D.C., R.J. Geider, and H.L. MacIntyre. 1996. Microphytobenthos: the ecological role of the "secret garden" of unvegetated, shallow-water marine habitats. II. Role in sediment stability and shallow-water food webs. *Estuaries* 19: 202–212.

Navarro, J.N. 1982. Marine diatoms associated with mangrove prop roots in the Indian River, Florida, USA. *Biblio. Phycol.* Band 61. Vaduz: J. Cramer.

Noël, D. 1982. Les diatomées des saumures des marais salants de Salin-de-Giraud (Sud de la France). *Géol. Méd.* 9: 413–446.

Oppenheim, D.R. 1988. The distribution of epipelic diatoms along an intertidal shore in relation to principal physical gradients. *Bot. mar.* 31: 65–72.

Oppenheim, D.R. 1991. Seasonal changes in epipelic diatoms along an intertidal shore, Berrow flats, Somerset. *J. Exp. Mar. Biol. Ecol.* 71: 579–596.

Paterson, D.M. 1995. Biogenic structure of early sediment fabric visualized by low temperature scanning electron microscopy. *J. Geol. Soc. London* 152: 131–140.

Poulin, M., L. Bérard-Therriault, and A. Cardinal. 1984. Les diatomées benthiques de substrats durs des eaux marines et saumâtres du Québec 3. Fragilarioideae (Fragilariales, Fragilariaceae). *Nat. Can.* 111: 349–367.

Poulin, M., L. Bérard-Therriault, and A. Cardinal. 1986. Les diatomées (Bacillariophyceae) benthiques de substrats durs des eaux marines et saumâtres du Québec 6. Naviculales: Cymbellaceae et Gomphonemaceae. *Nat. Can.* 113: 405–429.

Riaux, C. 1983. Structure d'un peuplement estuarien de diatomées épipéliques du Nord-Finistère. *Oceanol. Acta* 6: 173–183.

Riaux-Gobin, C. et al. 1998. Microphytobenthos de deux sédiments subtidaux de Nord-Bretagne. I: Biomasses pigmentaires, fluctuations saisonnières et gradients verticaux. *Ann. Inst. Océanogr.* 74: 29–41.

Riera, P. and P. Richard. 1996. Isotopic determination of food sources of *Crassostrea gigas* along trophic gradient in the estuarine bay of Marennes–Oleron. *Estuar. Coast. Shelf Sci.* 42: 347–360.

Shannon, C.E. and W. Weaver. 1963. *The Mathematical Theory of Communication*, Urbana: University of Illinois Press.

Simonsen, R. 1987. *Atlas and Catalogue of the Diatom Types of Friedrich Hustedt*. Berlin: J. Cramer.

Snoeijs, P. 1993. *Intercalibration and Distribution of Diatom Species in the Baltic Sea*, Vol. 1. Uppsala: Opulus Press.

Snoeijs, P. and S. Vilbaste. 1994. *Intercalibration and Distribution of Diatom Species in the Baltic Sea*, Vol. 2. Uppsala: Opulus Press.

Snoeijs, P. and M. Potapova. 1995. *Intercalibration and Distribution of Diatom Species in the Baltic Sea*, Vol. 3. Uppsala: Opulus Press.

Snoeijs, P. and J. Kasperoviciene. 1996. *Intercalibration and Distribution of Diatom Species in the Baltic Sea*, Vol. 4. Uppsala: Opulus Press.

Snoeijs, P. and N. Balashova. 1998. *Intercalibration and Distribution of Diatom Species in the Baltic Sea*. Vol. 5. Uppsala: Opulus Press.

Sylvestre, F. et al. 2001. Modern diatom distribution in a hypersaline coastal lagoon: the Lagoa de Araruama (R.J.) Brazil. *Hydrobiologia* 443: 213–231.

Sylvestre, F., Guiral, D., and J.P. Debenay. 2004. Modern diatom distribution in mangrove swamps from Kaw estuary (French Guiana). *Mar. Geol.* 208: 281–293.

Sournia, A. 1981. Morphological bases of competition and succession. *Can. Bull. Fish. Aquat. Sci.* 210: 339–346.

Taguchi, S. 1976. Relationship between photosynthesis and cell size of marine diatoms. *J. Phycol.* 12: 185–189.

Trenberth, K.E. 2007. Observations: surface and atmospheric climate change. In *Climate Change 2007: The Physical Science Basis*, Solomon, S. et al., Eds. Cambridge: Cambridge University Press.

Underwood, G.J.C. and J. Kromkamp. 1999. Primary production by phytoplankton and micro-phytobenthos in estuaries. *Adv. Ecol. Res.* 29: 93–153.

Underwood, G.J.C. 1994. Seasonal and spatial variation in epipelic diatom assemblages in the Severn Estuary. *Diatom Res.* 9: 451-472.

Williams, R.B. 1964. Division rates of salt marsh diatoms in relation to salinity and cell size. *Ecology* 45: 877–880.

Witkowski, A. 1994. *Recent and Fossil Diatom Flora of the Gulf of Gdansk, Southern Baltic Sea: Origin, Composition and Changes of Diatom Assemblages during the Holocene.* Stuttgart: J. Cramer.

Witkowski, A., H. Lange-Bertalot, and D. Metzelin 2000. Diatom flora of marine coasts I. In *Iconographia Diatomologica: Annoted Diatom Micrographs*, 7th ed. Lange-Bertalot, H., Ed. Ruggell: A.R.G. Gantner Verlag.

13 Foraminifera

Jean-Pierre Debenay

CONTENTS

The main objectives of this study were to (1) analyze benthic foraminiferal assemblages in two neighbouring estuaries; (2) analyze the information on the impact of pollution provided by these bioindicators; and (3) investigate the relationship between foraminiferans and the ragworm *Nereis diversicolor*, including predation. The first step consists of establishing an inventory of the species in relation to the positions of the samples in the estuarine ecosystems.

13.1 FORAMINIFERA: SHORT DESCRIPTION

Foraminifera are single-celled amoeboid protists with numerous thin reticulating pseudopods called reticulopodia. They typically produce a test (shell), the form and composition of which are the primary means by which Foraminifera are identified and classified (Loeblich and Tappan 1988). More than 50,000 species are recognized, both living and fossil, as far back as the Cambrian period, more than 500 million years ago. The test may be made of organic material (organic forms), mineral particles collected in sediment and cemented together (agglutinated forms), or secreted calcium carbonate (calcareous forms). A test may consist of one chamber or multiple chambers added during growth. The name Foraminifera is derived from the connecting hole (foramen) through the wall between each chamber. Openings that allow cytoplasm to flow from the test are called apertures. Fully grown individuals range from a few tens of micrometres to almost 15 cm and are usually smaller than 1 mm in size.

Modern Foraminifera are found in all marine and brackish environments where they may have planktonic or benthic modes of life. A few species have been reported to survive in fresh water.

Foraminifera feed on a wide range of food including dissolved organic matter, bacteria, diatoms and other single celled organisms, and even animals such as small crustaceans that they catch with their reticulopodia networks. Benthic and planktonic foraminiferans that inhabit the photic zone often bear unicellular symbiotic algae from diverse lineages such as dinoflagellates, diatoms, and chlorophytes inside their cell. Some smaller species retain chloroplasts from ingested algae to conduct photosynthesis (reviewed by Debenay et al. 1996a). The position of Foraminifera in the trophic web is not well known, but different organisms prey on them, including worms, crustaceans, gastropods, echinoderms, and fish. The life cycles of only about 20 among the approximately 5000 living species are known. The general characteristic is an alternation between an asexually produced haploid generation and a sexually produced diploid generation. Multiple cycles of asexual reproduction between sexual generations are common in benthic forms.

Because of their diversity, abundance, and evolutionary trends, fossil foraminiferans are useful for giving relative dates to rocks (biostratigraphy). Moreover, as their tests contain stable isotopes of carbon and oxygen from the past oceans in which they lived, they may be used for palaeoclimate reconstructions. By comparison with the environmental behavior of some living species, it is also possible to reconstruct past environments. Living foraminiferal assemblages are used as bioindicators for coastal environments including lagoons and estuaries. In estuarine environments, the composition of benthic foraminiferal assemblages reflects the complex interactions of biotic and abiotic parameters and their multiple changes over space and time.

The abundance of foraminiferans makes them particularly suitable for statistical approaches, allowing the quantification of environmental characteristics. Their short reproductive cycles (6 to 12 months) result in rapid responses to changes in environmental conditions, while their hard tests commonly preserved in sediments provide a picture of average environmental conditions, smoothing the great seasonal variabilities. When studied in cores, dead assemblages provide the possibility of studying historical environmental changes, such as increases of anthropogenic pressure. The biodiversity, density, and structure of foraminiferal assemblages may be used for biological monitoring along with analyzing the morphology and ultrastructure of tests or cytological characteristics.

13.2   METHODS FOR STUDYING FORAMINIFERANS

13.2.1   FIELD METHODS

Foraminifera may live as deep as 30 cm in sediment (Goldstein et al. 1995), but since the highest numbers of living foraminiferans are found in the 0- to 1-cm surface layer (review in Alve and Murray 2001), environmental studies based on foramin-iferal assemblages generally consider only surface sediments. In the present study, all the samples were collected at low tide in intertidal areas. At each station, in order to limit the bias due to small-scale patchiness (centimetre to decametre), the uppermost layer (0 to 1 cm) of sediment was scraped off randomly at several points over a surface of about 1 m². This procedure, categorized as *pseudoreplication* by sampling strategists (Hurlbert 1984), also minimizes bias resulting from the pertur-bations caused by repeated samplings at the same stations.

The subsamples were mixed together, homogenized, and stored in 96% ethyl alco-hol containing 1 g.l⁻¹ Rose Bengal stain (Walton 1952; Murray and Bowser 2000). When sediments are collected in subtidal areas, a grab sampler may be used. It must be hermetically closed to prevent washing away of the sediment surface during col-lection. In the boat, the sampler must be opened in a container in which the sediment is placed in its initial position. Generally, a diatom film covers the surface and indi-cates the absence of disturbance during collection. The surface sediment may then be collected following the procedure described for intertidal areas.

All the samples for this study were collected in intertidal areas (Figure 13.1). Sixteen samples were collected in February 2004 in selected environments, from channels to schorres (i.e., vegetated salt marshes) and from tidal flats to the water's edge in both the Authie and Seine estuaries (Table 13.1). To obtain a larger dataset, two additional samples were collected on the open tidal flat of the Somme estuary (15 km south of the Authie estuary). The stations studied by most participants in the PNETOX programme (see Figure 1.3 in Chapter 1) were sampled concomitantly for investigations on foraminiferans. Three stations (Authie Nord [AN] Authie Port [AP], and Pont de Normandie [PN]) were sampled seasonally in 2003 (February 20, May 19, August 26, November 27) and 2004 (February 8, May 17, August 31, November 24). One station was sampled at Honfleur (HON) in 2003 (May 19,

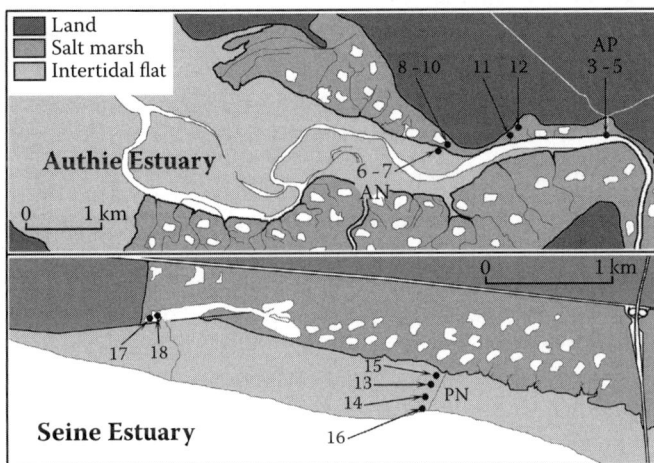

FIGURE 13.1 Location map for 18 samples collected in February 2004. AN, AP, and PN refer to stations selected for 2-year seasonal survey in the Authie estuary (AP = 5 and AN = 6) and in the Seine estuary (PN = 13).

TABLE 13.1

Main Characteristics of Stations Sampled in February

Station Number	Environmental Characteristics
1	Somme estuary: wide, poorly vegetated, intertidal sandflat with exposed sand
2	Somme estuary: wide, well vegetated, intertidal sandflat, in small depression
3	Authie estuary: slikke, close to main channel, near low water level
4	Authie estuary: slikke, close to main channel
5 (AP)	Authie estuary: slikke, close to main channel
6 (AN)	Authie estuary: slikke, close to main channel
7	Authie estuary: slikke, close to main channel
8	Authie estuary: schorre covered with dense halophilous vegetation
9	Authie estuary: schorre covered with dense halophilous vegetation
10	Authie estuary: schorre, depression with scarce halophilous vegetation
11	Authie estuary: tidal creek in schorre
12	Authie estuary: high marsh with dense grass vegetation
13 (PN)	Seine estuary: eroded mudflat
14	Seine estuary: eroded mudflat
15	Seine estuary: mudflat bounded by halophilous vegetation
16	Seine estuary: deeply eroded mudflat, near main channel
17	Seine estuary: exposed muddy sand in tidal embayment
18	Seine estuary: subtidal bottom of tidal embayment

Notes: See Figure 13.1 for localization. Schorre = vegetated salt marshes. Slikke = tidal flats or nonvegetated salt marshes.

August 26, November 27). Its survey was discontinued due to disturbances resulting from earthworks. Additionally, about ten ragworms *Nereis diversicolor* were collected in 2004 (May 17, August 31, November 24) at stations AN and PN to evaluate the predation of this worm on foraminiferans.

13.2.2 LABORATORY METHODS

Samples were kept for at least 3 days in Rose Bengal stain. This method of recognizing living specimens by staining the cytoplasm is questionable because cytoplasm may be preserved in the test several weeks after death (Boltovskoy and Lena 1970; Cann and De Dekker 1981). Other methods are proposed (review in Bernhard 2000), but most of them are logistically complicated, time consuming, and/or expensive, and are not suitable for environmental studies. Murray and Bowser (2000) noted that Rose Bengal staining is as reliable as other techniques. Only brightly stained foraminiferans, with all chambers except the last one stained, can be considered alive.

Benthic foraminiferans that live in fine estuarine sediments are very small. The first step for their extraction consists of washing a constant volume of sediment through a set of two nested sieves, a larger one with 500-μm openings and a smaller one with 50- or 63-μm openings (limit between nano- and microfossils and limit between silt-and-clay mixture and sand, respectively). The small sieve retains the foraminiferans.

The sediment is washed until the entire fine fraction has been removed (i.e., the liquid passing through the smaller sieve is clear). As foraminiferans may be rare in impacted estuarine environments, we recommend using 50 cm^3 of wet sediment. Samples with rich assemblages may be split into equal aliquots to avoid overcounting. Sufficient aliquots must be examined to obtain at least 100 specimens, which is significant for studying the main species (Fatela and Taborda 2002). When considering less abundant species, at least 300 specimens are needed (Phleger 1960). When the population density is <100 individuals in 50 cm^3, all the individuals are counted. After the counting is completed, the remaining sediment is checked for rare species not found during counting. In the tables, these species are reported with a frequency of 0%. The density (total number of specimens in 50 cm^3 of sediment) is estimated and the absolute and relative abundances of each species are calculated.

Samples with excess sand can be dried at 50°C and foraminiferans can be separated by flotation on a high-density liquid such as trichloroethylene (C_2HCl_3). When the density is about 1.5, sand grains settle, but owing to their empty chambers, foraminiferans float. After deposition of the heavy components, all floating particles including foraminiferans may be filtered. Density separation is a commonly used, long lasting process for investigating benthic foraminiferans (Murray 1969; Sen Gupta 1971). Generally a few more specimens are found after flotation than before flotation when they were hidden among sand grains. For muddy sediments, it may not be necessary to apply the flotation method because the entire terrigenous fraction is washed out through the smaller sieve.

After preparation, a fine dusting of dried sediment is scattered onto a black picking tray divided into squares. Foraminifera are counted and/or picked out with a fine brush while the tray is scanned under a microscope. Dried foraminiferans with thick tests must be humidified to detect Rose Bengal staining. If preservation is needed, the picked specimens can then be mounted onto single- or multi-celled cardboard slides with glass covers.

For samples with high organic contents, the recommendation is to examine the wet sediments in Petri dishes because the fragile species that live in organic-rich sediments may be destroyed during drying (Scott et al. 2001). If necessary, a wet splitter can be used to separate samples into equal aliquots (Scott and Hermelin 1993).

Living and dead individuals were counted separately. The suitability of counting total (living plus dead) assemblages is still debated (Murray 2000b; Debenay et al. 2001). We consider that it smoothes small-scale variability and is a way to circumvent the uncertainty of Rose Bengal staining. It does not change the information but allows the integration of this information over a longer period including the present. Moreover, the study of living populations in estuaries where they have pronounced seasonal cycles makes no sense unless at least a full year's monitoring is conducted (Debenay et al. 1996a). Consequently, in the present study, living assemblages were used for the seasonal survey, but total assemblages were used for the study of spatial distribution in February 2004.

Species diversity was calculated using Shannon–Weaver's formula ($H = -\sum p_i \log_2 p_i$) where p_i is the proportion of the ith species (%/100), and $\log_2$ the log base two of pi. Foraminifera also allow the calculation of a confinement index (Ic) that reflects the impact of the balance between fresh water and sea water on the benthos (review in Debenay and Guillou 2002). A disturbance in the natural gradient of Ic may also indicate the impact of pollution (Beck-Eichler et al. 1995).

Contrary to biocenotic indices such as diversity or dominance, Ic is calculated on the basis of individual identifications and counts and therefore takes into account the role of the species in the community and potential interest in the species as a bioindicator. Ic was first defined by statistical analyses in the restricted environments of West Africa (Debenay 1990). It may be transposed with slight modifications to Brazilian coasts (Debenay et al. 1996b), a Mediterranean lagoon (Debenay et al. 2005), and the Mekong delta (Debenay and Luan 2006). It can also be transposed to French coasts by changing some of the species, according to local characteristics of the microfauna. A first attempt was made in 2004 (Debenay 2004). The expression of Ic is:

$$Ic = (C/(C + B) - A/(A + B) + 1)/2$$

where A, B, and C are the relative percentages of three assemblages characteristic of decreasing marine influence. The assemblages defined for western French coasts are: A (strong marine influence): *Brizalina striatula, Brizalina variabilis, Cribroelphidium excavatum, Elphidium pulvereum, Quinqueloculina stelligera, Rosalina* spp.; B (moderate to high continental influence): *Ammonia tepida, Cribroelphidium gunteri, Haynesina germanica, Quinqueloculina seminula*; C (very high continental influence): *Jadammina macrescens, Miliammina* spp., *Paratrochammina* spp., *Trochammina* spp. The Ic index varies from 0 in marine environments or environments under strong marine influence to 1 under strong continental influence. Ic = 1 if assemblage C is the only one represented (A + B = 0). Ic = 0 if assemblage A is the only one represented (B + C = 0).

Hierarchical (cluster) analysis was performed on foraminiferan relative abundance data sets using a subset of the 17 taxa comprising of more than 5% of the assemblage in at least one sample. The relative abundance of these taxa was treated in Q-mode and R-mode hierarchical analyses, based on Euclidean distance correlation coefficients using Ward's merging criterion, carried out with Statlab for Macintosh (SLP Infoware).

13.3 RESULTS AND DISCUSSION

13.3.1 GENERAL DISTRIBUTION

The following results were obtained by collecting sediment at 18 stations in February 2004 (Figure 13.1; Table 13.1). Ninety taxa were identified, including 62 hyaline, 12 agglutinated, and 16 porcelaneous taxa, already reported from European coasts (review in Murray 2006). Twenty-seven contained living specimens. Counts are displayed in Appendix 13.1 at the end of this chapter. Generally, the tests were small and only a few specimens were stained with Rose Bengal. The density of total assemblages ranged from 200 to 45,000 individuals in 50 cm³ of sediment. The highest average density was in the Authie estuary (25,200, standard deviation [SD] = 19,000) and the lowest in the Seine estuary (1,200, SD = 1,550). The average density was 3,600 in the Somme estuary (SD = 4,800).

The species richness of total assemblages ranged between 7 and 56, with the maximum average on the tidal flat of the Somme estuary (43, SD = 18), the lowest value in the Seine estuary (20, SD = 7), and an average of 31 (SD = 13) in the Authie estuary. The species richness of living assemblages ranged between 0 and 11, with an average of 4 (SD = 3) in the Somme estuary, 6 (SD = 3) in the Authie estuary, and 2 (SD = 1) in the Seine estuary.

The dominant species in total assemblages (more than 10,000 tests of the 18 samples) were the hyaline calcareous species *Cribroelphidium magellanicum*, *Cribroelphidium williamsoni*, *Cribrononion gerthi*, *Haynesina germanica*, and *Cribroelphidium excavatum*, by order of decreasing abundance. *Cribroelphidium williamsoni* exhibited smaller abundance in the Seine estuary, but was abundant in the Somme estuary along with *C. excavatum*. Two species were somewhat less abundant but showed remarkable distributions: the hyaline species *Ammonia tepida* showed noticeable percentages only in the Seine estuary and the agglutinated *Jadammina macrescens* species was almost absent from the Seine estuary.

Cluster analysis (Figure 13.2) clearly groups all the schorre samples (cluster D) that contain a high proportion of *J. macrescens* and/or *H. germanica*. Another clear grouping is cluster B with two samples from the slikkes (tidal flats or non-vegetated salt marshes) of the Authie estuary, far from the main channel (6 and 7) and one in a neighbouring tidal creek (11). Cluster C groups highly eroded sites from the Grande Vasière and a sample from an exposed sandy mud flat in a tidal embayment, all three in the Seine estuary. The samples grouped in cluster A are from tidal flats of different origins, including the sandflats from the Somme estuary, the muddy slikke of the Authie estuary, and one sample from the mudflat of the Seine estuary. It shows that the grain size of the sediment does not have a major influence on the distribution of foraminiferans in this area. Sample 18 remains isolated because of its higher proportion of *Miliolinella subrotunda*. The fact that *A. tepida* was better represented in the Seine estuary than in the other sites does not seem to have a major influence on the clustering of the samples. R mode analysis resulted in the grouping of the species in two clusters, while the most abundant species *C. magellanicum* and the schorre species *J. macrescens* remain isolated.

The diversity was the lowest in the schorre samples of cluster D (1.47 to 2.32); see Appendix 13.1. In the other samples, diversity was irregular, the index ranging from 2.52 to 4.03. The confinement index Ic was highest in the schorre samples (0.47 to 0.86). It was lowest in the Somme estuary (0.05 and 0.16) and on the subtidal bottom of a tidal embayment (0.17) in the Seine estuary, showing significant marine influence at these sites. In other samples, the indices were irregular, ranging from 0.20 to 0.39 without a clear gradient of marine influence, which is not surprising because all the samples were collected in brackish areas. Under such conditions, Ic is related to vertical zonation, determining the duration of subaerial exposure, rather than horizontal zonation. This vertical succession may vary, depending on local characteristics (De Rijk and Troelstra 1997; Goldstein and Watkins 1998; Debenay et al. 2004).

Haynesina germanica
Ammonia tepida
Cribrononion gerthi
Cribroelphidium williamsoni
Cribroelphidium excavatum
Quinqueloculina seminula
Cribrostomoides jeffreysii
Neoconorbina nitida
Brizalina spp.
Brizalina variabilis
Buccella frigida
Bolivina pseudoplicata
Stainforthia fusiformis
Miliolinella subrotunda
Brizalina striatula
Jadammina macrescens
Cribroelphidium magellanicum

Relative abundance
%

■ > 40

● 20 < ≤ 40

◉ 10 < ≤ 20

○ 5 < ≤ 10

• 0 < ≤ 5

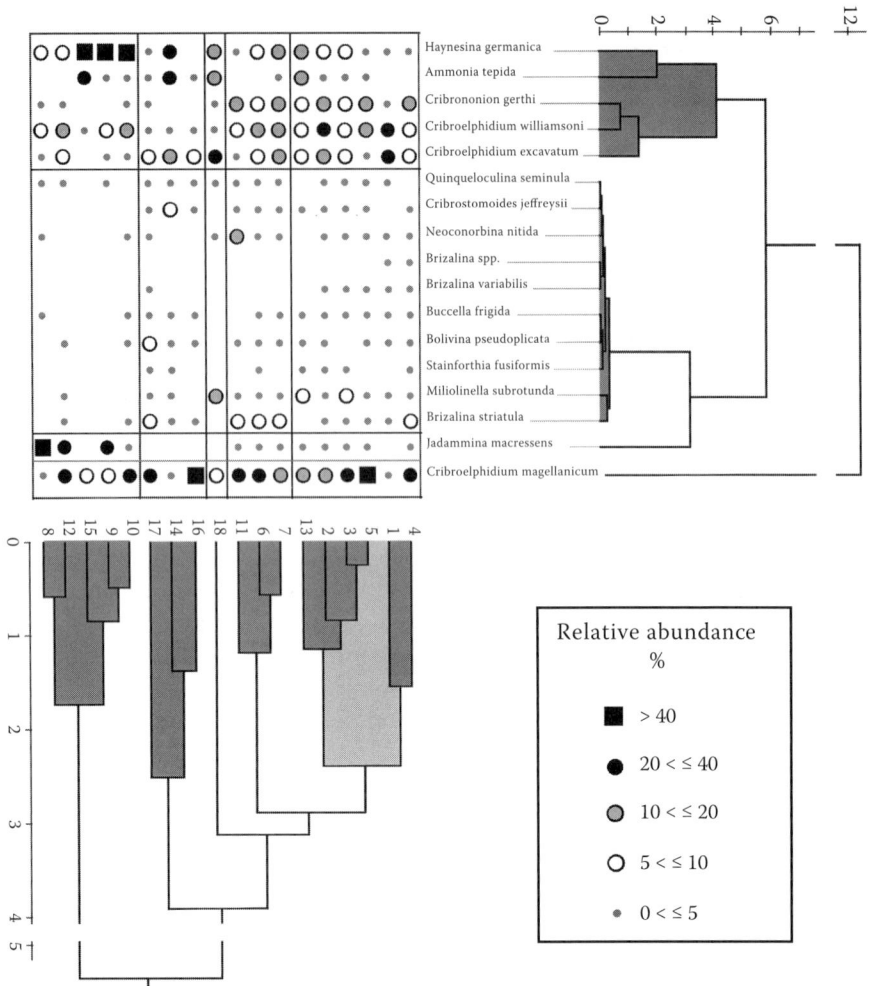

FIGURE 13.2 Q mode (bottom) and R mode (right) cluster analyses of assemblages collected in February 2004. The Q mode dendrogram defines clusters A, B, C, and D. The R mode dendrogram defines clusters I and II. Relative abundances are summarized in the chart.

13.3.2 Seasonal Changes in Assemblages

The following results were obtained by 2-year seasonal samplings at three stations (AN, AP, PN), and three samplings at station HON.

13.3.2.1 Assemblages

Totals of 109 taxa in dead assemblages and 42 in living assemblages were identified. In dead assemblages, the total number of taxa recorded over the eight sampling dates was slightly higher in the Authie estuary than in the Seine estuary: 83 at station AN; 88 at station AP; 79 at station PN; and 78 at station HON (Table 13.2). The difference

TABLE 13.2
Comparison of Assemblages Collected in the Authie and Seine Estuaries

	Authie Estuary		Seine Estuary	
	AN	AP	PN	HON
Total number of dead taxa over sampling period	83	88	79	78
Average species richness of dead assemblages	31	44	31	49
	(18)	(16)	(16)	(26)
Average density of dead assemblages	16,500	16,700	3,100	5,200
	(16,500)	(13,400)	(3,200)	(4,600)
Total number of living taxa over sampling period	32	25	13	15
Average species richness of living assemblages	9	6	4	8
	(7)	(5)	(3)	(4)
Average density of living assemblages	91	39	36	44
	(62)	(49)	(24)	(31)

Note: Standard deviation in parentheses; density = number of individuals in 50 cm^3 of sediment.

between the two estuaries was not perceptible when considering the average species richness that was highest at station HON (Table 13.2). This high value probably results from the fact that only three samples were collected there at the most favorable period. The temporal variations in species richness must be interpreted with care because they may result from the rarity and patchiness of rare species that are scattered around, and thus are not collected at each sampling period despite the pseudoreplication method used (Debenay et al. 2006).

The density of dead assemblages (number of individuals in 50 cm^3 of sediment) greatly changed with the seasons. The average density was more than three times higher in the Authie estuary than in the Seine estuary (Table 13.2). The general trend in the three stations sampled over the entire period (AN, AP, and PN) was an increase of species richness from February 2003 to November 2003 (Figure 13.3) that decreased in February 2004, reaching a minimum of eight species at station AN in May 2004. A slight increase was observed in November 2004, but the values remained much lower than in November 2003.

The density followed the same trend as the species richness in the Authie estuary, with a shift of the maximum in February 2004 instead of November 2003. This trend was less obvious at station PN (Figure 13.3).

Changes in density are directly related to reproduction patterns. The reproduction cycles of foraminiferans are still poorly understood, but many seasonal studies suggest that reproduction peaks responsible for higher densities occur once or few times a year, even if some species reproduce continuously (review in Alve 1999). For example, in the Vilaine estuary (France), *Cribroelphidium excavatum* exhibited reproduction peaks in winter (February) and autumn (September) and reproduced continuously in summer (Goubert 1997). In the lower part of the Vie estuary (France), two reproduction periods occur for most species in spring and early autumn (Debenay et al. 2006).

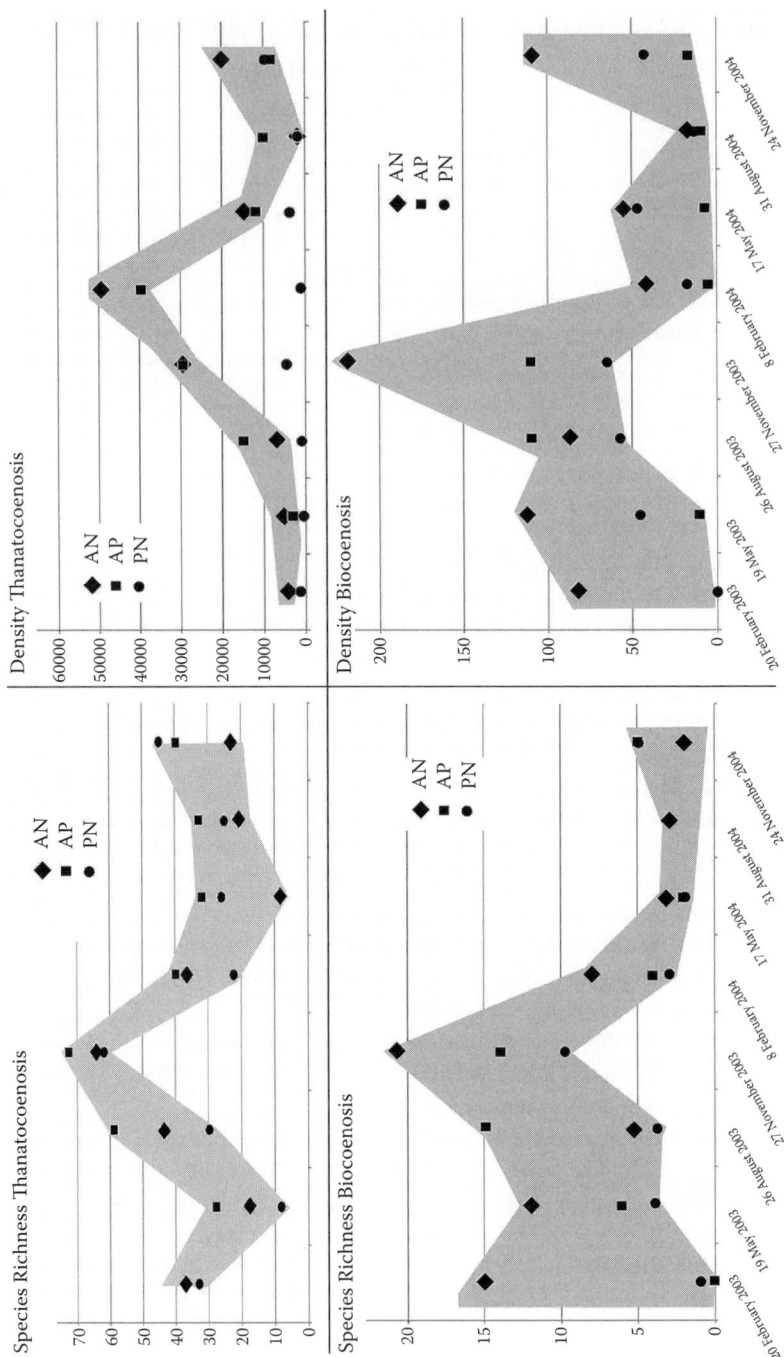

FIGURE 13.3 Changes in density (number of individuals in 50 cm³ sediment) and species richness (number of species) of dead and living foraminiferan assemblages over a 2-year seasonal survey in stations from the Authie (AN and AP) and Seine (PN) estuaries.

Reproduction peaks may be related to the abundance of food supply (benthic microflora or phytoplankton blooms; Walton 1955) but may also be related to temperature changes (Bradshaw 1961; Scott and Medioli 1980) and probably to a large set of environmental characteristics leading to complex reproduction patterns. For example, in the Tees estuary (UK), *Haynesina germanica* peaked in May to August and November to January on the low marsh; a single peak in July was noted on the tidal flat (Horton and Murray 2007). In the study area, it seems that the reproduction period of summer and autumn 2003 did not repeat itself with the same intensity in 2004.

During the study period, the dominant species of the dead assemblages were the same as those recorded in February 2004, in the same order. Except for two samples from station AP, these dominant species always comprised >60% of the assemblage in the Authie estuary and at station HON. *Ammonia tepida*, rare in the Authie estuary, was among the dominant species in the Seine estuary. In living assemblages, the order of the dominant species was somewhat changed with the highest proportion for *H. germanica*. It was impossible to show any trend in changes of diversity (Shannon–Weaver index) or Ic index. The only noticeable feature was a lower Ic index at station HON (<0.1) reflecting the proximity of the open sea.

The total number of living taxa collected during the 2 years was higher in the Authie estuary (32 at station AN, 25 at station AP) than in the Seine estuary (13 at station PN and 15 at station HON). This difference was less obvious when considering the average species richness (Table 13.2). The density of living assemblages greatly changed with the seasons, but was obviously higher at station AN than at other stations (Table 13.2). The general trend in the three stations was an increase of density and species richness from February 2003 to November 2003, with a temporary decrease at station AN in August (Figure 13.3). Both density and richness then decreased and remained low until November 2004, when a slight increase occurred, mainly due to *H. germanica*.

At station PN, a secondary peak occurred in May 2004. This behavior may have an impact on the populations of ragworms that lived at this station and showed glycogen maxima in May 2003 and 2004 (Amiard-Triquet, unpublished data). Such a correlation was not evident in the Authie estuary. The examination of both faeces released during transport to the laboratory and gut contents of the ragworms showed that they fed on foraminiferans only in May 2004, with different behaviors in the two estuaries. The number of foraminiferans ingested was not related to their availability in the sediment, and a selection of the species ingested may occur.

Seasonal changes in density and richness of living assemblages were different in 2003 and 2004. This is consistent with previous observations on the French Atlantic coast where Morvan et al. (2006) have shown that foraminiferal assemblages were not directly affected by seasonal cycles, but responded to more complex inter-annual patterns. In the present study, foraminiferal assemblages indicate that environmental conditions were less favourable in 2004 than in 2003. The delay between the maximum density of living assemblages in November 2003 and the maximum density of dead assemblages in February 2004 may be explained by the time necessary for the accumulation of empty tests after a reproduction episode (great production of living and dead specimens during reproduction). The concomitant decreases of living and

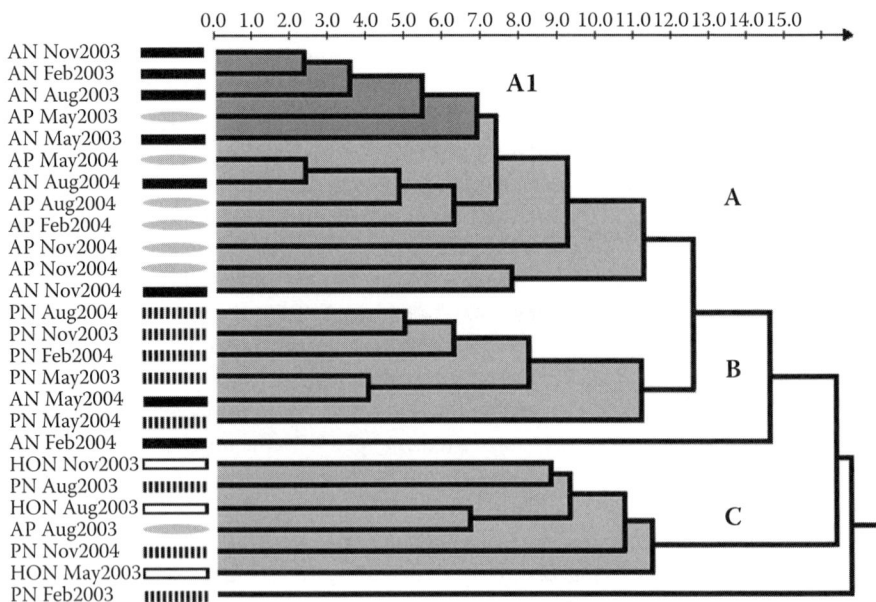

FIGURE 13.4 Q mode cluster analysis of assemblages collected during 2003 and 2004. Most samples from the Authie estuary are in cluster A. The separation between 2003 and 2004 may also be noted for both estuaries. Different symbols are used to simplify the distinction of AN, AP, PN, and HON stations.

dead assemblages at the beginning of 2004 seem to fit with a normal seasonal cycle, since the values reached in May 2004 are almost the same as those of May 2003.

However, the subsequent changes do not lead to an increase similar to what happened in 2003. Such complex inter-annual patterns in foraminiferal assemblages are well known and the seasonal pattern of variability is not necessarily repeated from year to year (Lutze 1968; Boltovskoy and Lena 1969; Scott and Medioli 1980; Basson and Murray 1995; Murray and Alve 2000; Murray 2000a; Buzas and Hayek 2000; Morvan et al. 2006). The cluster analysis of seasonal samplings reveals three clusters (Figure 13.4). Cluster A groups all but three samples from the Authie estuary, showing that the difference between the two estuaries is predominant over seasonal changes. The difference between 2003 and 2004 is demonstrated by the fact that most of the 2003 samples are grouped in sub-cluster A1 for the Authie estuary and in cluster C for the Seine estuary.

13.3.2.2 Species

The proportions of the dominant species at each sampling date are shown in Figure 13.5. *Cribroelphidium magellanicum* comprised >30% of the dead assemblages in all the samples but one from the Authie estuary (AN and AP). Its proportion was about three times lower at stations PN and HON of the Seine estuary. *Cribroelphidium williamsoni* was also better represented in the Authie estuary. In the Seine estuary, its only significant occurrence was at station PN in February 2004.

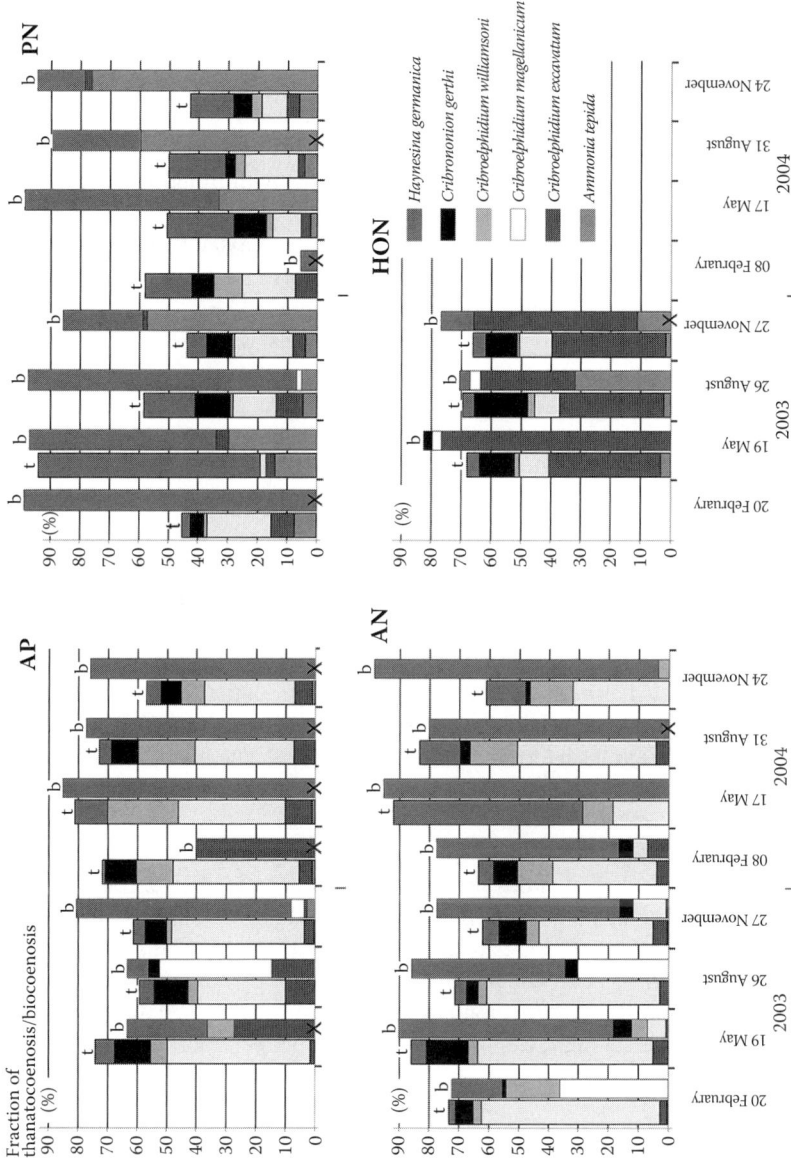

FIGURE 13.5 Proportions of dominant species at each sampling date in the dead assemblages (thanatocoenoses [t]) and living assemblages (biocoenoses [bl]). X indicates samples with only a few specimens, leading to uncertain results.

Cribrononion gerthi had about the same proportion at the four stations. It was not found in the Authie estuary in May 2004 and was absent from station PN in May 2003. *Haynesina germanica* was present in all the stations, but was irregularly represented. It generally exhibited a higher proportion at station PN throughout the study period and at station AN at the three last sampling dates.

Cribroelphidium excavatum was fairly well represented at station HON, where it represented >30% of the assemblages; it never exceeded 10% at the other stations. *Ammonia tepida* was poorly represented in the Authie estuary, never exceeding 1% of the assemblage. Present at all stations of the Seine estuary, it reached 14% of the assemblage. Beyond these dominant species, two showed interesting features. *Brizalina striatula* was present at all but three stations, with proportions reaching 10%. *Buccella frigida* was well represented at station PN, with proportions up to 22%. Despite these relatively high proportions in the dead assemblages, *B. frigida* was never found in living assemblages, possibly the result of an epiphytic or infaunal mode of life in which only superficial sediments are sampled.

In living assemblages (Figure 13.5), *C. magellanicum* was well represented at the Authie estuary, stations AN and AP in 2003. It was never recorded in 2004. *Cribroelphidium williamsoni* was relatively abundant in the Authie estuary in February 2003 and was very rare after that and found only at station AN. *Cribrononion gerthi* occurred in low proportions at AN, AP, and HON and was not recorded after February 2004. *Haynesina germanica* was the dominant species at AN, AP, and PN on most sampling dates. At station HON, it was replaced by *C. excavatum* that was also well represented at station AP until February 2004. *Ammonia tepida*, very rare in the Authie estuary, was well represented in the Seine estuary. It must be noted that only a few living species were collected after February 2004. The better represented were *H. germanica* at all stations and *A. tepida* at station PN.

Unfortunately, little has been published on the behavior of the North Atlantic species *C. magellanicum* and *C. gerthi*. Other species have been thoroughly studied and may provide information for interpreting our assemblages. The main characteristics of these species are shown in Appendix 13.2 at the end of this chapter.

13.3.2.3 Deformation of Test

The development of abnormalities in the tests of benthic foraminiferans was reported for a number of polluted settings (Bhalla and Nigam 1986; Sharifi et al. 1991; Jayaraju and Reddy 1996; Geslin et al. 1998; Yanko et al. 1998; Samir 2000; Coccioni 2000; Elberling et al. 2003; Burone et al. 2006). However, the relationship between pollution and test deformation has not yet been established and the quasi-absence of abnormal tests in the Seine estuary shows that pollution does not always induce abnormalities.

13.4 GENERAL DISCUSSION AND CONCLUSION

13.4.1 Comparison of Two Estuaries

The main differences between foraminiferal assemblages in the Seine and Authie estuaries were (1) a higher percentage of *Ammonia tepida* in the Seine estuary; the species was rare in the Authie estuary and (2) lower density and total number of taxa in living and dead assemblages in the Seine estuary than in the Authie estuary. These differences cannot be related to differences in elevation of the sampling stations (+6.5 m above the lowest sea level), or in sediment characteristics. Carbonate content was not significantly different between the two estuaries (25 to 50%); organic matter content was similar at each site, ranging from 9.5 to 19%. Little temporal variability in grain size characteristics was indicated, with a sand fraction (>200 μm) comprising 5 to 15% of the sediment on the Seine mudflat and <10% on the Authie mudflat (Deloffre et al. 2007). *A. tepida* may live in a wide range of sediment types and show a great tolerance to substrate variability (Alve and Murray 1999).

Lower density and total number of taxa of foraminiferal assemblages in the Seine estuary than in the Authie estuary should be partly related to lower salinity in the Seine estuary (average seasonal values between 15 and 20; Durou et al. 2007) than in the Authie estuary (25 to 30). However, as *Haynesina germanica* generally lives farther from the sea than *A. tepida* (review in Debenay et al. 2000), it should be favoured by lower salinity compared to *A. tepida*. Consequently, the behavior of these two species does not corroborate a salinity impact.

The literature indicates that *A. tepida* often becomes dominant in highly impacted environments (Nagy and Alve 1987; Alve 1991; Sharifi et al. 1991; Yanko et al. 1994; Samir and El-Din 2001; Vilela et al. 2004). Thus, even if we cannot exclude an impact of the difference in sedimentation processes with periods of erosion in the Seine estuary, we can infer, on the basis of lower density and higher proportion of *A. tepida*, that the main parameter leading to different foraminiferal assemblages in the two estuaries is pollution. This is in agreement with the chemical analyses of worms and sediments that allowed the designation of the Seine and Authie estuaries as polluted and relatively clean sites, respectively (see Chapter 3).

13.4.2 Changes in 2003 and 2004

Changes in foraminiferal assemblages and in the behavior of the ragworm *Nereis diversicolor* that fed on foraminiferans only at the first sampling date (May 2004) indicate changes in environmental conditions from 2003 to 2004, and more probably after spring 2004. The decreases of density and species richness suggest possible impacts of toxic elements since foraminiferal density and species richness decrease with increasing toxic element concentrations (Ferraro et al. 2006). However, the same trend occurred in both estuaries and affected the nonpolluted Authie. As a consequence, the deterioration of environmental conditions is probably related to a more general (climatic?) phenomenon.

The increase in the proportion of the tolerant species *Haynesina germanica*, in the Authie estuary, is a supplementary indication of adverse conditions. Conversely, little change in species proportions in the polluted Seine estuary shows that all the species are almost equally affected by this adverse phenomenon in this polluted estuary. This observation is consistent with the concept of Murray (2001) who considers that distribution patterns of benthic foraminiferans are controlled by environmental factors that have reached their critical thresholds.

The increase of the proportion of *Ammonia tepida* together with *H. germanica* in the sediment may be considered an alarm signal, showing the negative impact of pollution on the benthos. These two species are particularly well adapted for use as environmental indicators in estuaries because they are strongly euryhaline and tolerant of restricted conditions. Indeed, to be good indicators of pollution, taxa must live in areas where natural conditions are favourable to their maintenance and reproduction (Boltovskoy and Lena 1969; Schafer 1973).

REFERENCES

Alve, E. 1990. Variations in estuarine foraminiferal biofacies with diminishing oxygen conditions in Drammensfjord, SE Norway. In *Paleoecology, Biostratigraphy, Paleoceanography and Taxonomy of Agglutinated Foraminifera*, Hemleben, C. et al. Eds. NATO ASI Series C327, pp. 661–694.

Alve, E. 1991. Benthic foraminifera in sediment cores reflecting heavy metal pollution in Sorfjord, Western Norway. *J. Foramin. Res.* 21: 1–19.

Alve, E. 1994. Opportunistic features of the foraminifer *Stainforthia fusiformis* Williamson: evidence from Frierfjord, Norway. *J. Micropalaeontol.* 13: 24–28.

Alve, E. 1999. Colonization of new habitats by benthic foraminifera: a review. *Earth Sci. Rev.* 46: 167–185.

Alve, E. and J.W. Murray. 1994. Ecology and taphonomy of benthic foraminifera in a temperate mesotidal inlet. *J. Foramin. Res.* 24: 18–27.

Alve, E. and J.W. Murray. 1999. Marginal marine environments of the Skagerrak and Kattegat: a baseline study of living (stained) benthic foraminiferal ecology. *Palaeogeogr. Palaeocl.* 146: 171–193.

Alve., E. and J.W. Murray. 2001. Temporal variability in vertical distributions of live (stained) intertidal foraminifera, southern England. *J. Foramin. Res.* 31: 12–24.

Alve, E. and F. Olsgard. 1999. Benthic foraminiferal colonization in experiments with copper-contaminated sediments. *J. Foramin. Res.* 29: 186–195.

Amiard-Triquet, C. 2005. Rapport final, PNETOX 2, Septembre 2005. Unpublished.

Armynot du Châtelet, E., J.P. Debenay, and R. Soulard. 2004. Foraminiferal proxies for pollution monitoring in moderately polluted harbors. *Environ. Pollut.* 127: 27–40.

Armynot du Châtelet, E. et al. 2005. Utilisation des foraminifères benthiques comme indicateurs de paléo-niveaux marins? Etude du cas de l'anse de l'Aiguillon: *C. R. Palevol.* 4: 209–223.

Barmawidjaja, D.M. et al. 1995. One hundred fifty years of eutrophication in the northern Adriatic Sea: evidence from a benthic foraminiferal record. *Mar. Geol.* 122: 367–384.

Basson, P.W. and J.W. Murray. 1995. Temporal variations in four species of intertidal Foraminifera, Bahrain, Arabian Gulf. *Micropaleontol.* 41: 69–76.

Bates, J.M. and R.S. Spencer. 1979. Modification of foraminiferal trends by the Chesapeake–Elizabeth sewage outfall, Virginia Beach, Virginia. *J. Foramin. Res.* 9: 125–140.

Beck-Eichler, B. et al. 1995. Répartition des foraminifères benthiques dans la zone Sud-Ouest du système laguno-estuarien d'Iguape-Cananeia (Brésil). *Bol. Inst. Oceanogr. USP São Paulo* 43: 1–17.

Bergin, F. et al. 2006. The response of benthic foraminifera and ostracoda to heavy metal pollution in Gulf of Izmir (Eastern Aegean Sea). *Estuar. Coast. Shelf Sci.* 66: 368–386.

Bernhard, J.M. 1986. Characteristic assemblages and morphologies of benthic foraminifera from anoxic, organic-rich deposits: Jurassic through Holocene. *J. Foramin. Res.* 16: 207–215.

Bernhard, J.M. 2000. Distinguishing live from dead foraminifera: methods review and proper applications. *Micropaleontol.* 46: 38–46.

Bhalla, S.N. and R. Nigam. 1986. Recent foraminifera from polluted marine environment of Velsao Beach, South Goa, India. *Rev. Paléobiol.* 5: 43–6.

Boltovskoy, E. and H. Lena. 1969. Seasonal occurrences, standing crop and production in benthic Foraminifera of Puerto Deseado. *Contr. Cushman Found. Foraminif. Res.* 20: 87–95.

Boltovskoy, E. and H. Lena. 1970. On the decomposition of the protoplasm and the sinking velocity of the planktonic foraminifers. *Int. Rev. Ces. Hydrobiol.* 55: 797–804.

Bradshaw, J.S. 1961. Laboratory experiments on the ecology of foraminifera. *Contr. Cushman Found. Foraminif. Res.* 12: 87–106.

Brasier, M.D. 1981. Microfossil transport in the tidal Humber basin. In *Microfossils from Recent and Fossil Shelf Seas*, Neale, P. and Brasier, M.D., Eds. Chichester: Ellis Horwood Ltd., pp. 314–322.

Buckley, D.E. et al. 1974. Canso strait and Chedabucto Bay: a multidisciplinary study of the impact of man on the marine environment. *Geol. Surv. Can. Paper* 1: 74–30.

Burone, L. et al. 2006. Foraminiferal responses to polluted sediments in the Montevideo coastal zone, Uruguay. *Mar. Pollut. Bull.* 52: 61–73.

Buzas, M.A. and L.C. Hayek. 2000. A case for long-term monitoring of the Indian River Lagoon, Florida: foraminiferal densities, 1977–1996. *Bull. Mar. Sci.* 67: 805–814.

Cann, J.H. and P. de Deckker. 1981. Fossil quaternary and living foraminifera from athalassic (non-marine) saline lakes, Southern Australia. *J. Paleontol.* 55: 660–670.

Cearreta, A. 1988. Population dynamics of benthic foraminifera in the Santona Estuary, Spain. *Rev. Paléobiol.* Spec. Vol. 2: 721–724.

Cearreta, A. et al. 2002. Modern foraminiferal record of alternating open and restricted environmental conditions in the Santo Andre lagoon, SW Portugal. *Hydrobiologia* 475: 21–27.

Coccioni, R. 2000. Benthic foraminifera as bioindicators of heavy metal pollution: case study from the Goro Lagoon (Italy). In *Environmental Micropaleontology: The Application of Microfossils to Environmental Geology*, Martin, R.E., Ed. New York: Kluwer, pp. 71–103.

Debenay, J.P. 1990. Recent foraminiferal assemblages and their distribution related to environmental stress in the paralic environments of West Africa (Cape Timiris to Ebrie Lagoon). *J. Foramin. Res.* 20: 267–282.

Debenay, J.P. 2004. A foraminifera-based biological index for the characterization of estuaries of the French Atlantic coast. *38th Symposium of Estuarine and Coastal Science Association*, Rouen, 13–17 Septembre 2004. Abstract 143.

Debenay, J.P. and J.J. Guillou. 2002. Ecological transitions indicated by foraminiferal assemblages in paralic environments. *Estuaries* 25: 1107–1120.

Debenay, J.P. and B.T. Luan. 2006. Foraminiferal assemblages and the confinement index as tools for assessment of saline intrusion and human impact in the Mekong delta. *Rev. Micropaléontol.* 49: 74–85.

Debenay, J.P., J. Pawlowski, and D. Decrouez. 1996a. *Les foraminifères actuels*. Paris: Masson.

Debenay, J.P. et al. 1996b. Les foraminifères paraliques des côtes d'Afrique et d'Amérique du Sud, de part et d'autre de l'Atlantique: comparaison, discussion. In *Géologie de l'Afrique et de l'Atlantique sud*, Jardiné, S., De Klazs, I., and Debenay, J.P., Eds. C.R. Colloq. Géol. d'Angers, 16–20 juillet 1994, Elf-Aquitaine, Pau, pp. 463–471.

Debenay, J.P. et al. 2000. Distribution trends of foraminiferal assemblages in paralic environments: a base for using foraminifera as bioindicators. In *Environmental Micropaleontology: The Application of Microfossils to Environmental Geology*, Martin, R.E., Ed. New York: Kluwer, pp. 39–67.

Debenay, J.P. et al. 2001. The influence of pollution on the distribution of foraminiferal assemblages in a harbour: Port Joinville, Ile d'Yeu, France *Mar. Micropaleontol.* 43: 75–118.

Debenay, J.P., D. Guiral, and M. Parra. 2004. Behaviour and taphonomic loss in foraminiferal
assemblages of mangrove swamps of French Guiana. *Mar. Geol.* 208: 295–314.

Debenay, J.P., B. Millet, and M. Angelidis. 2005. Relationships between foraminiferal
assemblages and hydrodynamics in the Gulf of Kalloni (Greece). *J. Foramin. Res.* 35:
327–343.

Debenay, J.P. et al. 2006. Relationships between spatio-temporal distribution of benthic foraminifera and the dynamic of the Vie Estuary (Vendée, W France). *Estuar. Coast. Shelf
Sci.* 67: 181–197.

Deloffre, J. et al. 2007. Sedimentation on intertidal mudflats in the lower part of macrotidal
estuaries: sedimentation rhythms and their preservation. *Mar. Geol.* 241: 19–32.

Duchemin, G. et al. 2005. Foraminiferal microhabitats in Plougoumelen high marsh (Morbihan,
France). *Palaeogeogr. Palaeocl.* 226: 167–185.

Durou, C. et al. 2007. Biomonitoring in a clean and a multi-contaminated estuary based on biomarkers and chemical analyses in the endobenthic worm *Nereis diversicolor. Environ.
Pollut.* 148: 445–458.

De Rijk, S. and S.R. Troelstra. 1997. Salt marsh foraminifera from the Great Marshes,
Massachusetts: environmental control. *Palaeogeogr. Palaeocl.* 130: 81–112.

Elberling, B. et al. 2003. Applying foraminiferal stratigraphy as a biomarker for heavy metal
contamination and mining impact in a fiord in west Greenland. *Mar. Environ. Res.* 55:
235–256.

Fatela, F. and R. Taborda. 2002. Confidence limits of species proportions in microfossil assemblages. *Mar. Micropaleontol.* 45: 169–174.

Ferraro, L. et al. 2006. Benthic foraminifera and heavy metals distribution: a case study from
the Naples Harbour (Tyrrhenian Sea, Southern Italy). *Environ. Pollut.* 142: 274–287.

Frontalini, F. and. R. Coccioni. 2008. Benthic foraminifera for heavy metal pollution monitoring: a case study from the central Adriatic Sea coast of Italy. *Estuar. Coast. Shelf Sci.*
76: 404–417.

Geslin, E., J.P. Debenay, and M. Lesourd. 1998. Abnormal textures in the wall of deformed
tests of *Ammonia* (hyaline foraminifer). *J. Foramin. Res.* 28: 148–156.

Goldstein, S.T. and G.T. Watkins. 1998. Elevation and the distribution of salt marsh foraminifera, St. Catherine's Island, Georgia: a taphonomic approach. *Palaios* 13: 570–580.

Goldstein, S.T., G.T. Watkins, and R.M. Kuhn. 1995. Microhabitats of salt marsh foraminifera:
St. Catherine's Island, Georgia, USA. *Mar. Micropaleontol.* 26: 17–29.

González-Regalado, M.L. et al. 2001. Total benthic foraminifera assemblages in the southwestern Spanish estuaries. *Geobios* 34: 39–51.

Goubert, E. 1997. *Elphidium excavatum* (Terquem), benthic Foraminifera living in the Vilaine
Bay (France, Britanny) from October 1992 to September 1996: morphology, population
dynamics and environmental relations. Reflexions about methodology, evolutive line
and use in palaeoecology. PhD Thesis, Nantes Univ., Nantes (France), 186 p.

Haslett, S.K., P. Davies, and F. Strawbridge. 1998. Reconstructing Holocene sea-level change
in the Severn and Somerset levels: the Foraminifera connection. *Archaeol. Severn
Estuary* 8: 29–40.

Haslett, S.K. et al. 2001. Vertical salt marsh accretion and its relationship to sea level in the
Severn Estuary, UK: an investigation using foraminifera as tidal indicators. *Estuar.
Coast. Shelf Sci.* 52: 143–153.

Hayward, B.W. and C.J. Hollis. 1994. Brackish foraminifera in New Zealand: a taxonomic and
ecologic review. *Micropaleontology* 40: 185–222.

Hayward, B.W. et al. 1999. Recent New Zealand shallow water benthic foraminifera: taxonomy, ecologic distribution, biogeography, and use in paleoenvironmental assessment. Monograph 21. Lower Hutt, NZ: Institute of Geological and Nuclear Sciences.

Horton, B.P. and J.W. Murray. 2007. The roles of elevation and salinity as primary controls on living foraminiferal distributions: Cowpen Marsh, Tees Estuary, UK. *Mar. Micropaleontol.* 63: 169–186.

Hurlbert, S.J. 1984. Pseudoreplication and the design of ecological experiments. *Ecol. Monogr.* 54: 187–211.

Jayaraju, N. and K.R. Reddi. 1996. Impact of pollution on coastal zone monitoring with benthic foraminifera of Tuticorin, southeast coast of India. *Ind. J. Mar. Sci.* 25: 376–378.

Le Campion, J. 1970. Contribution à l'étude des foraminifères du Bassin d'Arcachon et du proche océan. *Bull. Inst. Géol. Bassin Aquitaine* 8: 3–98.

Loeblich, A.R. Jr. and H. Tappan. 1988. *Foraminiferal Genera and Their Classification.* New York: Van Nostrand Reinhold.

Lutze, G.E. 1968. Jahresgang der Foraminiferen Fauna in der Bottsand Lagune (westlich Ostee). *Meyniana* 18: 13–30.

Morvan, J. et al. 2006. Patchiness and life cycle of intertidal foraminifera: implication for environmental and paleoenvironmental interpretation. *Mar. Micropaleontol.* 61: 131–154.

Mullins, H.T. et al. 1985. Oxygen minimum zone edge effects: evidence from the central California coastal upwelling system. *Geology* 13: 491–494.

Murray, J.W. 1969. Recent foraminifers from the Atlantic continental shelf of the United States: *Micropaleontology* 15: 401–419.

Murray, J.W. 1971. Living foraminiferids of tidal marshes: a review. *J. Foramin. Res.* 1: 153–161.

Murray, J.W. 1991. *Ecology and Palaeoecology of Benthic Foraminifera.* Harlow, UK: Longman.

Murray, J.W. 2000a. When does environmental variability becomes environmental change? The proxy record of benthic foraminifera. In *Environmental Micropaleontology: The Application of Microfossils to Environmental Geology,* Martin, E.R., Ed. New York: Kluwer, pp. 7–37.

Murray, J.W. 2000b. The enigma of the continued use of total assemblages in ecological studies of benthic foraminifera. *J. Foramin. Res.* 30: 244–245.

Murray, J.W. 2001. The niche of benthic foraminifera, critical thresholds and proxies. *Mar. Micropaleontol.* 41: 1–7.

Murray, J.W. 2006. *Ecology and Application of Benthic Foraminifera.* Cambridge: Cambridge University Press.

Murray, J.W. and E. Alve. 1999. Taphonomic experiments on marginal marine foraminiferal assemblages: how much ecological information is preserved? *Palaeogeogr. Palaeocl.* 149: 183–197.

Murray, J.W. and E. Alve. 2000. Major aspects of foraminiferal variability (standing crop and biomass) on a monthly scale in an intertidal zone. *J. Foramin. Res.* 30: 177–191.

Murray, J.W. and S.S. Bowser. 2000. Mortality, protoplasm decay rate, and reliability of staining techniques to recognize "living" foraminifera: a review. *J. Foramin. Res.* 30: 66–70.

Nagy, J. and E. Alve. 1987. Temporal changes in foraminiferal faunas and impact of pollution in Sandebukta, Oslo Fjord. *Mar. Micropaleontol.* 12: 109–128.

Ozarko, D.L., R.T. Patterson, and H.F.L. Williams. 1997. Marsh foraminifera from Nanaimo British Columbia (Canada): implications of infaunal habitat and taphonomic biasing. *J. Foramin. Res.* 27: 51–68.

Pascual, A. et al. 2002. Late Holocene pollution in the Gernika estuary (southern Bay of Biscay) evidenced by the study of Foraminifera and Ostracoda. *Hydrobiologia* 475: 477–491.

Phleger, F.B. 1960. *Ecology and Distribution of Recent Foraminifera*. Baltimore: Johns Hopkins University Press.

Phleger, F.B. 1970. Foraminifera population and marine marsh processes. *Limnol. Oceanogr.* 15: 34–52.

Phleger, F.B. and A. Soutar. 1973. Production of benthic foraminifera in three East Pacific oxygen minima. *Micropaleontology* 19: 110–115.

Poag, C.W. 1984. Distribution and ecology of deep water benthic foraminifera in the Gulf of Mexico. *Palaeogeogr. Palaeocl.* 48: 25–37.

Samir, A.M. 2000. The response of benthic foraminifera and ostracods to various pollution sources: a study from two lagoons in Egypt. *J. Foramin. Res.* 30: 83–98.

Samir, A.M. and A.B. El-Din. 2001. Benthic foraminiferal assemblages and morphological abnormalities as pollution proxies in two Egyptian bays. *Mar. Micropaleontol.* 41: 193–227.

Schafer, C.T. 1973. Distribution of foraminifera near pollution sources in Chaleur Bay. *Water Air Soil Poll.* 2: 219–233.

Schafer, C.T., E.S. Collins, and J.N. Smith. 1991. Relationship of foraminifera and the camoebian distributions to sediments contaminated by pulp mill effluent, Saguenay Fjord, Quebec, Canada. *Mar. Micropaleontol.* 17: 255–283.

Scott, D.B. 1976. Brackish water foraminifera from southern California and description of *Polysaccammina ipohalina* n. gen., n. sp. *J. Foramin. Res.* 6: 312–321.

Scott, D.B. and F.S. Medioli. 1978. Vertical zonations of marsh foraminifera as accurate indicators of former sea levels. *Nature* 272: 528–531.

Scott, D.B. and F.S. Medioli. 1980. Quantitative studies of marsh foraminiferal distributions in Nova Scotia: implications for sea level studies. *Contr. Cushman Found. Foraminif. Res.* 17: 1–58.

Scott, D.B. and J.O.R. Hermelin. 1993. A device for precision splitting of micropaleontological samples in liquid suspension. *J. Paleontol.* 67: 151–154.

Scott, D.B., F.S. Medioli, and C.T. Schafer. 2001. *Monitoring in Coastal Environments Using Foraminifera and the Camoebian Indicators*. Cambridge: Cambridge University Press.

Seiglie, G.A. 1975. Foraminifers of Guayanilla bay and their use as environmental indicators. *Rev. Esp. Micropaleontol.* 7: 453–487.

Sen Gupta, B.K. 1971. The benthonic foraminifera of the tail of the Grand Banks. *Micropaleontology* 17: 69–98.

Setty, M.G.A.P. 1976. The relative sensitivity of benthonic foraminifera in the polluted marine environment of Cola Bay, Goa. *Proceedings of VI Indian Colloquium of Micropaleontology and Stratigraphy*, pp. 225–234.

Setty, M.G.A.P. and R. Nigam. 1984. Benthic foraminifera as pollution indices in the marine environment of west coast of India. *Riv. Ital. Paleontol.* S89: 421–436.

Sharifi, A.R., T.W. Croudace, and R.L. Austin. 1991. Benthonic foraminiferids as pollution indicators in Southampton waters, Southern England. *J. Micropaleontol.* 10: 109–113.

Stubbles, S.J. 1993. Recent benthic Foraminiferida as indicators of pollution in Restronguet Creek, Cornwall. *Proc. Ussher Soc.* 8: 200–204.

Stubbles, S.J. et al. 1996. Responses of foraminifera to presence of heavy metal contamination and acidic mine drainage. Minerals, Metals, and the Environment II Conference, Institution of Mining and Metallurgy, Prague, 3–6 September, pp. 217–235.

Vilela, C. et al. 2004. Benthic foraminifera distribution in high polluted sediments from Niteroi Harbor (Guanabara Bay), Rio de Janeiro, Brazil. *Ann. Acad. Bras. Cienc.* 76: 161–171.

Walton, W.R. 1952. Techniques for recognition of living foraminifera. *Contr. Cushman Found. Foraminif. Res.* 3: 56–60.

Walton, W.R. 1955. Ecology of living benthonic Foraminifera, Todos Santos Bay, Baja California. *J. Paleontol.* 29: 952–1018.

Yanko, V. and A. Flexer. 1991. Foraminiferal benthonic assemblages as indicators of pollu-
tion: an example of northwestern shelf of the Black Sea. *Proceedings of Third Annual Symposium on the Mediterranean Margin of Israel*, Haifa.

Yanko, V., J. Kronfeld, and A. Flexer. 1994. Response of benthic foraminifera to various pollution sources: implications for pollution monitoring. *J. Foramin. Res.* 24: 1–17.

Yanko, V., M. Ahmad, and M. Kaminski. 1998. Morphological deformities of benthic foraminiferal tests in response to pollution by heavy metals; implications for pollution monitoring. *J. Foramin. Res.* 28: 177–200.

APPENDIX 13.1

Relative Abundances of All Taxa Identified in Thanatocoenoses of Samples Collected in February 2004. The density (number of tests in 50 cm³ of sediment), species richness, Shannon–Weaver diversity index, and Ic index are indicated at the bottom of the table. Asterisks indicate the presence of stained (living) individuals.

The last three columns (Somme, Authie, Seine) are grouped under the heading "Average Nb per 50 cm³".

Sample No.	1	2	3	4	5	6	7	8	9	10	11	12	13	14	15	16	17	18	Somme	Authie	Seine
Adelosina longirostra	0	0				0	0												0	1	
Adelosina sp.		0	1																	36	
Ammonia tepida		1	2	1		2		1	1				18*	31*	31*	3	3	10*	26	127	105
Angulogerina angulosa	2		1			1	1												2	85	
Astacolus crepidulus		0																	0		
Aubignyna planidorsa	2	1		1															27	44	
Bolivina pseudoplicata	3	0		1	1	0	3*			2	4*	1	2	1		1	6		3	235	45
Brizalina pseudopunctata	1		0		0	1	0						0			1			26	48	2
Brizalina spathulata															1					1	
Brizalina striatula	1	0*	0*	8*	1*	9*	5*			4	6*	2		2		1	6		1	1069	42
Brizalina variabilis	2	0	2	3	2												1		3	271	7
Brizalina spp.	1		1																1	44	
Buccella frigida	3	1	2	1	2	1	3	2*		1			5	4		3	4		29	252	41
Bulimina elegans		2	1	1	0					1			1			1			77	88	8
Bulimina marginata															1					2	
Buliminella elegantissima	1	1	1	1*		1*			2	1	0	1	1	2			4		27	170	29
Cassidulina crassa	1	0	2	1												1			1	124	7
Cibicides refulgens	1																		1		
Cribroelphidium cuvilieri	2	1	1				1						1					2	54	38	3
Cribroelphidium excavatum	23*	15*	9*	9*	4*	6*	10*	2	1	1	3*	6	7	18		6	7	36	538	1462	106
Cribroelphidium gunteri																1					1
Cribroelphidium magellanicum	5	18*	36*	39	42	33*	16	2	7	23	34*	28	16	1	5	61	30	5	649	8379	336
Cribroelphidium excavatum	27*	23*	6	5	12	11	15	6*	5*	10*	8	11*	8	3	1	1	2	3	825	2132	34
Cribrononion gerthi	6	10	8*	13	11	8*	10*	2		3	12	3	7			3	4		341	2187	34
Cribrostomoides jeffreysii		1*	1*	1	2*	1	3				3*		1	5		3	3	36	26	290	29
Cyclogira sp.			0*			1	1*						2			3	1			51	15
Edentostomina sp.			0		0	0	0				0	0				1			2	7	
Eggerelloides scabrus		0															2		0		2
Elphidium macellum													1								0
Elphidium pulvereum	2	3	1										1			1	0		105	34	7
Eoeponidella pulchella	1																		1		
Favulina hexagona					0															0	
Favulina lineata		0		1		0				0						1			0	45	7
Favulina melo	1		0																1	0	
Favulina squamosa		0	1																0	34	
Fissurina lucida	2	1				0			1	0			1					2	54	7	3
Fissurina spp.		0		1	1					0	1			1	1				0	107	3
Gavelinopsis praegeri		0	0	0*	1	2													0	131	
Glabratella cf. G.baccata		0														1	0		0		7
Globulina gibba			0														0		0	0	0
Homalohedra williamsoni	2	2																	2	68	
Hopkinsina pacifica	1	1*				0	0*					1							1	49	
Haynesina depressula		1	1		2	4				1	1		3	2		2	1	1	26	332	16
Haynesina germanica	2	7*	8*	1	5*	10*	8*	52*	50*		1	6*	14*	21*	60	4		14*	260	1734	121
Hyalinonetrion sp.			0	0	0						0					1	0		1		2
Jadammina macrescens		0	1	0	1	1*	0	72*	28*	0	1	32*	0						0	875	
Lachlanella sp.		0																	0		
Lagena laevis		0																	0		
Lagena semistriata		0		0															0	0	
Lagena striata			0		0											3*			1		20
Lagenosolenia lagenoides			0								0									1	
Lamarckina haliotidea		0		0		0	2				3		1	1					0	91	2
Lenticulina rotulata			1		1	0							1						69	0	
Lepidodeuterammina ochracea			1*	3	1*	2	2*				2	1		1	3	3			374	26	
Lobatula lobatula	2	3	1				1						3	1		1	1	3	105	38	16
Miliolinella obliquinodus						1	1												83		
Miliolinella subrotunda	2	4	5	4*	3	1	1				1	2	6	1		1	10		157	594	25
Milliammina fusca												0*	0*						0		
Neoconorbina nitida	1	1	2	1	2	3	2	1		1	10					1	1		52	646	7
Nonion pauperatum	1	1	1	0							1	1			1				27	77	2
Palliolatella orbignyana			0				1								1					4	2
Parafissurina sp.		0		0								0							0	1	
Paratrochammina cf. P.haynesi	0	0	1	0	1	0					1	1							0	135	2
Paratrochammina sp.	1																			28	
Patellina corrugata		0	0	0	0						1	0		1					0	5	2
Planorbulina mediterranensis		0	2		0		1												0	72	

(continued on next page)

Sample No.	1	2	3	4	5	6	7	8	9	10	11	12	13	14	15	16	17	18	Average Nb per 50 cm³ Somme	Authie	Seine
Polymorphina sp.		1	0		1	0	0												26	36	
Pseudononion atlanticum		0	1																0	34	
Pyrgo sp.						0														0	
Quinqueloculina laevigata			0	0	1		1*			1			1							50	2
Quinqueloculina lamarckiana	1		0				0							1		1			1	0	7
Quinqueloculina lata			0		1				1				1							55	0
Quinqueloculina seminula	1	1	1		3	3*	2	2	2*	3	1		4		4	1	5		27	402	19
Quinqueloculina spp.		1		2	2		1*		0	1		2	2					2	26	190	5
Quinqueloculina stelligera	1	0	1	0	1		3				1								1	109	
Remaneica plicata		0	0*	0	1	0	0				2								0	91	
Reophax nana		0	1																0	44	
Rosalina anglica						1														48	
Rosalina cf. *vilardeboana*		1	1	0*	0	3	1			0		1							26	196	
Rosalina globularis		0			1												1		0	35	1
Spirillina vivipara		0	0	0		0	1*				3								0	88	
Spiroloculina dilatata		0				0													0	0	
Spirophtalmidium sp.				0							0									1	
Spiroplectinella wrighti						1														4	
Stainforthia fusiformis		1	0	0		0							1	1		8			26	1	55
Svratkina tuberculata	1	0																	1		
Textulara earlandi			0																	0	
Textularia truncata		1														1			26		7
Triloculina trigonula			1																	34	
Trochammina inflata		0	0	0	0		0	4*	2*	1	1	1*	1						0	67	2

n°speciments/50cm3	200	7000	40000	45000	40000	50000	4000	300	8000	10000	30000	15000	1000	300	500	1000	4000	500			
Number of species	30	56	53	40	40	37	40	10	11	19	32	22	22	22	7	21	29	20			
Number of living species	2	6	10	6	4	8	11	5	5	2	6	5	2	2	1	0	1	2			
Diversity (Shannon-Weaver)	3.7	3.6	3.6	3.3	3.3	3.6	4.0	1.7	2.0	2.3	3.5	2.9	3.7	3.1	1.5	2.5	4.0	3.1			
Ic x 100	5	16	26	28	25	25	22	86	67	47	39	64	37	37	50	31	17	20			

APPENDIX 13.2

Main Characteristics of Foraminiferan Species

AMMONIA TEPIDA

In the literature, *Ammonia tepida* is described as a strongly euryhaline species that may experience salinity ranging from ~10 to ~100 (Bradshaw 1961; Brasier 1981; Debenay 1990; review in Murray 2006). This tolerance allows it to live worldwide in coastal and paralic environments, where it is often associated with *Haynesina germanica* (review in Debenay and Guillou 2002). In paralic environments, its distribution is related to the prevailing sea-to-freshwater gradient (Murray 1971; Scott 1976; Hayward and Hollis 1994; Hayward et al. 1999; review in Debenay and Guillou 2002). It generally grows in an intermediate position between *C. excavatum* in a lower estuary and *H. germanica* that is more tolerant of restricted conditions (Alve and Murray 1994; review in Debenay and Guillou 2002). It is commonly encountered worldwide in restricted environments under pollution stress, even capable of surviving high concentrations of heavy metals (Seiglie 1975; Setty 1976; Setty and Nigam 1984; Yanko and Flexer 1991; Sharifi et al. 1991; Stubbles et al. 1996; Coccioni 2000; Armynot du Châtelet et al. 2004; Bergin et al. 2006; Ferraro et al. 2006; Frontalini and. Coccioni 2007).

In highly impacted environments, it often becomes dominant (Nagy and Alve 1987; Alve 1991; Sharifi et al. 1991; Yanko et al. 1994; Samir and El-Din 2001; Vilela et al. 2004). Under such conditions, it is often associated with *H. germanica* in upper estuaries and with *Cribroelphidium excavatum* in lower estuaries and coastal environments. In polluted normal marine environments, *A. tepida* may be totally replaced by *C. excavatum* (Debenay et al. 2001). In moderately polluted environments, *A. tepida* and *Haynesina germanica* show a positive correlation with pollutants that is stronger than their correlation with the sea-to-freshwater gradient (Armynot du Châtelet et al. 2004). Several authors have suggested that this species may be used as a heavy metal pollution bioindicator (Yanko et al. 1994; Ferraro et al. 2006).

HAYNESINA GERMANICA

Haynesina germanica is an extremely euryhaline species tolerant of very restricted conditions (Brasier 1981; Murray 1991; Alve and Murray 1999; Pascual et al.2002; review in Debenay and Guillou 2002), allowing it to live in upper intertidal areas (Cearreta 1988; Murray 1991; Debenay et al. 2000; Cearreta et al. 2002). It generally lives farther from the sea and in upper topographic conditions than *Ammonia tepida* (review in Debenay et al. 2000), but an inverse distribution may be observed (Haslett et al. 2001; Debenay et al. 2001; Horton and Murray 2007). *Haynesina germanica* may be tolerant of high inputs of organic matter (Alve and Murray 1994), heavy metal (Stubbles 1993; Stubbles et al. 1996), and hydrocarbon pollution (Armynot du Châtelet et al. 2004).

CRIBROELPHIDIUM EXCAVATUM

Cribroelphidium excavatum is able to grow successfully in polluted, near-shore environments (Schafer 1973; Buckley et al. 1974; Bates and Spencer 1979; Setty and

Nigam 1984; Alve 1991; Schafer et al. 1991). Sharifi et al. (1991) established that it is the most tolerant species to heavy metal pollution, followed by *H. germanica* and *A. beccari* [*tepida*] in that order. Debenay et al. (2001) already considered this species as a pollution bioindicator, especially in open ocean harbors. Because it can survive only moderate salinity variations, it cannot be used as a pollution indicator in upper estuaries where the impact of low salinity may be stronger than the impact of pollution for this species. Its relative rarity in our samples may be due to the fact that it is common only in subtidal areas although it is present in low abundance in the intertidal zone (Alve and Murray 1999).

JADAMMINA MACRESCENS

Jadammina macrescens, often associated with *Trochammina inflata*, is generally considered a typical high marsh species (Scott and Medioli 1978). However, it is not confined to marshes. It was found living at the bottom of a 30-cm deep eutrophic basin (Debenay et al. 2001) and Murray and Alve (1999) reported that it can dominate in subtidal areas down to 1 m depth. It may be epifaunal (Goldstein et al. 1995), infaunal (Ozarko et al. 1997), and may migrate down to 30 cm in sediments in response to increasing or decreasing water content (Duchemin et al. 2005). Alve and Murray (1999) reported that it is a typical epiphytal species. Whether it lives within or above sediment, it seems to be able to flourish only in vegetated environments. *Jadammina macrescens* may also be influenced by salinity, and a positive correlation was established between its abundance and mean salinity (De Rijk and Troelstra 1997). In the southwestern Spanish estuaries, *J. macrescens* prefers the more marine habitats of protected salt marshes with higher salinities (González-Regalado et al. 2001).

BOLIVINIDS

Bolivinids, including the genera *Bolivina* and *Brizalina*, are known to survive in oxygen-deficient environments (Bernhard 1986; Murray 1991; Phleger and Soutar 1973; Poag 1984; Mullins et al. 1985). In marine harbors, high proportions of bolivinids are found in fine sediments where they may resist metal and other chemical pollution (Debenay et al 2001, Armynot du Châtelet et al. 2004).

STAINFORTHIA FUSIFORMIS

In temperate environments, *Stainforthia fusiformis* can take advantage of an excess of organic matter in fine sediment, even impoverished in oxygen (Alve 1990; Barmawidjaja et al. 1995), but it is sensitive to copper pollution (Alve and Olsgard 1999) and its tolerance to subaerial exposure is low (Armynot du Châtelet et al. 2005). It is an opportunistic species that may be a successful colonizer of previously anoxic habitats (Alve 1994).

CRIBROELPHIDIUM WILLIAMSONI

Cribroelphidium williamsoni has been reported in marshes of the Atlantic seaboard of Europe (Le Campion 1970; Phleger 1970; review in Murray 1991; Haslett et al.

1998). However, it seems that the distribution of this species may change greatly depending on the region. Alve and Murray (1999) reported it as characteristic of nonmarsh areas but sometimes present at the seaward edges of marshes. Horton and Murray (2007) consider that it is among the species that characterize the low marsh and tidal flat subenvironments.

14 Patterns of Abundance, Diversity, and Genus Assemblage Structure of Meiofaunal Nematodes in the Seine (Pont de Normandie) and Authie (Authie Port) Estuaries

Timothy J. Ferrero

CONTENTS

The meiofauna or meiobenthos are classified as animals of intermediate size that can pass though a sieve with a mesh aperture size of 1000 to 500 μm and be retained on a mesh of 32 to 63 μm. This classification was originally intended as a means to differentiate the meiobenthos from the macrobenthos and microbenthos (Mare 1942) but also reflects separate ecological and evolutionary histories (Warwick 1989). Of the 34 recognised phyla comprising the Kingdom Animalia, 20 are mainly meiofaunal or have meiofaunal representatives, and 5 are only known as meiofaunal organisms (Higgins and Thiel 1988; Giere 1993).

Many meiofaunal taxa are relatively constrained to sandy and coarse sediments and comprise a specialist interstitial fauna (Swedmark 1964), living in the spaces between the particles of benthic sediments. However, other taxa, notably nematodes, are commonly found at high densities in muddier sediments and in brackish and estuarine systems and regularly occur at densities from 1 to 15×10^6 m^{-2} and higher. The maximum density recorded (27×10^6 m^{-2}) was recorded at the Lynher estuary in the United Kingdom (Warwick and Price 1979).

14.1   MEIOFAUNA IN ECOTOXICOLOGICAL STUDIES

Many authors have promoted the meiofauna as suitable model organisms for the assessment of pollution impacts, both in field and laboratory studies (Heip 1980; Platt and Warwick 1980; Moore and Bett 1989). This recommendation rests largely on intrinsic features such as the distribution, abundance, diversity, and life histories of meiofaunal organisms. These features are shared by many meiofaunal taxa, but are exemplified by the Copepoda and Nematoda in terms of their utility for pollution studies. Specifically, these taxa are relatively abundant compared with macrofaunal taxa, and can thus be sampled adequately with small samples, often employing relatively simple sampling equipment and requiring minimum investigator impact within the study site or area.

Nematodes and copepods also tend to exhibit higher diversity than macrofaunal taxa, which may increase the range of potential responses to pollutants and thus the overall sensitivity of the fauna in detecting their impacts. These two factors mean that data obtained have high information content and statistical analysis is often facilitated. In addition, the short generation times of most meiofaunal taxa mean that populations can respond rapidly to environmental perturbations, while their generally conservative reproductive strategies, characterised by relatively low gamete numbers and direct benthic development without planktonic larval dispersal, increase the likelihood that significant changes in populations reflect some change in biotic or abiotic environmental conditions.

The low dispersal characteristics of meiofaunal taxa also mean that most individuals within a population complete their entire life cycles in intimate contact with the sediment and pore water, and thus any contaminants they contain. Some taxa, particularly the nematodes, include species with very high tolerances to certain pollutants. Thus, nematodes are often the last metazoans to persist in the most grossly polluted systems and can be the only model organisms to study under such conditions.

Studies intended to investigate the effects of pollutants on meiofauna have employed a wide range of approaches (*in vitro* toxicity testing, microcosm and mesocosm studies of selected taxa and assemblage responses, and field studies). The early studies have been extensively reviewed (Vincx and Heip 1987; Moore and Bett 1989; Coull and Chandler 1992). In studies of assemblage responses, workers have attempted to use meiofaunal major taxa as a relatively simple means of assessing environmental impacts, avoiding the perceived taxonomic difficulties associated with working at genus or species level. However, these studies have produced variable results as the abundance of an entire taxon has not been found to be particularly sensitive to pollution or produce reliable results because of confounding variables that may affect abundance and interact with pollution effects (Coull and Chandler 1992).

In practice, the most reliable and informative approaches involved work with the two most abundant and ubiquitous meiofaunal taxa, the nematodes and copepods and studies in which the fauna have been identified to species (or genus) level, allowing either single-species toxicological/autecological approaches or multi-species synecological assemblage studies.

As will be seen, this study focused almost exclusively on the nematode fauna of the Seine and Authie estuaries and the following introductory comments refer for the most part to this taxon.

Of the pollutants present in the Seine estuary and forming the focus of the PNETOX programme, the effects of trace metals on nematodes have received particular attention (see Coull and Chandler 1992). Laboratory toxicity testing has shown that mortality generally increases with increasing metal concentration, but that nematodes tend to be tolerant of rather high concentrations of some metals, cadmium for example (Vranken et al. 1985), while copper and mercury often cause higher toxicity. Interestingly, when paired metals were tested on the opportunistic nematode *Monhystera disjuncta* (Vranken et al. 1988), the recorded effects appeared to be antagonistic rather than additive. However, most authors acknowledge that mortality is a rather poor measure of a pollutant's likely impact in natural situations. Perhaps more important for the interpretation of ecological results has been the discovery of significant effects on reproductive success, development, and fecundity (Bogaert et al. 1984; Verschraegen et al. 1986; Vranken and Heip 1986).

Field studies have shown a range of responses to trace metal contamination. Lorenzen (1974) found no impacts associated with industrial waste disposal in the German Bight of the North Sea and similarly Tietjen (1977) found no significant impact of mixed metal contamination in a study of subtidal sediments in Long Island Sound, USA. However, the same author found significantly depressed nematode diversity in medium sands contaminated with several metals in the nearby New York Bight (Tietjen 1980). Somerfield et al. (1994) investigated the effects of a suite of metals derived from mining activities in a series of creeks in the Fal estuary, UK, and found nematodes to be more sensitive to sediment metals than copepods, showing reduced diversity in the more polluted creeks and significant changes in species assemblage structure.

These findings were supported by Austen and Somerfield (1997) who achieved similar results in a mesocosm study modelled on the Fal system, and by Millward and Grant (1995) who tested the toxicity of copper to nematode assemblages from the Fal system and found that several species from the most polluted sediments were more tolerant to the metal. A similar observation was made by Howell (1984), who found that specimens of the predatory nematode *Enoplus brevis* from a polluted site were less sensitive to a suite of metals than those from an unpolluted site. Gyedu-Ababio and Baird (2006) also used microcosms to assess the effects of heavy metals on meiofaunal assemblages and again demonstrated suppression of nematode abundance and diversity with corresponding changes in genus assemblage structure, as well as differences in responses to metals and organic enrichments.

The micro/mesocosm approach has often proved useful in separating the effects of pollutants from other factors. Schratzberger et al. (2002) demonstrated effects on nematode diversity and assemblage structure due to the presence of tributyltin (TBT) in test sediments, but also demonstrated the importance of sediment burial by deposition of the contaminant. Similarly, Lee and Correa (2007) demonstrated toxic effects of copper mine tailings, but also distinct effects caused solely by the physical impact of the deposition of fine particles onto a coarser sediment, resulting in reduction in pore size. Beyrem et al. (2007) demonstrated an important synergistic relationship between trace metals and oil pollution, finding a significant impact of combined cadmium and diesel oil contamination compared with no significant impact of the same contaminants in isolation. Nematode assemblages treated with combined contaminants showed significant reduction in both density and diversity.

These examples highlight the range of potential effects of trace metals that are significant pollutants in the Seine estuary system, compared with the comparatively clean Authie, but it is important to remember that in a system like the Seine, the fauna are exposed to a range of pollutants at different and variable levels. This may be termed chronic pollution and is typical of many large estuaries within Europe. Under these conditions, the fauna are subject to the continuous presence of a mixture of contaminants, rather than the considerably higher levels of specific contaminants associated with point sources discharges into the receiving system.

Studies of estuarine meiofauna have generally not tried to assess specifically the effects of chronic pollution, although general meiofaunal surveys have often been located in estuarine systems having many different pollution sources (Bouwman et al. 1984; Moore 1987). The usual model has been to attempt to study the impact of a single pollutant (trace metal, oil, organic waste) or pollution source (contaminant spill or discharge point such as a sewage outfall). In these studies, the meiofauna have often been shown to be sensitive to such spatially or temporally discrete pollution impacts (Heip et al. 1985; Somerfield et al. 1994; Coull and Chandler 1992). In these cases, both organic and chemical pollutants were shown to exert measurable, though not always predictable, impacts on meiofaunal abundance, diversity, and assemblage structure.

Millward et al. (2004) demonstrated that different types of pollutants (in this case trace metals and diesel oil) may interact to produce contrasting impacts on both abundance and species composition of copepod assemblages, though not on nematodes, in microcosm studies. Similarly, detecting the response to reduction of pollution can also produce contrasting results. Essink (2003) found that a reduction in the organic loading of the Ems estuary over more than 10 years produced a distinct pattern of recovery in the fauna. However, in a similar study, removal of trace metal-contaminated sediments from an area of the Hudson River estuary led to significant differences between the "restored" site and unpolluted sites 6 years after the initial clean-up, which the authors ascribed to the impact of anthropogenic disturbance caused by the removal of contaminated sediments (Kelaher et al. 2003).

In this study, a field study approach was adopted to assess the abundance, diversity, and assemblage structure of the nematode fauna at two comparable locations within the multi-polluted Seine and relatively clean Authie estuary systems over a series of seasonal sampling occasions to assess any differences attributable to the

impact of chronic, multisource pollution typical of many large estuaries in industrialised areas.

14.2 MATERIALS AND METHODS

At two sampling sites (Pont de Normandie [PN] in the Seine estuary and Authie
Port [AP] in the Authie estuary; see Figure 1.3 in Chapter 1) six randomly positioned
replicate samples were taken on each of the five sampling occasions (February, May,
August, and November 2003 and February 2004) to a depth of 10 cm using a Perspex
hand coring tube of 15 cm^2 sample surface area. Each sample was fixed in the field
with a 4% solution of buffered formaldehyde prepared with filtered (<45 µm) local
estuary or sea water.

Meiofauna were extracted by a combination of five successive decantations in filtered tap water over a 45-µm mesh sieve followed by five repeated centrifugations at
3500 g of the material retained on the sieve in a solution of Ludox-TM (1.15 specific
gravity) colloidal silica. Despite these combined treatments, the resulting extract
was, in all cases, still contaminated by significant quantities of detritus and further
volumetric sample splitting was employed prior to enumeration of the meiofauna in
Petri dishes under a dissecting microscope. During enumeration, the first 120 nematodes encountered were picked into a solution of 5% alcohol, 5% glycerol, and 90%
distilled water. They were then evaporated to anhydrous glycerol and mounted on
slides for identification to genus. The few harpacticoid copepods encountered were
transferred to cavity slides containing lactic acid before identification.

Nematode genus abundance data were used to calculate univariate measures of
diversity (G = genus number, H′ = Shannon–Weiner diversity, J′ = Pielou's evenness)
and conduct multivariate analyses (cluster analysis and nonmetric multidimensional
scaling [nMDS]) using PRIMER 5 for Windows version 5.2.9. ANOVA with Tukey's
post hoc test, and graphs of univariate measures were produced using GraphPad
PRISM version 4.00.

14.3 RESULTS

14.3.1 FAUNAL COMPOSITION

A total of seven meiofaunal taxa were found at the two sampling locations: Nematoda,
Copepoda, Turbellaria, Ostracoda, Tardigrada, Halacarida and Oligochaeta. However, despite the occurrence of other meiofaunal taxa, the meiofaunal assemblages
at both sites were dominated by the Nematoda, accounting for >95% of all individuals. Of the few specimens of copepods encountered, most were nauplii or copepodid I and II stage juveniles that could not be identified. However, specimens of
Microarthridion fallax were identified from the Pont de Normandie site and a single
specimen of *Tachidius discipes* from the Authie Port site. Copepods showed no
peaks in abundance at either site and, unfortunately, their occurrence was so low and
sporadic that they did not prove to be useful candidates for biological monitoring at
these sites. Similarly, the other meiofaunal taxa occurred either in very low numbers
or sporadically, thus making further analysis uninformative.

The high numerical abundance of Nematoda resulted in their selection as the only meiofaunal taxon for which full univariate and multivariate analyses were possible. One hundred individuals were identified from each of the 60 sample cores, mostly at the genus level of taxonomic discrimination. The total number of nematode taxa at the two sites was 58, of which 43 were recorded on the Seine at Pont de Normandie and 53 at Authie Port. Of these 58 taxa, 38 were found at both sites, representing 66% of the total. The 5 taxa found only at Pont de Normandie all occurred at very low abundance, represented by <5 (≤0.05% total) individuals identified in the raw data. However, of the 15 taxa recorded only at Authie Port, one genus, *Anoplostoma*, was one of the ten most abundant genera recorded over the entire survey, accounting for 2.75% of all individuals identified.

Table 14.1 shows the ten most abundant nematode genera at each site for the five sampling campaigns. In general, the most dominant taxa tended to remain represented in the lists across the entire study period, but notable changes in relative dominance and composition of subdominant taxa indicating seasonal change were seen.

At Pont de Normandie, *Calomicrolaimus* and *Halalaimus* were the first and second most dominant genera, accounting for 40 to 45% of all individuals, in February and May 2003 and again in February 2004. In the late summer and autumn of 2003 (August and November), their ranking was reversed and *Halalaimus* was the most abundant genus. Their combined numbers still accounted for approximately 40% of the total in August 2003, but in November 2003 and February 2004, these two genera accounted for over 50% of the total fauna. Between February and May 2003, only five taxa were consistently present in the top ten listings, but as the study progressed, less taxon turnover occurred, with eight taxa remaining common to the top ten in all subsequent surveys.

At Authie Port, *Dichromadora* and *Leptolaimus* were the first and second most dominant genera, similarly abundant and accounting for a somewhat higher proportion (approximately 55 to 65%) of all individuals in February and May 2003. This ranking was the same in February 2004, with the exception that numbers of *Dichromadora* were clearly higher than those of *Leptolaimus*. In the late summer and autumn of 2003 (August and November) seasonal changes similar to those found at Pont de Normandie occurred; *Leptolaimus* became the most abundant genus. However, in August 2003, although the numbers of both *Leptolaimus* and *Dichromadora* increased from their May 2003 levels and their individual percentage abundances were not greatly different, there was a dramatic peak in the numbers of *Daptonema*, with the result that only three genera represented over 75% of all individuals in the samples. In contrast to the findings at Pont de Normandie, taxon turnover was stable from February to May 2003 and from May to August 2003. Eight taxa in the top ten listing were common in both cases, but higher during the latter half of the study with only six taxa remaining common on all subsequent listings.

14.3.2 NEMATODE ABUNDANCE AND DIVERSITY

Mean and 95% confidence interval values of univariate measures of nematode abundance and diversity are shown in Figure 14.1. ANOVA of the abundance data showed significant differences at both sampling sites over the study period (Table 14.2), but

TABLE 14.1
Summary of Mean and Percentage Abundance of the Ten Most
Abundant Nematode Genera at Pont de Normandie and Authie Port

Pont de Normandie			Baie de l'Authie		
Genus	N 10 cm⁻²	%	Genus	N 10 cm⁻²	%
February 2003			**February 2003**		
Calomicrolaimus	1259.6	21.5	*Dichromadora*	1797.9	29
Halalaimus	1223.7	20.8	*Leptolaimus*	1773.4	28.7
Ptycholaimellus	564.4	9.6	*Calomicrolaimus*	1148	18.5
Thallassomonhystera	537.7	9.2	*Anoplostoma*	418.8	6.8
Paracanthonchus	448.9	7.6	*Daptonema*	169.2	2.7
Praecanthonchus	396.1	6.7	*Thallassomonhystera*	164.1	2.7
Tripyloides	241.9	4.1	*Axonolaimus*	130.6	2.1
Leptolaimus	241.2	4.1	*Halalaimus*	89.5	1.4
Deontolaimus	218.8	3.7	*Eudiplogaster*	67.7	1.1
Camacolaimus	107.6	1.8	*Theristus*	54.9	0.9
May 2003			**May 2003**		
Calomicrolaimus	1419.7	33.5	*Dichromadora*	2635.5	29.1
Halalaimus	454.4	10.7	*Leptolaimus*	2208.1	24.3
Spilophorella	453.6	10.7	*Daptonema*	1891.6	20.9
Ptycholaimellus	438.7	10.3	*Calomicrolaimus*	888.6	9.8
Daptonema	426.3	10	*Anoplostoma*	316.7	3.5
Diplolaimella	226.5	5.3	*Diplolaimella*	239.5	2.6
Deontolaimus	138.9	3.3	*Thallassomonhystera*	121	1.3
Calyptronema	134.6	3.2	*Desmolaimus*	86.7	1
Camacolaimus	122.1	2.9	*Eudiplogaster*	81.7	0.9
Cyatholaimidae juvs	74.8	1.8	*Axonolaimus*	66	0.7
August 2003			**August 2003**		
Halalaimus	632	24.4	*Leptolaimus*	4023.5	26.4
Calomicrolaimus	415.1	16	*Daptonema*	4010.7	26.3
Camacolaimus	262.8	10.1	*Dichromadora*	3761.2	24.7
Ptycholaimellus	240	9.3	*Anoplostoma*	1505.7	9.9
Calyptronema	205.2	7.9	*Calomicrolaimus*	805.4	5.3
Spilophorella	174.6	6.7	*Diplolaimella*	337.7	2.2
Xyalidae gen (LT)	114.3	4.4	*Thallassomonhystera*	195.7	1.3
Diplolaimella	108.3	4.2	*Eudiplogaster*	104.6	0.7
Daptonema	97.2	3.8	*Paracanthonchus*	77.6	0.5
Antomicron	51.6	2	*Theristus*	50.3	0.3

(continued on next page)

TABLE 14.1 (continued)
Summary of Mean and Percentage Abundance of the Ten Most Abundant Nematode Genera at Pont de Normandie and Authie Port

Pont de Normandie			Baie de l'Authie		
Genus	N 10 cm^{-2}	%	Genus	N 10 cm^{-2}	%
November 2003			**November 2003**		
Halalaimus	1031.3	29	Leptolaimus	667.1	25
Calomicrolaimus	923.8	26	Dichromadora	432.1	16.2
Spilophorella	370.5	10.4	Calomicrolaimus	350.8	13.2
Ptycholaimellus	250.3	7	Daptonema	181.5	6.8
Camacolaimus	147	4.1	Thallassomonhystera	165.1	6.2
Diplolaimella	111	3.1	Anoplostoma	132	5
Calyptronema	104.5	2.9	Axonolaimus	95.7	3.6
Tripyloides	95.1	2.7	Oncholaimus	79.7	3
Paracanthonchus	86.2	2.4	Tripyloides	55.3	2.1
Daptonema	85.1	2.4	Praecanthonchus	52.3	2
February 2004			**February 2004**		
Calomicrolaimus	1165.8	27	Dichromadora	2071.6	41.5
Halalaimus	1087.3	25.2	Leptolaimus	1081	21.6
Spilophorella	459.4	10.6	Daptonema	499.8	10
Ptycholaimellus	252.2	5.8	Calomicrolaimus	325.4	6.5
Paracanthonchus	212.3	4.9	Thallassomonhystera	222.7	4.5
Xyalidae gen	140	3.2	Anoplostoma	190.1	3.8
Diplolaimella	135.6	3.1	Haliplectus	66.8	1.3
Daptonema	117.7	2.7	Diplolaimella	54.3	1.1
Calyptronema	117.2	2.7	Syringolaimus	49.4	1
Atrochromadora	106.8	2.5	Camacolaimus	48.6	1

the pattern of change in mean nematode abundance also showed a distinct difference between the two estuaries (Figure 14.1A). At Authie Port, abundance increased steadily to a peak of >15 × 10^6 individuals per square meter in August 2003. Abundance then crashed to its lowest value at that location in November 2003, before recovering in February 2004 to a level similar to that recorded in February 2003.

Tukey's *post hoc* test revealed abundance in August 2003 to be significantly higher than at all other times, and abundance in May 2003 to be significantly higher than in November 2003 and February 2004. Nematode abundance at Pont de Normandie was similar to that found at Authie Port in February 2003, November 2003, and February 2004. The difference between the two sites was that nematode abundance was generally much more stable at Pont de Normandie. Instead of the peak in abundance recorded at Authie Port in late spring (May 2003) and late summer (August 2003), abundance at Pont de Normandie appeared to be suppressed in August 2003. Tukey's *post hoc* test revealed abundance in both August and November 2003 to be significantly lower than in February 2003.

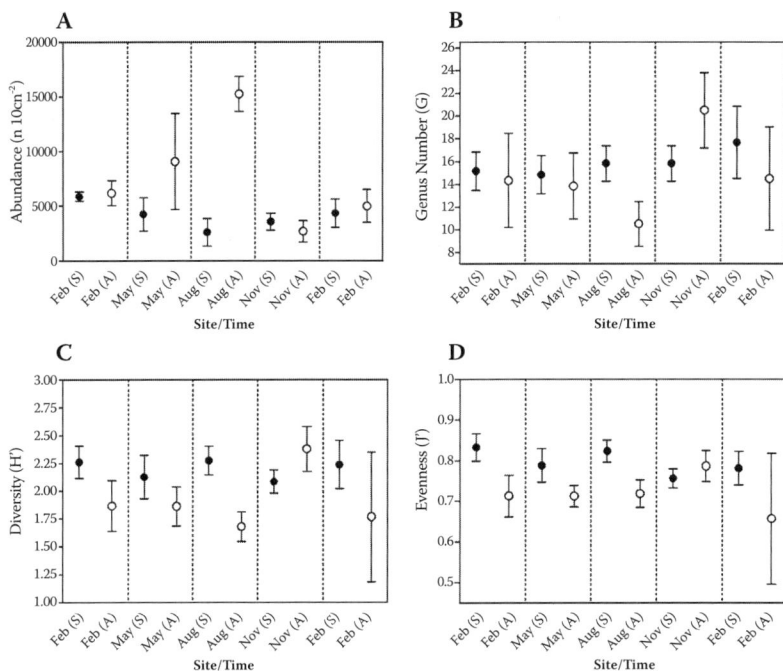

FIGURE 14.1 Nematoda: Univariate measures of diversity at Pont de Normandie (black circles/(S)), and Authie Port (open circles/(A)) from the February, May, August, November 2003, and February 2004 surveys. A = Abundance (N 10 cm^{-2}). B = genus number (G). C = Diversity (Shannon H′). D = Evenness (Pielou J′). Values are means (n = 6) with 95% confidence intervals.

TABLE 14.2
Results of ANOVA of Nematode Abundance and Univariate Measures of Diversity on Five Sampling Occasions at Pont de Normandie and Authie Port

	Pont de Normandie					Authie Port			
	F	R^2	p	sig		F	R^2	p	sig
N	7.44	0.545	0.0004	**	N	29.21	0.824	<0.0001	***
G	1.94	0.237	0.1346	n.s.	S	7.08	0.531	0.0006	**
H′	1.77	0.221	0.1658	n.s.	H′	5.10	0.449	0.0038	*
J′	5.67	0.476	0.0022	**	J′	2.19	0.260	0.0992	n.s.

Note: N = abundance. G = genus number. H′ = Shannon–Weiner diversity. J′ = evenness. Significance levels (sig): NS = not significant (p >0.05). * p = <0.05. ** p = <0.001. *** p = <0.0001.

Patterns of nematode diversity measured by mean genus number (G), Shannon–Weiner diversity index (H'), and Pielou's evenness index (J') also revealed differences between the two estuaries. At Pont de Normandie, diversity measures (H' and G) were rather stable, showing no distinct pattern and ANOVA confirmed the absence of significant differences. As in the case of nematode abundance values, a more dynamic pattern was observed at Authie Port and ANOVA confirmed significant differences in both G and H'. Both parameters were somewhat lower than at Pont de Normandie on the first three sampling campaigns (February, May, and August 2003), particularly in August 2003 when the peak in numbers, particularly of *Daptonema* as noted above, led to a drop in diversity. It is important to remember that at least some of this drop may have been due purely to mathematical artefacts of diversity indices; when identifying only 100 individuals from a sample dominated by a few taxa, it is statistically less likely that rarer species that may well be present will be encountered. The crash in nematode abundance at Authie Port in November 2003 was accompanied by an increase in diversity such that on this occasion, diversity was actually somewhat higher than at Pont de Normandie.

It is also notable that in November 2003, the highest genus number of all the surveys was recorded at Authie Port, indicating that beta diversity may be higher at this location than is usually represented by measures of alpha/sample diversity. However, by February 2004, diversity at Authie Port had again fallen back to a level similar to that in February 2003. Tukey's *post hoc* test revealed that G was significantly higher in November 2003 than at all other sampling times and that H' was significantly higher in November 2003 than in August 2003 and February 2004.

14.3.3   Genus Assemblage Structure

Figures 14.2 and 14.3 show the results of multivariate analyses of both untransformed and root–root transformed nematode genus data. In both analyses, the results were broadly similar, indicating a strong multivariate signal in both the dominant genera (untransformed data) and the less abundant genera (root–root-transformed data).

Clearly, as indicated by the differences in relative abundances of the most numerous nematode taxa, the results of all multivariate analyses indicated that the nematode assemblages of the Pont de Normandie and Authie Port were distinctly different. This could be seen as a clear separation of samples from the two estuaries in both the dendrograms and nMDS ordinations. ANOSIM tests showed this difference to be highly statistically significant (global R values = 0.927 and 0.959; p = 0.001 and 0.001 for untransformed and root–root-transfomed data, respectively).

In the analysis of untransformed data (Figures 14.2a and b) in which the relative abundance of the most dominant genera made the greatest contribution to the resulting dendrogram and ordination, the samples from Pont de Normandie demonstrated a reasonable degree of separation according to the sampling season, although the samples from May and August 2003 showed a rather high degree of overlapping variability, and those from the two February sampling campaigns (2003 and 2004) did not seem to show any great differences. ANOSIM results confirmed these observations, indicating that at Pont de Normandie, the nematode assemblages on each

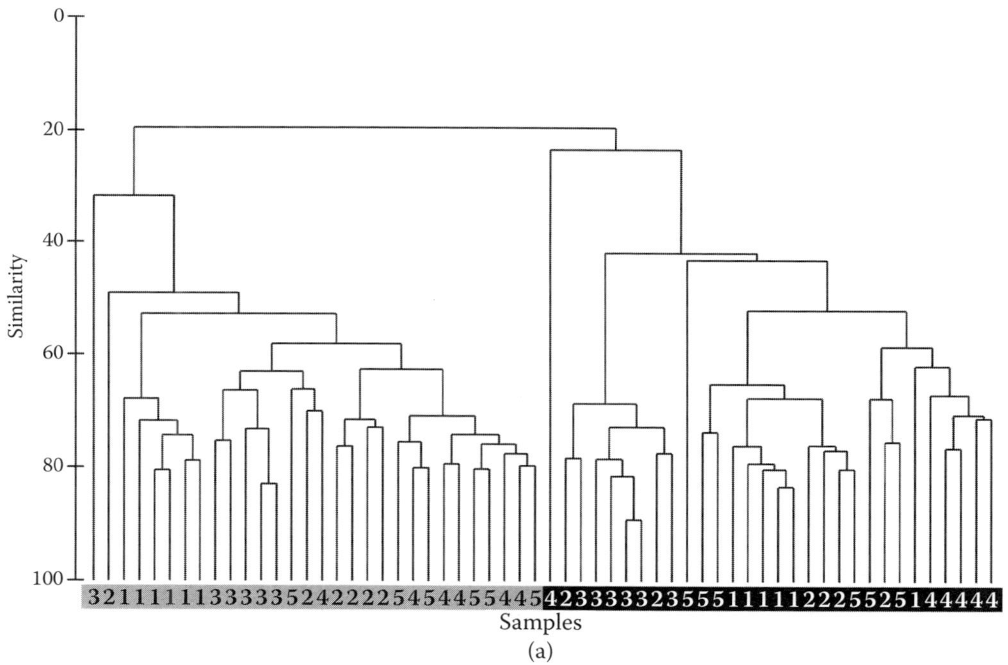

FIGURE 14.2 Multivariate analysis of untransformed nematode genus abundance data. (a) Group average clustering dendrogram of Bray Curtis similarities. (b) Two-dimensional nonmetric multidimensional scaling (nMDS) ordination. Coding Key: 1 = February 2003, 2 = May 2003, 3 = August 2003, 4 = November 2003, 5 = February 2004. Black shading = Pont de Normandie. Grey shading = Authie Port.

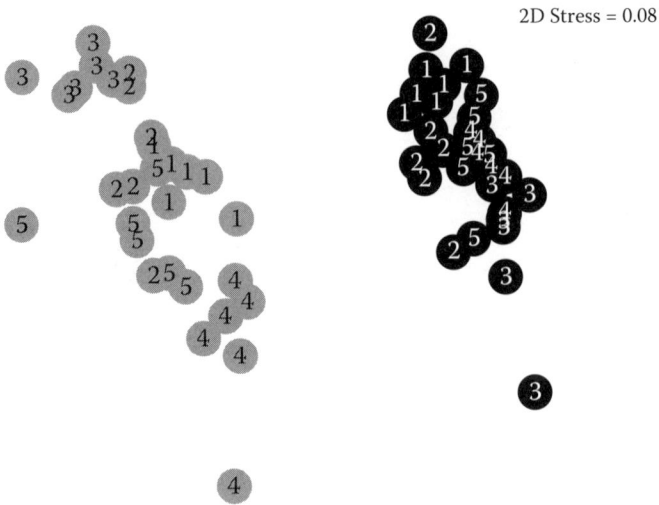

(b)

FIGURE 14.2   (continued).

sampling occasion were significantly different (global $R = 0.647$; $p = 0.001$) with the exception of the last two sample series taken in November 2003 and February 2004.

At Authie Port, the results were similar. Samples showed relatively good grouping according to sampling season. Some variability overlapping was noted, particularly in the samples from May 2003. However, in contrast, the samples from the last two campaigns showed a greater degree of separation; this observation was confirmed by the results of ANOSIM analysis, which showed significantly different nematode assemblages on each of the five sampling seasons (global $R = 0.661$; $p = 0.001$). The results for the Authie Port also indicated a greater degree of assemblage change over the entire study. This may be seen most clearly in the nMDS ordination in which samples appeared to occupy a greater "multivariate space" than those from the Pont de Normandie. The effect of the late summer (August 2003) peak, and the subsequent population crash and assemblage change in late autumn (November 2003), may also be seen in the great distance in the ordination between the samples from these surveys.

In the analysis of root–root transformed data (Figures 14.3a and b) in which the contribution of the dominant species was reduced and patterns of abundance of the rarer species could influence the outcome, the results were broadly similar but with some differences in detail. In general, the analyses showed slightly fewer clear differences between samples from the different campaigns and the stress value of the nMDS ordination was rather high, indicating a significant degree of distortion. This was particularly notable in the analysis of the Authie Port samples in which fewer distinct patterns of separation could be found. However, the overall variation remained higher than that at Pont de Normandie and some of the distortion was

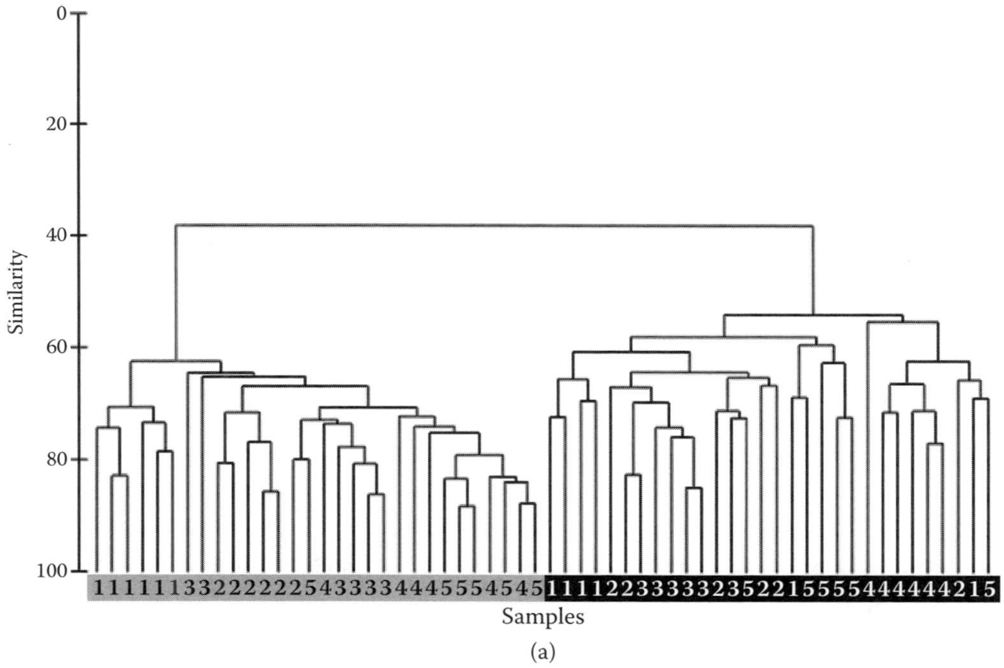

FIGURE 14.3 Multivariate analysis of root–root transformed nematode genus abundance data. (a) Group average clustering dendrogram of Bray Curtis similarities; (b) Two-dimensional nonmetric multidimensional scaling (nMDS) ordination. Coding Key: 1 = February 2003, 2 = May 2003, 3 = August 2003, 4 = November 2003, 5 = February 2004. Black shading = Pont de Normandie. Grey shading = Authie Port.

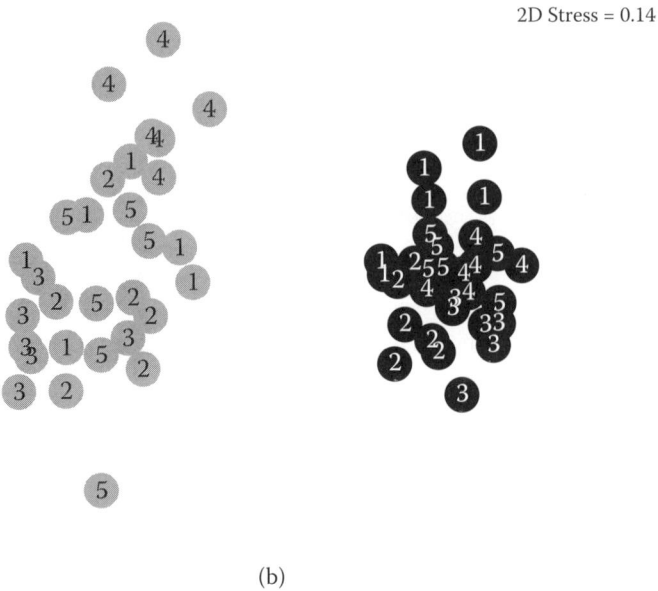

(b)

FIGURE 14.3 (continued).

misleading since subsequent ANOSIM indicated that most nematode assemblages from the different sampling occasions were, in fact, statistically significantly different (global R = 0.498; p = 0.001).

The exception was the high assemblage variability in samples from the May 2003 survey, which was also apparent in the analysis of untransformed data; the result was no significant difference between these samples and those from the August and November 2003 samplings. The fact that the nematode assemblages from Authie Port were generally more dominated by a few species may be significant. As there were fewer individuals remaining from the 100 identified specimens from each sample, the multivariate signal contribution from the rarer species may have been low in information content.

The samples from Pont de Normandie again showed less overall variability, but in contrast to the results for the Authie Port samples, the root–root transformation appeared to have led to further discrimination of sampling seasons as subsequent ANOSIM analysis showed that the nematode assemblages were significantly different on each of the five sampling seasons (global R = 0.647; p = 0.001).

14.3.4 GENUS ABUNDANCE PATTERNS

Figure 14.4 shows the abundance of results for the ten most abundant nematode genera recorded over the entire study and provides a useful illustration of seasonal assemblage changes at the two locations during the study. The overall difference between Pont de Normandie and Authie Port was illustrated by differences in the

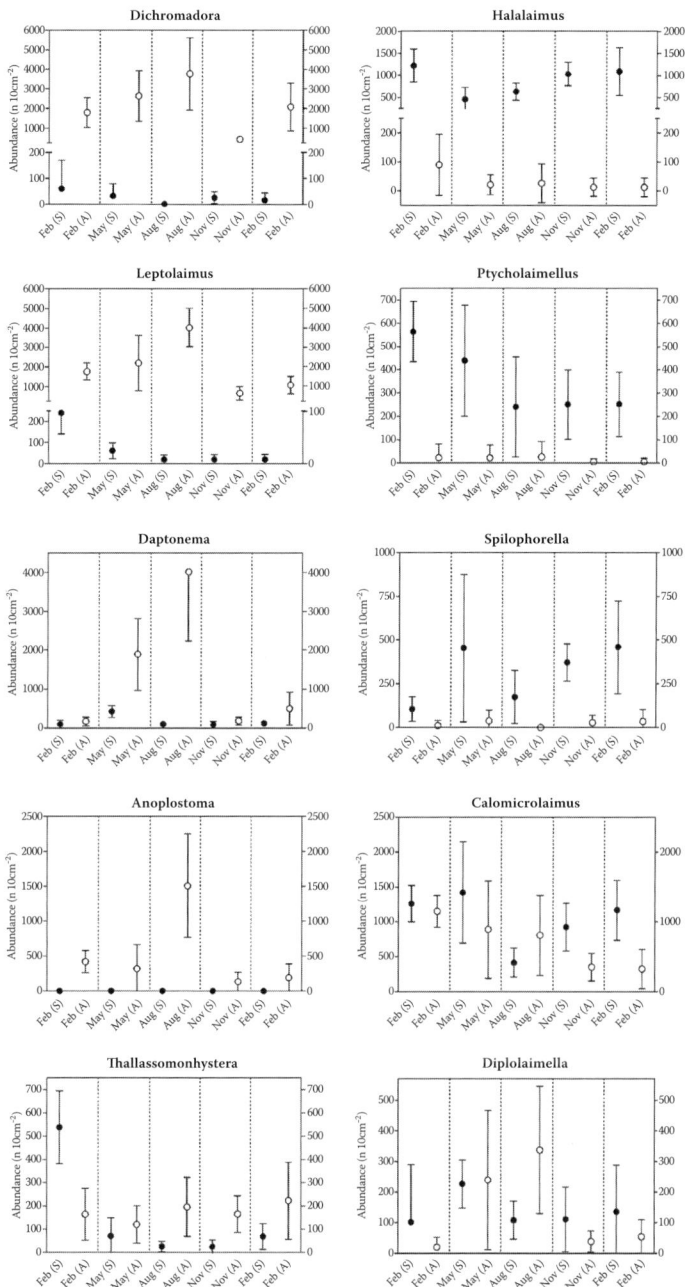

FIGURE 14.4 Mean abundance 10 cm^{-2} (n = 6) and 95% confidence intervals of the ten most abundant nematode genera encountered over the entire PNETOX study on five sampling occasions (February, May, August, November 2003, and February 2004). Black circles/(S) = Pont de Normandie. Open circles/(A) = Authie Port.

abundance of *Dichromadora, Leptolaimus, Daptonema,* and *Anoplostoma,* all genera significantly more abundant at Authie Port.

These four genera also showed a distinct summer peak in numbers that was not observed at Pont de Normandie. Three genera, *Halalaimus, Ptycholaimellus,* and *Spilophorella,* were significantly more abundant at Pont de Normandie, but their patterns of abundance showed no sign of a summer peak, which again demonstrated the difference between the two locations.

Even in genera in which abundance values were more similar at both locations (*Calomicrolaimus, Thalassomonhystera,* and *Diplolaimella*) no genus showed evidence of summer population increases at Pont de Normandie; even genera such as *Diplolaimella,* which although not dominant at Authie Port, still showed evidence of a summer increase in abundance. Overall abundance at Pont de Normandie exhibited a pattern of slight depression in the summer (Figure 14.1A), but this pattern was not clearly shown by many of the genera, with the exception of *Calomicrolaimus,* which showed a significant depression in abundance in August 2003. However, relatively low summer abundance occurred commonly in many Pont de Normandie genera, even if individual patterns were not clear.

14.4 DISCUSSION

The absence of significant numbers of meiofaunal taxa other than the Nematoda and to a lesser degree, the Copepoda, from many areas within estuarine systems limits the range of analyses that may be performed and the usefulness of any conclusions drawn from data based on the simple enumeration of meiofaunal major taxa. In practice then, for meiofauna to be employed for bioassessment or biological monitoring, it is necessary to commit to working with nematodes or copepods, and to work at the highest taxonomic resolution that can be achieved: at least the family level and preferably at the genus or species level.

The results of this series of surveys have shown that only the nematodes occurred at high abundance and diversity, and thus provided data with information content sufficient for meaningful statistical analysis. It is possible that data on other meiofaunal taxa may refine the conclusions cited here, but it is most likely that different collecting and processing strategies would be required to ensure that enough specimens were encountered. In view of the extra effort required, it seems that in most estuarine systems, studies focused on the nematode fauna may provide the best return in terms of their utility for ecological monitoring for the required input of sampling and processing effort (Coull 1999).

The aim of this study was to use patterns of nematode abundance, diversity, and genus assemblage structure to assess the potential impacts of conditions in the multi-polluted Seine estuary in comparison with the relatively unpolluted Authie estuary. This aim rested on the assumption that the site at Authie Port would represent an adequate "control" for comparison with the "impact" site at Pont de Normandie. In terms of two of the principal abiotic parameters, salinity and sediment grain size, the similar values recorded from both estuaries (Table 1.1 in Chapter 1) suggest that this criterion had been substantially met. However, as Underwood (1992) pointed out in his analysis of the BACI (before–after control–impact) model for the detection of environmental impacts, the adoption of a single control site always presents limitations as one can never be certain that an unaccounted or stochastic factor may be responsible for any observed effects or even mask true effects.

Underwood (1992) recommends the use of multiple control sites, but this is not always possible for purely pragmatic reasons of cost and time. In this study, although a single control site was used, the model was considerably strengthened by the adoption of a series of replicated sampling occasions over five seasonally scaled time periods. This design significantly reduces the likelihood that a single stochastic event could compromise the interpretation of the results.

Patterns of nematode abundance and diversity were significantly different over the study period. Authie Port was characterised by more variable abundance due to a strong summer abundance peak and more variable diversity characterised by generally lower sample diversity and higher dominance. However, the highest alpha diversity values over the entire study period recorded at Authie Port (November 2003), and the greater number of taxa recorded at Authie Port than at Pont de Normandie (53 versus 43) indicate potentially higher alpha diversity at Authie Port.

The nematode assemblages at the two study sites were composed from essentially the same suite of common estuarine genera. However, patterns of individual genus abundance and assemblage structure were significantly different, and multivariate analysis confirmed that each location maintained a distinct series of nematode assemblages throughout the study period. The results also indicated significant seasonal differences in nematode assemblage structure. Once again, multivariate analysis revealed that changes in assemblage structure between sampling occasions were generally greater at Authie Port and that the samples occupied more "multivariate space" than did those from Pont de Normandie. These results led to a general conclusion that a more dynamic system existed at Authie Port during the study period.

Certainly, peaks in nematode abundance have been recorded frequently (see review in Heip et al. 1985), although the timing and number of peaks in a year appear to be variable. For example, Skoolmun and Gerlach (1971) found distinct peaks in abundance in May and June and another in October, while Coull (1985), as part of an 11-year study, found a spring peak in average nematode abundance in muddy sediments, but not in sandy sediments. Bouwman et al. (1984) found that the timing and number of abundance peaks differed along a transect of sites subject to different levels of organic enrichment. The temporal resolution of the present study was probably not fine enough to accurately determine the exact time of the peak in abundance recorded at Authie Port since abundance appeared elevated in May as well as in August when the highest values were recorded. However, the pattern is consistent with a late spring–summer peak and is in contrast to the pattern shown at Pont de Normandie, where abundance was rather stable and somewhat suppressed in the middle of the year.

If pollution loading was a significant factor in causing the observed differences in nematode assemblages between Pont de Normandie and Authie Port, perhaps the natural tendency for nematode populations to peak and crash was suppressed, particularly during the warmer months when chronic pollution effects may be enhanced as metabolic rates rise. This may explain why nematode assemblages were more dominated and variable at Authie Port, such a condition representing the "natural" state for an estuary.

Estuaries are generally dynamic environments where food is not limited, and spatial and temporal variability can be extreme on a number of scales. It is possible that higher levels of chronic pollution may serve to suppress this variability while not causing extreme effects. However, these results do not clearly indicate that the Pont de Normandie site is under particularly strong direct pollution stress. With the exception of the abundance peak at Authie Port, patterns of abundance were not greatly different, and diversity, although more variable at Authie Port, did not indicate that one site was distinctly more diverse than the other.

The difference in assemblage structure between the Seine and Authie sites is compelling as evidence that the two systems are subject to very different ecological pressures despite similarities in salinity, sediment granulometry, and organic loading. It is possible that this reflects the combined effects of the chronic levels of multiple pollutants found in the Seine. However, it remains possible that some other environmental or stochastic factor was operating. It is clear that the Pont de Normandie site, situated on the much larger Seine estuary, is subject to a far more energetic and variable hydrodynamic regime, coupled with significant and rapid periods of sediment erosion and deposition (Chapter 2). Such a regime may represent physical disturbance for the nematode assemblages and result in significant impacts on population growth and diversity (Huston, 1994).

Some important biological differences between the two sampling sites were also noted and may have contributed to some of the differences in the nematode fauna. During sampling, the abundance of *Nereis diversicolor* appeared significantly higher at Authie Port than at Pont de Normandie. This observation was confirmed by Gillet et al. (Chapter 8) who recorded significantly higher *N. diversicolor* abundance within the Authie system. Reise (1979) recorded significant predation effects on meiofauna due to the presence of *N. diversicolor*, and Tita et al. (2000) found in a mesocosm study that the related polychaete *N. virens* caused a significant reduction in nematode abundance in the surface 0 to 1 cm of sediment that they ascribed to disturbance and predation resulting from the mainly nonselective feeding behaviour of *N. virens*.

The authors also found some effects on diversity, mainly in the upper sediment layers, and assemblage structure. Ferrero (1992) conducted similar experiments, employing an *in situ* field caging experiment during the winter and summer, using *N. diversicolor*. Suppression of nematode abundance was not as great as found by Tita et al. (2000) and was most significant in the 1- to 2-cm sediment depth layer. However, Ferrero (1992) also found that *N. diversicolor* could influence both nematode diversity and assemblage structure. In contrast, Reise (1981) found that the subsurface tubes of polychaetes could significantly promote meiofaunal numbers through the creation of micro-oxic regions in anoxic sediment layers, and Kennedy (1993) found no significant predation effects on meiofauna in a field experimental study.

Coull (1999) concluded that, although macrofauna prey on meiofauna, their effect is seldom sufficient to affect population numbers. The patterns of *N. diversicolor* abundance found at the Authie Port do not seem to suggest that they are particularly responsible for a suppression of nematode abundance since the greatest populations of both *N. diversicolor* and nematodes coincided in August 2003. If predation was a significant factor, what is perhaps more significant is that the comparative absence of *N. diversicolor* at Pont de Normandie did not seem to be associated with increased nematode abundance. However, if the nematode populations were significantly stimulated by subsurface burrow structures, then the effect at Pont de Normandie could be reduced oxygenation of anoxic sediments during the warmer summer months, with a potential suppressive effect on nematode abundance. As the core samples employed in this study were not sectioned, further studies would have to be conducted to explore these interactions.

The abundance of benthic diatoms (Chapter 12) was also found to be higher at Authie Port than at Pont de Normandie during the May and August 2003 sampling occasions when nematode abundance was also elevated. Diatoms are important as food sources for many nematodes and higher food availability and abundance of some genera may be linked. Of the genera that contributed significantly to the peak in abundance seen over May and August 2003, two are of particular interest in this context. Species of *Dichromadora* (Figure 14.4) are generally classified as group 2A epigrowth (diatom) feeders under the scheme of Wieser (1953), based on their possession of small teeth in the buccal cavity. Species of *Daptonema* (Figure 14.4) are generally recorded as belonging to group 1B nonselective deposit feeders under the same scheme, but the relation of nematode buccal cavity structure to food source is now known to be far more complex (see Moens and Vincx 1997), and in fact many species of *Daptonema* can be observed with diatoms packing the gut after being ingested whole (personal observation).

The degree to which these abiotic and biological differences between the two estuaries may be responsible for the observed differences in nematode assemblages, compared with effects that may be directly attributable to pollution, cannot be quantified without further study, and it is likely that differences in biological parameters (*N. diversicolor* and diatom populations) are also affected by pollution.

The results of this study demonstrate the challenges inherent in attempting to study the relatively low level, multi-source chronic pollution to which most major estuaries in Europe are affected to a greater or lesser extent. However, they also demonstrate the sensitivity of nematode assemblages to both natural and anthropogenically modified environmental change and the range of different responses this important component of estuarine fauna can display.

ACKNOWLEDGMENTS

The author wishes to acknowledge the important contributions made to this work by the following people, through their assistance with sampling, sample processing, meiofaunal enumeration, and mounting of nematodes: Natalie Barnes, Graham Bennell, Kat Birch, Lisa King, Andreia Salvador, Brian Smith, and Richard Williams. Thanks also to Phil Rainbow for helpful comments on the manuscript.

REFERENCES

Austen, M.C. and P.J. Somerfield. 1997. A community level sediment bioassay applied to an estuarine heavy metal gradient. *Mar. Environ. Res.* 43: 315–328.

Beyrem, H. et al. 2007. Individual and combined effects of cadmium and diesel on a nematode community in a laboratory microcosm experiment. *Ecotox. Environ. Saf.* 68: 412–418.

Bogaert, T., M.R. Samoiloff, and G. Persoone. 1984. Determination of the toxicity of four heavy metal compounds and three carcinogens using two marine nematode species, *Monhystera microphthalma* and *Diplolaimelloides bruciei*. In *Proceedings of International Symposium on Ecotoxicological Testing for the Marine Environment, Ghent, Belgium,* September 1983, Vol. 2, Persoone, G. et al., Eds. Gent, Belgium: State University, pp. 21–30.

Bouwman, L.A., K. Romeijn, and W. Admiraal. 1984. On the ecology of meiofauna in an organically polluted estuarine mudflat. *Estuar. Coast. Shelf Sci.* 19: 633–653.

Coull, B.C. 1985. Long-term variability of estuarine meiobenthos: an 11-year study. *Mar. Ecol. Prog. Ser.* 24: 205–218.

Coull, B.C. and G.T. Chandler. 1992. Pollution and meiofauna: field laboratory and mesocosm studies. *Oceanogr. Mar. Biol. Ann. Rev.* 30: 191–271.

Coull, B. C. 1999. Role of meiofauna in estuarine soft-bottom habitats. *Aust. J. Ecol.* 24: 327–343.

Essink, K. 2003. Response of an estuarine ecosystem to reduced organic waste discharge. *Aquat. Ecol.* 37: 65–76.

Ferrero, T.J. 1992. The effect of biological activities and biogenic structures on free-living estuarine nematode assemblages. Ph.D. thesis, University of Bristol.

Giere, O. 1993. *Meiobenthology: The Microscopic Fauna in Aquatic Sediments.* Berlin: Springer-Verlag.

Gyedu-Ababio, T.K. and D. Baird. 2006. Response of meiofauna and nematode communities to increased levels of contaminants in a laboratory microcosm experiment. *Ecotox. Environ. Saf.* 63: 443–450.

Heip, C. 1980. Meiobenthos as a tool in the assessment of marine environmental quality. *Rapp. P. Réun. Cons. Int. Explor. Mer* 179: 182–187.

Heip, C., M. Vincx, and G. Vranken. 1985. The ecology of marine nematodes. *Oceanogr. Mar. Biol. Ann. Rev.* 23: 399–489.

Howell, R. 1984. Acute toxicity of heavy metals to two species of marine nematodes. *Mar. Environ. Res.* 11: 153–161.

Higgins, R.P. and H. Thiel. 1988. *Introduction to the Study of Meiofauna.* Washington: Smithsonian Institution Press.

Huston, M.A. 1994. *Biological Diversity: The Coexistence of Species on Changing Landscapes.* Cambridge: Cambridge University Press.

Kelaher, B.P. et al. 2003. Changes in benthos following the clean-up of a severely metal-polluted cove in the Hudson River estuary: environmental restoration or ecological disturbance? *Estuaries* 26: 1505–1516.

Kennedy, A.D. 1993. Minimal predation upon meiofauna by endobenthic macrofauna in the Exe Estuary, south west England. *Mar. Biol.* 117: 311–319.

Lee, M.R. and J.A. Correa. 2007. An assessment of the impact of copper mine tailings disposal on meiofaunal assemblages using microcosm bioassays. *Mar. Environ. Res.* 64: 1–20.

Lorenzen, S. 1974. Die Nematodenfauna der sublitoralen Region der Deutschen Bucht, insbesondere im Titan-Abwassergebiet bei Helgoland. *Veröff. Inst. Meeresforsch. Bremerh.* 14: 305–327.

Mare, M.F. 1942. A study of a marine benthic community with special reference to the microorganisms. *J. Mar. Biol. Assoc. U.K.* 25: 517–554.

Millward, R.N. et al. 2004. Mixtures of metals and hydrocarbons elicit complex responses by a benthic invertebrate community. *J. Exp. Mar. Biol. Ecol.* 310: 115–130.

Millward, R.N. and A. Grant. 1995. Assessing the impact of copper on nematode communities from a chronically metal-enriched estuary using pollution-induced community tolerance. *Mar. Pollut. Bull.* 30: 701–706.

Moens, T. and M. Vincx. 1997. Observations on the feeding ecology of estuarine nematodes. *J. Mar. Biol. Assoc. U.K.* 77: 211–227.

Moore, C.G. 1987. Meiofauna of the industrialised estuary and Firth of Forth, Scotland. *Proc. R. Soc. Edinburgh* 93B: 415–430.

Moore, C.G. and B.J. Bett. 1989. The use of meiofauna in marine pollution impact assessment. *Zool. J. Linn. Soc.* 96: 263–280.

Platt, H.M., and R.M. Warwick. 1980. The significance of free-living nematodes to the littoral ecosystem. In *The Shore Environment: Ecosystems*, Price, J.H. et al., Eds. New York: Academic Press, pp. 729–759.

Reise, K. 1979. Moderate predation on meiofauna by the macrobenthos of the Wadden Sea. *Helgol. Wissensch. Meeresunters.* 32: 453–465.

Reise, K. 1981. High abundance of small zoobenthos around biogenic structures in tidal sediments of the Wadden Sea. *Helgol. Wissensch. Meeresunters.* 34: 413–425.

Schratzberger, M. et al. 2002. Effects of paint-derived tributyltin on structure of estuarine nematode assemblages in experimental microcosm. *J. Exp. Mar. Biol. Ecol.* 272: 217–235.

Skoolmun, P. and S.A. Gerlach. 1971. Jahreszeitliche Fluktuationen der nematodenfauna im gezeitenbereich des Weser-Ästuars (Deutsche Bucht). *Veröff. Inst. Meeresforsch. Bremerh.* 13: 119–138.

Somerfield, P.J., J.M. Gee, and R.M. Warwick. 1994. Soft sediment meiofaunal community structure in relation to a long-term heavy metal gradient in the Fal Estuary. *Mar. Ecol. Prog. Ser.* 105: 79–88.

Swedmark, B. 1964. The interstitial fauna of marine sand. *Biol. Rev.* 39: 1–42.

Tietjen, J.H. 1977. Population distribution and structure of the free-living nematodes of Long Island Sound. *Mar. Biol.* 43: 123–136.

Tietjen, J.H. 1980. Population structure and species composition of the free-living nematodes inhabiting sands of the New York Bight apex. *Estuar. Coast. Mar. Sci.* 10: 61–73.

Tita, G. et al. 2000. Predation and sediment disturbance effects of the intertidal polychaete *Nereis virens* (Sars) on associated meiofaunal assemblages. *J. Exp. Mar. Biol. Ecol.* 243: 261–282.

Underwood, A.J. 1992. Beyond BACI: the detection of environmental impacts on populations in the real, but variable, world. *J. Exp. Mar. Biol. Ecol.* 161: 145–178.

Verschraegen, K. et al. 1986. Comparison of demographic (life and fecundity table analysis) and biochemical (ATP and AEC) characteristics as sublethal pollution indices in the marine nematode *Monhystera disjuncta*, In *Ecology, Ecotoxicology and Systematics of Marine Benthos*, Heip, C. and Coomans, A., Eds. Gent: Laboratorium voor Morfologie en Systematiek der Dieren Sektie Mariene Biologie Rijksuniversiteit Gent, pp. 145–172.

Vincx, M. and C. Heip. 1987. The use of meiobenthos in pollution monitoring studies. A review. *ICES CM E* 33L: 1–18.

Vranken, G. and C. Heip. 1986. Toxicity of copper, mercury and lead to a marine nematode. *Mar. Pollut. Bull.* 17: 453–457.

Vranken, G., C. Tire, and C. Heip. 1988. The toxicity of paired metal mixtures to the nematode *Monhystera disjuncta* (Bastian, 1865). *Mar. Environ. Res.* 26: 161–179.

Vranken, G., R. Vanderhaeghen, and C. Heip. 1985. Toxicity of cadmium to free-living marine and brackish water nematodes (*Monhystera micropthalma, Monhystera disjuncta, Pelliodotis marina*). *Dis. Aquat. Org.* 1: 49–58.

Warwick, R.M. 1989. The role of meiofauna in the marine ecosystem: evolutionary considerations. *Zool. J. Linn. Soc.* 96: 229–241.

Warwick, R.M. and R. Price. 1979. Ecological and metabolic studies on free-living nematodes from an estuarine mudflat. *Estuar. Coast. Mar. Sci.* 9: 257–271.

Wieser, W. 1953. Die Beziehung zwischen Mundhöhlengestalt Ernährungsweise und Vorkommen bei freilebenden marinen Nematoden. *Ark. Zool.* 4: 439–484.

15 Dynamic Diagenetic Modelling and Impacts of Biota

Lionel Denis, Dominique Boust, Bénedicte Thouvenin, Pierre Le Hir, Julien Deloffre, Jean-Louis Gonzalez, and Patrick Gillet

CONTENTS

Particulate and dissolved exchanges between pelagic and benthic compartments have served as the foci of numerous studies to enhance our understanding of the global ocean cycling of carbon and associated elements. The importance of benthic–pelagic coupling has been widely recognized (Graf 1992) since the empirical model developed by Hargrave (1973) and the work of Rowe et al. (1975), Hinga et al. (1979), and Graf et al. (1982). Since the late 1970s, understanding of the processes of mineralization of organic matter in surficial sediments (the so-called diagenetic sequence, Froelich et al. 1979) has led to the development of several models, a domain primarily investigated by Berner (1980).

From simple early formulations, models have become increasingly more complicated in various ways. For instance, the composition of total organic matter has commonly been considered to be partitioned into multiple (usually two or three) reactivity classes that are respired at contrasting rates: for example, fast decaying organic matter, slow decaying organic matter, and refractory organic matter (multi G models, Jørgensen 1978; Westrich and Berner 1984; Boudreau 1996; Soetaert et al. 1996). Some models, however, have also considered a continuum between fast decaying and refractory organic matter (Middelburg 1989; Boudreau and Ruddick, 1991).

Numerous modelling studies have described applications of the models identified above (Boudreau et al. 1998; Herman et al. 2001); other papers reported additions or simplifications of these models (Luff et al. 2000; Wijsman et al. 2002).

Such development of relevant models has resulted from the increasing accuracy of field measurements as a consequence of methodological developments, from sediment sampling or processing using innovative techniques, to improvements of analytical precision (Viollier et al. 2003). As a consequence, spatial and temporal variabilities have been described at finer and finer scales. For deep-sea sediments, temporal changes are generally much reduced, and barely taken into account due to the spatial heterogeneity of sedimentation processes. However, for coastal environments, temporal variability is one of the main drivers, especially in areas subjected to numerous sources of hydrodynamic forcing (tidal currents, waves, fresh water inputs).

As a consequence, the assumption of steady state commonly adopted in diagenetic models (Dhakar and Burdige 1996; Van Cappellen and Wang 1996) may no longer be valid; measurements now must be considered as transient states between two hypothetical steady states, i.e., equilibrium between environmental forcing, inputs, outputs, and sedimentary processes (hypothetical because steady states may not actually be reached). To our knowledge, few diagenetic models (Boudreau 1996; Soetaert et al. 1996 and their derivatives) are dynamic and calculate a succession of steady states while a temporal recording of organic matter input is used as a forcing function.

In this context, this chapter describes the first stages of the development of a fully dynamic diagenetic model including the temporal variability of sedimentary processes (accumulation, resuspension) as a forcing function. The model was developed on the basis of a water column and sediment model, which first aims to reproduce hydrological parameters and their consequences on surficial muddy sediments. The first calibrations presented in this chapter were performed in the Seine estuary, where numerous previous studies conducted under the Seine Aval project provided a large dataset, in an environment where tidal currents and fresh water inputs constitute the major drivers of the system. More specifically, field data presented here were obtained from the largest intertidal part of the northern mudflat in the Seine estuary (3.2 km²) that was studied extensively by the DYVA project (Lafite et al. 2002). After a comparison of the outputs of the model with field data, sensitivity tests of bioturbation intensities are proposed and discussed.

15.1   GENERAL FUNCTIONING OF THE MODEL

The simulation advection multivariable model (SiAM-1DV) used in this work was developed by IFREMER and represents the sedimentary part (consolidation model) of SiAM-3D applied in the Seine estuary (Le Hir et al. 2000 and 2001; Cugier and Le Hir 2002). This is a non-steady state model suited to reproduce the consequences of erosion and accumulation events in surficial muddy sediments. The SiAM-1DV version used in this study focuses on processes in sediments (consolidation, diffusion and dispersion, chemical reactions) and exchanges at the interface with the overlying water. The sediment column is split into thin layers (typically 2 mm thick) and consolidation is deduced from a mass exchange between layers due to specific rates of sedimentation. This settling rate is assumed to depend only on local porosity, according to a (series of) power laws (Equation 15.1):

$$\omega_s = Xv_1 \times \phi^{Xv_2} \qquad (15.1)$$

where Xv_1 and Xv_2 are coefficients that depend on porosity. As the model is adapted for muddy sediments, compaction is neglected if the porosity is <0.65. From the resulting change of concentration in each layer, an upward porewater advection is computed. Then the sediment model is able to simulate the vertical transfer of any particulate or dissolved constituent. The former is advected by settling and possibly diffused by bioturbation, whereas the latter is transported with advecting porewater and diffused by molecular movement or by bioturbation. The diffusion of dissolved species in the interstitial water is simulated; diffusion coefficients of dissolved species are from Schulz and Zabel (2006). Bioturbation processes are described as diffusive processes only (D_b coefficient, see below).

Density and porosity are linked by the relationship (Equation 15.2) that involves grain density:

$$\phi = \left(1 - \frac{C_s}{\rho_s} \right) \qquad (15.2)$$

where C_s is sediment weight per volume and ρ_s, grain density.

When erosion or deposition occurs, corresponding exchanges of particulate or dissolved constituents are accounted for, as any exchange of sediment is accompanied by porewater (depositing material has a given density and thus a given porosity that should be considered as a parameter). It should be noted that due to consolidation and erosion, the upper layer can become thinner or disappear. On the other hand, deposition induces a thickening of the upper layer or the creation of a new one when the layer thickness exceeds a given value that defines the vertical discretisation of the sediment model.

Diagenetic processes have been integrated in the sediment but processes in the water column are not considered in this work. Horizontal transport is actually very strong in this area (tidal dynamics) and a one-dimensional vertical (1DV) model may not be able to reproduce vertical sedimentary processes in the water column such as erosion or deposition cycles without a good estimation of horizontal concentration gradients. Consequently, deposition and erosion events have been artificially introduced into the model and forced from continuous measurements (Altus data, see below).

Starting from an "artificial" sediment column of 20 cm depth, the model was systematically run over 6 months (183 days) with the measured sedimentary inputs or erosion events to set the initial conditions. Then, the run was performed again during the whole period to be modeled, from July 2001 until February 2002, respectively, corresponding to the beginning of Altus measurements and to the last field campaign (DYVA 5; 378 days of run = 183 days for initial conditions + 195 days).

15.2   BIOGEOCHEMICAL REACTIONS

The idealized diagenetic reactions calculated in the model follow the general sequence of Froelich et al. (1979), and primary reactions are consecutive to the input of organic matter (calculated as a constant fraction of particulate input) from the overlying water column. Particulate organic matter (POM) composition follows the Redfield ratios $(CH_2O)_{106}(NH_3)_{16}(H_3PO_4)_1$, and two fractions of organic matter are considered and characterized by different labilities. POM_1 is considered to be fast decaying organic matter; POM_2 is considered to be slow decaying organic matter. Primary diagenetic equations are listed in Table 15.1A and applied to POM_1 and POM_2 with different rate constants. Mineralization processes follow Monod kinetics according to the equations in Table 15.1A. Several secondary reactions (Table 15.1B) are also indirectly implied in organic matter degradation processes. Corresponding state variables calculated in the model are listed in Table 15.2.

Briefly, oxic mineralization produces ammonium (R1, Table 15.1A) which may be further oxidized to nitrate via the nitrification pathway (R9, Table 15.1B). In the suboxic or anoxic zone of the sediment, denitrification (R2, Table 15.1A) may follow and lead to ammonium and nitrogen gas; the latter is not modelled and is considered to be lost from the benthic system. In the field, several studies demonstrated the tight coupling of nitrification and denitrification processes, but such a coupling is not imposed in this model.

Mineralization of organic matter via the reduction of manganese oxides (R3, Table 15.1A) is associated with several secondary reactions, such as the reoxidation of reduced manganese (R7, Table 15.1B) or coupled to the cycling of iron and sulfur, manganese oxides used to reoxidize reduced iron (R10, Table 15.1B) or hydrogen sulfide (R11, Table 15.1B). In a similar way, the reduction of iron oxides (R4, Table 15.1A) is associated with the reoxidation of reduced iron (R6, Table 15.1B) or coupled to the cycling of manganese oxides (R10, Table 15.1B), hydrogen sulfide (R12, Table 15.1B), or iron sulfide (R13, Table 15.1B).

15.3   MAIN FORCING OF RESUSPENSION AND ACCUMULATION PROCESSES

In the case of a fully dynamic model (i.e., no steady state assumption) in estuarine areas, the major driver is the erosion or the accumulation of sedimentary particles that severely modifies the functioning of surficial sediments. On the north mudflat of the Seine estuary, previous studies demonstrated that up to 10 cm of sediment may be eroded within a few days (Deloffre et al. 2005), but accumulation processes may also be very rapid. With the aim of calibrating the sedimentary routines of the model, we worked on a sedimentary sequence as recorded by means of the Altus system (Jestin et al. 1998) between July 2001 and August 2002.

We observed drastic topographic variations in time (Figure 15.1): the north mudflat is marked by sharp accumulation events bringing up to 8 cm of sediment in a few days, as in October 2001. During this survey, erosion processes occurred more regularly and led to more moderate topographic variations. Such variations are directly linked to the complex interactions in the estuary of the highly variable

TABLE 15.1
Primary Reactions Directly Involving Particulate Organic Matter Degradation and Secondary Reactions Calculated in the SiAM 1DV Diagenetic Model

A Primary Reactions Involving Particulate Organic Matter Degradation

	Rate Constant (s^{-1})	
	POM_1	POM_2

R1

Oxic mineralization

$POM_i + (c + 2.25 \, n) \, O_2 \Rightarrow c \, CO_2 + n \, NH_4^+ + p \, H_3PO_4^{2-} + (c + 1.5 \, n) \, H_2O$

$$Sw(R1) = \frac{O2}{(K_{O2} + O2)} = Sw_1^{O2}$$

1.40E-06	7.00E-08

R2

Denitrification

$POM_i + (4 \, c/5) \, NO_3^- \Rightarrow (c/5) \, CO_2 + (4 \, c/5) \, HCO_3 + (2 \, c/5) \, N_2 + nNH_3 + pH_3PO_4^{2-} + (3 \, c/5) \, H_2O$

$$Sw(R2) = \frac{K'_{O2}}{(K'_{O2} + O2)} \frac{NO_3}{\left(K_{NO3} + NO3\right)} = Sw_2^{O2} Sw_1^{NO3}$$

2.80E-08	1.40E-09

R3

Manganese reduction

$POM_i + (2 \, c) \, MnO_2 + (3 \, c) \, CO_2 + cH_2O \Rightarrow (4 \, c) \, HCO_3^- + (2 \, c)Mn^{2-} + nNH_3 + pH_3PO_4^{2-}$

$$Sw(R3) = \frac{K'_{O2}}{(K'_{O2} + O2)} \frac{K'_{O3}}{(K'_{O3} + NO3)} \frac{MnO2}{\left(K_{MnO2} + MnO2\right)} = Sw_2^{O2} Sw_2^{NO3} Sw_1^{MnO2}$$

2.80E-08	1.40E-09

R4

Iron reduction

$POM_i + (4 \, c) \, Fe(OH)_3 + (7 \, c) \, CO_2 \Rightarrow (8 \, c) \, HCO_3^- + (4 \, c) \, Fe^{2+} + nNH_3 + pH_3PO_4^{2-} + (3 \, c) \, H_2O$

$$Sw(R4) = \frac{K'_{O2}}{(K'_{O2} + O2)} \frac{K'_{O3}}{(K'_{O3} + NO3)} \frac{K'_{MnO2}}{(K'_{MnO2} + MnO2)} \frac{Fe(OH)3}{\left(K_{Fe(OH)3} + Fe(OH)3\right)}$$

$$= Sw_2^{O2} Sw_2^{NO3} Sw_2^{MnO2} Sw_1^{Fe(OH)3}$$

5.60E-09	2.80E-10

R5

Sulfate reduction

$POM_i + (0.5 \, c) \, SO_4^{2-} \Rightarrow cHCO_3^- + (0.5 \, c) \, H_2S + nNH_3 + pH_3PO_4^{2-}$

$$Sw(R5) = \frac{K'_{O2}}{(K'_{O2} + O2)} \frac{K'_{NO3}}{(K'_{NO3} + NO3)} \frac{K'_{MnO2}}{(K'_{MnO2} + MnO2)} \frac{K'_{Fe(OH)3}}{(K'_{Fe(OH)3} + Fe(OH)3)} \frac{SO4^{2-}}{\left(K_{SO4} + SO4^{2-}\right)}$$

$$= Sw_2^{O2} Sw_2^{NO3} Sw_2^{MnO2} Sw_2^{Fe(OH)3} Sw_1^{SO4}$$

7.00E-08	3.50E-09

(continued on next page)

TABLE 15.1 (continued)
Primary Reactions Directly Involving Particulate Organic Matter Degradation and Secondary Reactions Calculated in the SiAM 1DV Diagenetic Model

B Secondary Reactions calculated in SiAM 1DV diagenetic model

	Rate constant (mole^{-1}.s^{-1})
R6	
$4Fe^{2+} + O_2 + 8HCO_3^- + 2H_2O \Rightarrow 4Fe(OH)_3 + 8CO_2$	6.34E-02
R7	
$2Mn^{2+} + O_2 + 4HCO_3^- \Rightarrow 2MnO_2 + 4CO_2 + 2H_2O$	6.34E-02
R8	
$H_2S + 2O_2 + 2HCO_3^- \Rightarrow SO_4^{2-} + 2CO_2 + 2H_2O$	1.90E-10
R9	
$NH_4^+ + 2O_2 + 2HCO_3^- \Rightarrow NO_3^- + 2CO_2 + 3H_2O$	4.76E-04
R10	
$MnO_2 + 2Fe^{2+} + 2HCO_3^- + 2H_2O \Rightarrow 2Fe(OH)_3 + 2CO_2 + Mn^{2+}$	3.17E-07
R11	
$MnO_2 + H_2S + 2CO_2 \Rightarrow Mn^{2+} + S^0 + 2HCO_3^-$	3.17E-07
R12	
$H_2S + 2Fe(OH)_3 + 4CO_2 \Rightarrow 2Fe^{2+} + S^0 + 4HCO_3^- + 2H_2O$	3.17E-07
R13	
$FeS + 2Fe(OH)_3 + 6CO_2 \Rightarrow 3Fe^{2+} + S^0 + 6HCO_3^-$	2.80E-09

Note: Two fractions of particulate organic matter (POM$_i$, with i = 1 to 2) were considered and followed the Redfield stoichiometry [$(CH_2O)_c(NH_3)_n(H_3PO_4)_p$; c, n, and p = 106, 16, and 1, respectively]. Rate constants used in the model are given for primary (with POM$_1$ and POM$_2$) and secondary reactions. R13 is reversible (iron sulfide precipitation and dissolution).
Reactions rates POM$_1$ (fast decaying organic matter).
Reactions rates POM$_2$ (slow decaying organic matter).

river discharge (from 100 m^3.s^{-1} during low-water summer periods up to 1500 to 2000 m^3.s^{-1} during winter floods), storms, and tidal flow (maximal tidal range 7 m). Experimental data from Deloffre et al. (2006) have been corrected for dewatering processes formulated according to a logarithmic law (Terzaghi and Peck 1967). Deposition fluxes and erosion events were deduced from the temporal variations of sediment topography issued from the Altus dataset and were used as constraints for the consolidation model SiAM-1DV. The model simulates sediment concentrations, porosity profiles, and variations of sediment topography (plotted in Figure 15.1 with the Altus dataset), and discrepancies between raw topographic data and modelled sediment thickness remain low.

TABLE 15.2
State Variables Calculated in the Model and Corresponding Unit

Name	Unit	Description
Inorganic matter	$kg.m^{-3}$	Mass density of inorganic particles in sediment
POM_1	$mmol.kg^{-1}$	Concentration of fast decaying particulate organic matter
POM_2	$mmol.kg^{-1}$	Concentration of slow decaying particulate organic matter
MnO_2	$mmol.kg^{-1}$	Concentration of manganese oxide
$Fe(OH)_3$	$mmol.kg^{-1}$	Concentration of iron oxide
O_2	$mmol.l^{-1}$	Concentration of oxygen
ΣCO_2	$mmol.l^{-1}$	Concentration of carbonate ($\Sigma CO_2 = CO_2 + HCO_3^- + CO_3^{2-}$)
NO_3^-	$mmol.l^{-1}$	Concentration of nitrate
SO_4^{2-}	$mmol.l^{-1}$	Concentration of sulfate
Fe^{2+}	$mmol.l^{-1}$	Concentration of dissolved iron
ΣH_2S	$mmol.l^{-1}$	Concentration of dihydrogen sulfide ($\Sigma H_2S = H_2S + HS^+$)
Mn^{2+}	$mmol.l^{-1}$	Concentration of manganese
NH_4^+	$mmol.l^{-1}$	Concentration of ammonium

15.4 INITIAL CONDITIONS AND ORGANIC MATTER INPUTS AT THE SEDIMENT–WATER INTERFACE

Field data of particulate organic carbon concentrations in surficial sediments average ($\pm$ standard deviation) 1700 ± 400 mmol.kg wet sediment^{-1} and around 10% is considered as fast or slowly decaying; the remaining fraction considered as refractory (Garnier et al. 2005). In this study, as in the biogeochemistry model in the Seine estuary (Even and Thouvenin 2004; Garnier et al. 2005), we chose an input of organic matter with a ratio of 0.25 between fast and slow decaying organic matter. Hence, concentrations of POM_1 and POM_2 in the sediment are ~30 mmol.kg^{-1} wet sediment and 130 mmol.kg^{-1} wet sediment, respectively.

At the beginning of the simulation, neither POM_1 nor POM_2 was present in surficial sediments. For initial conditions, the reactive fractions of manganese and iron oxides in the sediment were estimated to be 1.5 mmol.kg^{-1} and 95 mmol.kg^{-1}, respectively. Initial concentrations of NO_3, NH_4, H_2S, Fe^{2+}, and Mn^{2+} in the bottom water and interstitial water were set to zero. In the bottom water, oxygen and alkalinity concentrations were respectively set to 0.275 *mM* and 2.2 *mM* while they were set to zero in the interstitial water (Table 15.3).

Initial concentrations of SO_4^{2-} in the interstitial water were also set to zero while bottom water concentrations were calculated from salinity data. Temporal variations of salinity have been calculated with the SiAM 3D model as described in Cugier and Le Hir (2002), and this calculation has been validated against field profiles of sodium in surficial sediments.

15.5 MODEL ADJUSTMENT TO FIELD DATA

To test the validity of the model, we calculated dissolved and particulate profiles obtained on 4 February 2002 (195 days after the starting point on 24 July 2001)

a)

b)

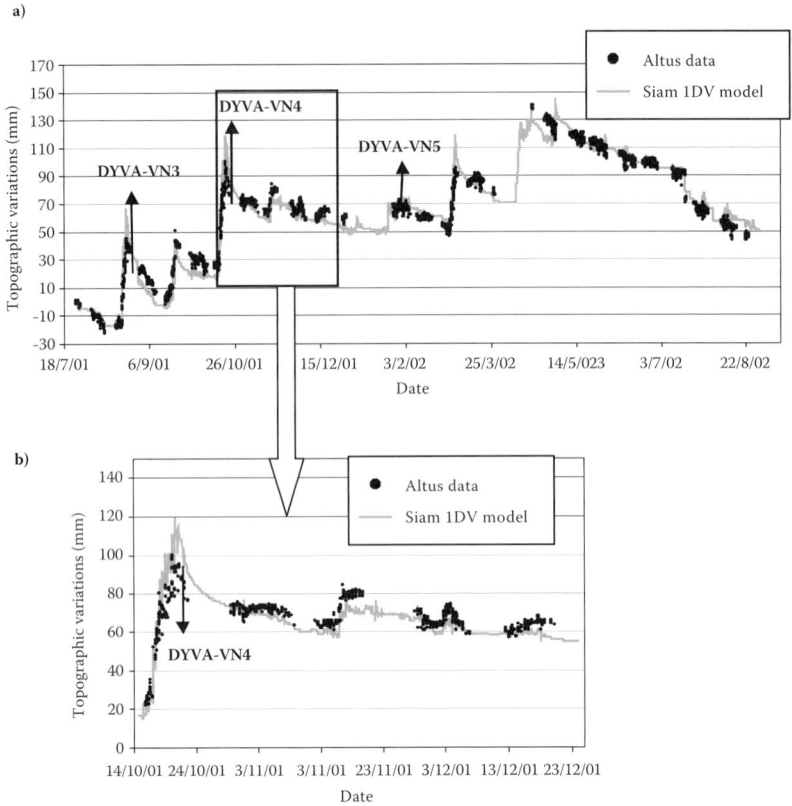

FIGURE 15.1 Temporal variations recorded with the Altus system (corrected for dewatering, temperature, and salinity) and topographic values produced by the Siam 1DV model (a) over the whole period modelled (July 2001 and August 2002) and (b) detailed view of short-term variations after a sedimentation event. The sampling cruises achieved during this period are mentioned.

and compared these results with field data obtained during the DYVA campaigns (Figure 15.2; Boust 1999; Lafite et al. 2002; Bally et al. 2004; Ouddane et al. 2008). After several tests, the best fit was obtained and corresponding parameters are reported in Table 15.3. We adopted reaction rates estimated in the Seine estuary by Garnier et al. (2005) for POM_1 and POM_2, respectively, 1.4×10^{-6} s^{-1} (0.12 d^{-1}) and 7×10^{-8} s^{-1} (0.006 d^{-1}). The same ratio of 20 between fast and slow decaying organic matter was preserved for other mineralization pathways (Table 15.1).

This ratio is close to the one Wijsman et al. (2002) chose (ratio of 25: 8.71×10^{-7} s^{-1} and 3.47×10^{-8} s^{-1}) but these authors used the same reaction rates for all mineralization pathways. Primary production, poorly documented on the site, was not taken into account, and bioturbation intensity was maximal (1 $cm^2.y^{-1}$) down to 10 cm depth, exponentially decreasing down to 20 cm depth, and considered negligible below 20 cm (see next section for details).

TABLE 15.3
Parameters Used in the SiAM 1DV Model

Parameter	Value	Unit
Initial sediment height	0.2	M
Computing time step	300	S
Maximal thickness of surficial layer	0.002	M
Grain density	2600	$kg.m^{-3}$
Initial inorganic matter concentration in bottom water	0.3	$kg.m^{-3}$
Initial inorganic matter concentration in sediments	900	$kg.m^{-3}$
Temperature	10	°C
Minimal depth of Db decrease	0.1	M
Total depth of bioturbation	0.2	M
Exponential coefficient for depth decrease of Db	6	—
POM1 concentration in bottom water	0.01	$mmol.l^{-1}$
POM2 concentration in bottom water	0.04	$mmol.l^{-1}$
$Fe(OH)_3$ concentration in bottom water	0.0058	$mmol.l^{-1}$
O_2 concentration in bottom water	0.275	$mmol.l^{-1}$
ΣCO_2 concentration in bottom water	2.2	$mmol.l^{-1}$
NO_3^- concentration in bottom water	0.01	$mmol.l^{-1}$
Half saturation constant for O_2 limitation in oxic mineralization (R1)	0.008	$mmol.l^{-1}$
Half saturation constant for NO_3^- limitation in denitrification (R2)	0.03	$mmol.l^{-1}$
Half saturation constant for O_2 inhibition in denitrification (R2)	0.008	$mmol.l^{-1}$
Half saturation constant for MnO_2 limitation in manganese reduction (R3)	1	$mmol.l^{-1}$
Half saturation constant for O_2 inhibition in manganese reduction (R3)	0.008	$mmol.l^{-1}$
Half saturation constant for NO_3^- inhibition in manganese reduction (R3)	0.03	$mmol.l^{-1}$
Half saturation constant for $Fe(OH)_3$ limitation in iron reduction (R4)	1	$mmol.l^{-1}$
Half saturation constant for O_2 inhibition in iron reduction (R4)	0.008	$mmol.l^{-1}$
Half saturation constant for NO_3^- inhibition in iron reduction (R4)	0.001	$mmol.l^{-1}$
Half saturation constant for MnO_2 inhibition in iron reduction (R4)	10	$mmol.l^{-1}$
Half saturation constant for SO_4^{2-} limitation in sulfate reduction (R5)	1	$mmol.l^{-1}$
Half saturation constant for O_2 inhibition in sulfate reduction (R5)	0.008	$mmol.l^{-1}$
Half saturation constant for NO_3^- inhibition in sulfate reduction (R5)	0.001	$mmol.l^{-1}$
Half saturation constant for MnO_2 inhibition in sulfate reduction (R5)	10	$mmol.l^{-1}$
Half saturation constant for $Fe(OH)_3$ inhibition in sulfate reduction (R5)	10	$mmol.l^{-1}$

Note: Bioturbation parameters correspond to the best fit with field data (Figure 15.2) and were modified for tests of bioturbation intensity (see text and Figure 15.4).

Because no POM was present in the sediment at the beginning of the run, the entire amount of POM present in profiles obtained after 378 days originates from inputs from the water column. The plot of inorganic matter concentration (Figure 15.2) illustrates the compaction processes, but impacts on the porosity profile remain limited. The sum of POM_1 and POM_2 in the sediment is in good agreement with field data, and variations of POM_1 are probably consecutive to the deposition and erosion events.

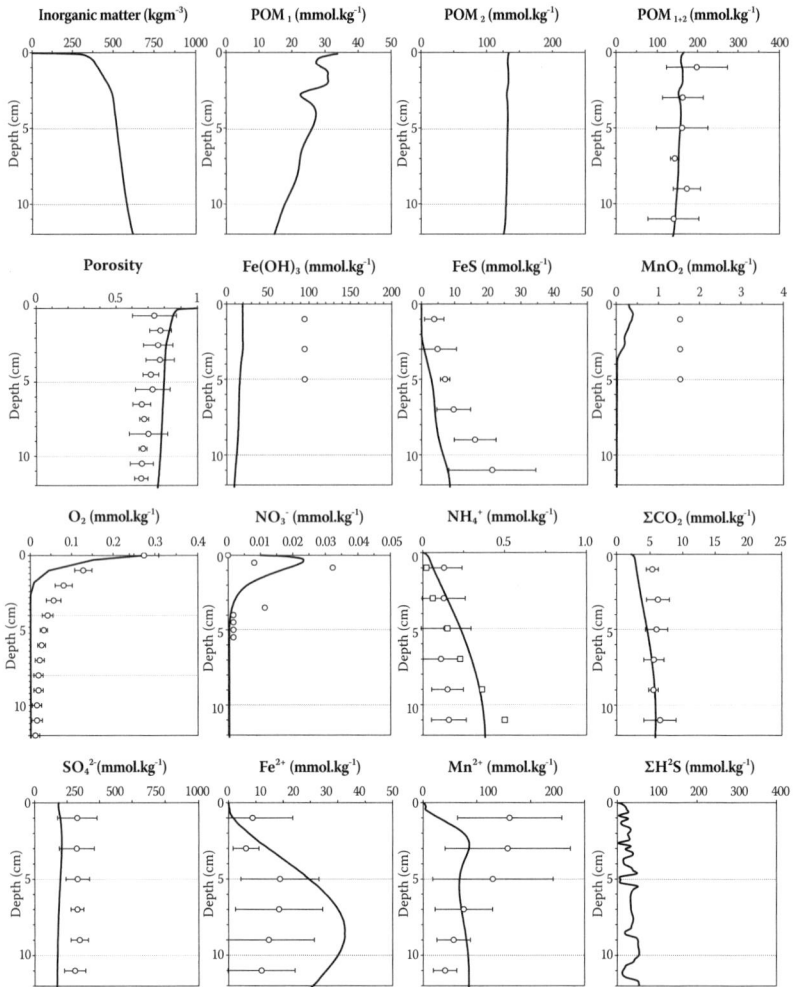

FIGURE 15.2 Vertical profiles of particulate and dissolved compounds obtained with the SiAM1DV model after 378 days of run (curves), compared with available field data from the DYVA sampling campaign (average values of three replicates ± standard deviation). The parameters used for this run are listed in Tables 15.1 and 15.3. Additional duplicate measurements of NH_4^+ concentrations are also plotted (squares).

As expected, the fast-decaying POM_1 pool significantly decreases at depth, while the POM_2 pool remains almost constant on the vertical scale. The oxygen penetration depth modelled (0.7 mm) is lower than the measured one (more than 2 mm), but additional tests (data not shown) demonstrated that the short-term temporal variability of modelled oxygen profiles was extreme, as a consequence of the continuous deposition and erosion cycles in the first millimetres of surficial sediments. Punctual field measurements and a single modelled profile may not agree because *in situ* oxygen profiles are directly consecutive to the short-term (minutes, hours) history of surficial sediments, characterized by high spatial heterogeneity.

Nitrogen cycling seems to be correctly represented in the model, despite the absence of several secondary reactions such as NH_4^+ adsorption on particles (Laima et al. 1999) and anaerobic ammonium oxidation (anammox; see review by Dalsgaard et al. 2005).

Ammonium and nitrate profiles are in good agreement with measured concentrations. Total alkalinity (ΣCO_2) also agrees relatively well with modelled concentrations. When quantifying the diagenetic pathways mainly implied in ammonium and carbonate production, we observed that more than 80% of each dissolved compound originates from sulfate reduction. Inputs of $Fe(OH)_3$ had been set to a value of 5.8 × 10^{-3} mM of the total inorganic inputs (from the water column at a concentration of 0.3 g.L^{-1}), corresponding to about 1.1g.kg^{-1} in the particulate fraction.

Despite this high input, modelled interstitial $Fe(OH)_3$ concentrations are significantly lower than field data, and iron sulfide concentrations remain lower than measured acid-volatile sulfides (AVS); however, the iron sulfide produced in the model may represent only a fraction of the measured AVS. On the other hand, Fe^{2+} concentrations are in good agreement with average field data. Despite MnO_2 inputs of 1.8 × 10^{-4} mM at the sediment–water interface (equivalent to 80 mg.kg^{-1} of the particulate fraction), MnO_2 concentrations in the sediments remain an order of magnitude lower than field data, while the modelled Mn^{2+} profile is close to the average measured one. Modelled interstitial SO_4^{2-} concentrations are slightly lower than field data, but since they did not limit sulfate reduction, inputs at the sediment–water interface (calculated from salinity data) were not modified. Interstitial sulfide concentrations are very low, because sulphides are efficiently converted to FeS.

We are aware that several other secondary reactions may be implemented in the model and may help produce a better fit of field data. For instance, manganese precipitation with carbonates was not taken into account and may seriously modify the Mn^{2+} profile at depth. The distinction between FeS and FeS_2 is noted in several papers (Berg et al. 2003). But despite the original structure of this diagenetic model, mainly driven by high frequency erosion–deposition events, concentrations obtained after more than 6 months of calculation are in relatively good agreement with field data.

15.6 SENSITIVITY TO THE IMPACTS OF BIOTURBATION

Based on the parameters and forcing described earlier, we tested the influence of biota on the biogeochemistry of surficial sediments. Fauchald and Jumars (1979) developed classifications of benthic invertebrates that advocate a subdivision of the trophic functional groups traditionally recognized in the benthos (herbivores, suspension feeders, detritus feeders, carnivores, and omnivores) that relate to detailed information available on the feeding behaviour and morphology of the species.

The classification of Pearson and Rosenberg (1987) used four modes of functional traits (feeding, mobility, degree of mobility, and feeding habit) aimed to link marine benthic community structure and benthic habitat. Consequently, bioturbation can be considered the direct or indirect result of macrofaunal and meiofaunal activities on surficial sediments. Sediment reworking depends on the types and densities of organisms implied in the mixing processes. From a qualitative view, bioturbators are generally classified in several groups, depending on their reworking mode: diffusers, gallery diffusers, upward or downward conveyor belts, and regenerators. Further details of the classification of bioturbators can be found in François et al. (1997 and 2002) or Michaud et al. (2005).

In this study, only quantitative sediment reworking by macrofauna was considered. Classically, the quantitative impact of biota on early diagenesis is characterized by the apparent bioturbation coefficient D_b. This coefficient is apparent, because it is not a direct measurement of bioturbation processes (i.e., movement of organisms), but rather an indirect measurement of the effects on particulate or dissolved vertical transfer (in the case of a 1DV model). The estimation of the apparent bioturbation coefficient can be obtained by using fluorescent inert tracers (François et al. 1997; Gilbert et al. 2007) or by steady state modelling of radionuclides (Lecroart et al. 2007; Schmidt et al. 2007) and the D_b coefficient (in square centimetres per year) is derived from the general diffusion equation (Equation 15.3) of a tracer in surficial sediments (Berner, 1980):

$$\frac{\partial C}{\partial t} = Db \times \frac{\partial^2 C}{\partial z^2}$$

(15.3)

where C is tracer concentration, t is time, and z is the depth in the sediment. Generally, bioturbation coefficients measured in intertidal sandy–muddy to muddy areas for the whole benthic community range between 0.7 and 50 $cm^2.y^{-1}$ (Osaki et al. 1997; Rusch et al. 2000).

Unfortunately, the dataset available for the Seine estuary does not include such measurements of bioturbation coefficients. Although no global measurement of macrofaunal reworking of surficial sediments was performed during the PNETOX sampling collections, estimations of the sediment reworked by the ragworm *Nereis diversicolor* were reported by Gillet et al. (2008). It is well recognized that in estuarine systems, this species plays a key role in the physical, chemical, and biological properties of the marine water–sediment interface and it has been extensively studied (Fernandes et al. 2006; Gillet et al. 2008).

Moreover, *N. diversicolor* is considered to be a good indicator of organic pollution (Scaps and Borot 2000; Scaps 2001) and its tolerance to metals and other stressors in contaminated estuaries has been widely documented (Durou et al. 2005 and 2007). The influence of polychaetes on biogeochemical processes has mainly been attributed to their sediment reworking and bioirrigation activities (Gillet et al. 2008) that are directly connected to worm density and biomass. *N. diversicolor* constructs Y-shaped burrows (Koretsky et al. 2002) irrigated through lateral body undulations (Christensen et al. 2000).

Previous studies have demonstrated that *N. diversicolor* may reduce or inhibit anaerobic metabolism and alter sediment sulfur cycles, although to different degrees (Banta et al. 1999). Hence, the abundance and the biomass of the population of *N. diversicolor* may partly control the amount of sediment reworked in the Seine estuary. During the PNETOX sampling collections, the estimations of the sediment

(a)

(b)

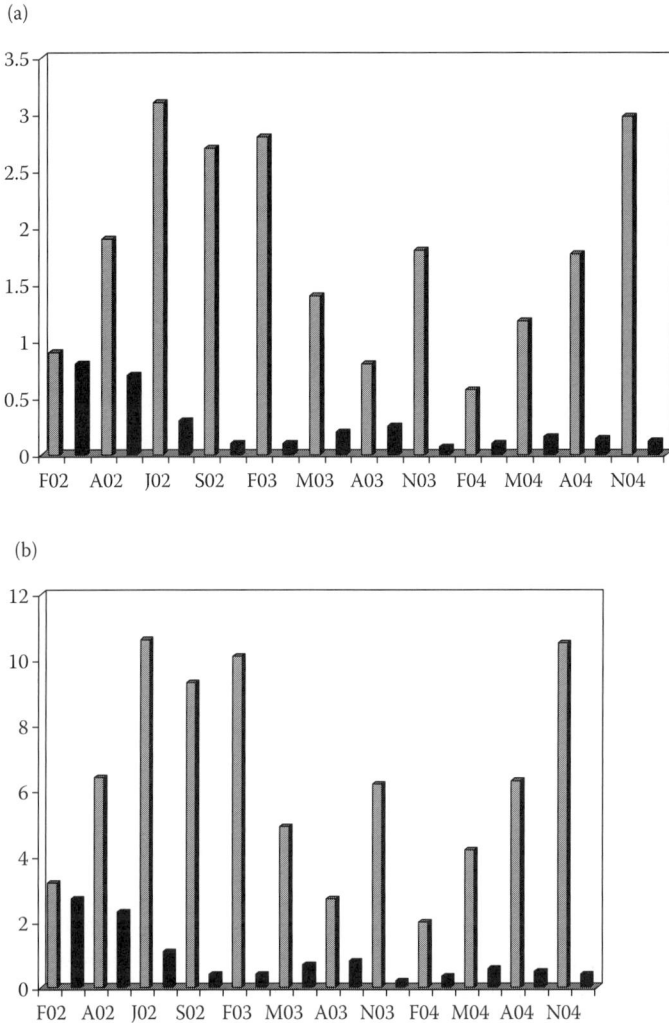

FIGURE 15.3 (a) Intersite and temporal differences in surface areas of burrow walls (in $m^2.m^{-2}$ mudflat surface) and (b) in pumping rate ($L.day^{-1}.m^{-2}$ mudflat surface) of ragworms *Nereis diversicolor* originating from the Authie estuary (grey) and the Seine estuary (black) in 2002, 2003, and 2004.

reworked by *N. diversicolor* varied in the range of 0.2 to 0.4 $cm^3.g^{-1}$ dw.d^{-1} in the Seine estuary (Gillet et al. 2008), while burrow wall surface and pumping rates ranged from 0.13 to 0.47 $m^2.m^{-2}$ and from 0.45 to 1.6 $L.d^{-1}.m^{-2}$, respectively. Similar measurements performed by these authors in the Authie estuary demonstrated higher densities and bioturbation activities than in the Seine estuary along with different population structures (Figure 15.3). The observed differences were attributed to the effect of pollution in the Seine estuary, while the Authie is generally considered a noncontaminated estuary.

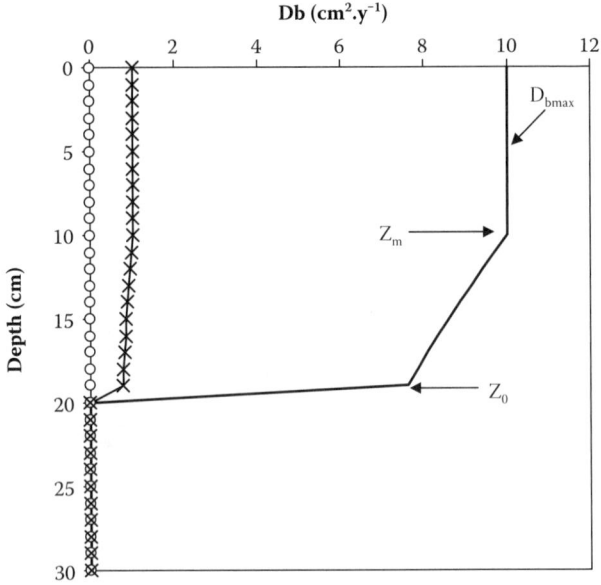

FIGURE 15.4 Vertical profiles of apparent bioturbation coefficient D_b used in model with parameters $z_m = 10$ cm, $z_0 = 20$ cm, and $k = 6$ (see text for details). The apparent bioturbation coefficients tested were $D_{bmax} = 0$ cm^2.y^{-1} (curve with open circles), $D_{bmax} = 1$ cm^2.y^{-1} (curve with crosses), and $D_{bmax} = 10$ cm^2.y^{-1} (bold curve). The z_m, z_0, and D_{bmax} parameters are shown on the bold curve.

All these data provide only partial information about the vertical profiles of apparent bioturbation coefficients and such observations may vary greatly over time. Decreases of sediment reworking, burrow wall surface, and pumping rate of *N. diversicolor* were observed between 2002 and 2004 for both sites. Therefore, based on the range of density measured in the Seine estuary (90 to 920 individuals per square meter), we estimated a bioturbation profile from literature data on *N. diversicolor* (Fernandes et al. 2006; Gillet et al. 2008) that may range from 0.7 to 10 cm^2.y^{-1}. In the present model, the bioturbation coefficient D_b (treated as a diffusion coefficient for particulate and dissolved constituents) at a given depth z (positive downward) is described in three different layers (Figure 15.4):

$$\text{If } z < z_m, \text{ then } D_{bz} = D_{b\,max} \tag{1}$$

$$\text{If } z_m < z < z_0, \text{ then } \qquad D_{bz} = D_{bmax} \times e^{\left(-k\frac{(z-z_m)}{(z_0/z_m)}\right)} \tag{2}$$

$$\text{If } z_0 < z, \text{ then } D_{bz} = 0 \tag{3}$$

In the upper layer (1), the apparent bioturbation coefficient is constant and maximal ($D_{b\ max}$), to mimic the presence of living organisms near the sediment–water interface for most of the time, down to a given depth z_m. In the next layer (2), the bioturbation coefficient decreases exponentially downward according to a coefficient k, down to a depth z_0. Deeper down (3), the bioturbation is considered negligible and the bioturbation coefficient is set to zero.

First, the quantitative impact of bioturbation was tested and three different situations (absence of bioturbation $D_{bmax} = 0$; a moderate bioturbation rate $D_{bmax} = 1\ cm^2.y^{-1}$; and a high bioturbation rate $D_{bmax} = 10\ cm^2.y^{-1}$) were considered, while the parameters z_m, z_0, and k were systematically set to 10 cm, 20 cm, and 6 respectively.

No differences were observed for dissolved oxygen, alkalinity, nitrate, and sulfate concentrations in surficial sediments; particulate variables were only slightly affected (Figure 15.5). The higher bioturbation rate tended to smooth the vertical profiles of particulate organic matter (POM) whereas they were initially characterized by numerous peaks consecutive to pulse inputs at the sediment–water interface. Other particulate profiles were also smoothed. The depth penetration of POM was increased, but mineralization processes remained only slightly affected, except the Fe^{2+} profile characterized by lower concentrations in the interstitial water of the sediment due to active reoxidation when the bioturbation rate was higher.

From a global view, these results demonstrate that the intensity of bioturbation in the surficial layers had a limited impact on the profiles calculated by the model (Figure 15.5). The intensities of mineralization pathways after 378 days (DYVA 5) remain very close to each other, despite the wide range of bioturbation tested.

Further tests were also conducted on the depth of maximal bioturbation rate (z_m), the maximal depth of bioturbation (z_0), and the exponential coefficient of bioturbation decrease (k) (data not shown), but they also resulted in similar conclusions. Hence, despite high bioturbation rates tested, the sedimentary dynamics (accumulation, resuspension, dewatering) largely dominated the biogeochemical characterization of surficial sediments. Considering that only the expected activity of *N. diversicolor* was taken into account in the model, and the low densities observed, it is not surprising to observe a low impact of bioturbation processes.

We are aware that nonlocal exchanges were not included in the model and may have modified the geochemical processes, but the range of bioturbation tested (from zero up to a D_{bmax} of 10 $cm^2.y^{-1}$ on a depth of 20 cm) is large. We consider that the absence of nonlocal exchanges was largely counterbalanced by the intensity of "local" bioturbation that was applied.

Despite high levels of metal and organic contamination, several studies have shown that the Seine estuary continues to support a very high benthic biomass and remains productive (Dauvin 2008), but responses to contaminants vary greatly among species and assemblages. Gillet et al. (2008) hypothesized that the low densities of small-sized *N. diversicolor* in the Seine estuary comparable with densities from polluted parts of the Tees estuary (Gray 1976) may result from pollution that affects the recruitment of juveniles, the number of cohorts, and individual growth. Such a contaminated environment may exert direct consequences on the density and the average size of *N. diversicolor* individuals, hence limiting their bioturbation efficiency.

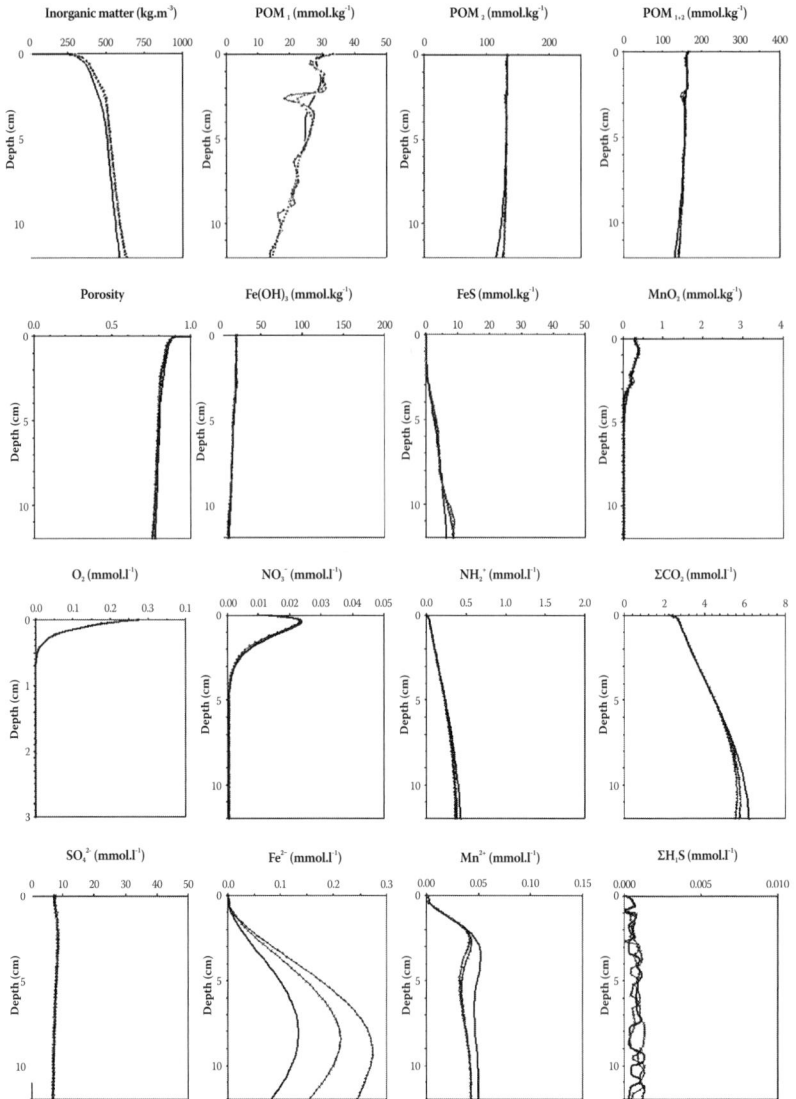

FIGURE 15.5 Vertical profiles of particulate and dissolved compounds obtained with the SiAM1DV model after 378 days of run with different bioturbation intensities: D_{bmax}, = $0\,cm^2.y^{-1}$ (curve with open circles), $D_{bmax} = 1\,cm^2.y^{-1}$ (curve with crosses), and $D_{bmax} = 10\,cm^2.y^{-1}$ (bold curve). The characteristics of apparent bioturbation coefficients appear in Figure 15.4 and are detailed in the text; other parameters are listed in Tables 15.1 and 15.3.

However, in the case of this estuarine mudflat, the influence of even very high bioturbation rates ($10\ cm^2.y^{-1}$) on surficial biogeochemistry is limited. Thus, the high-frequency erosion–deposition cycle caused by complex hydrodynamic forcing in the Seine estuary may explain the low influence of bioturbation in this unsteady sedimentary system. Previous studies have focused on the main causes of such short-term changes in surficial sediments: waves, tidal currents, high river flows (Lesourd et al. 2002; Deloffre et al. 2006; Verney et al. 2006).

This modelling work clearly demonstrates the small influence of bioturbation on geochemical processes in surficial sediments, but further tests based on direct measurements of apparent bioturbation coefficients and their temporal variability are needed. This work also illustrates the need for a full model of sediment erosion and deposition, taking into account organic matter and nutrients in water in order to improve the results of a diagenetic model of this coastal system.

15.7 CONCLUSIONS

The original approach of this model is to integrate sediment accumulation and resuspension in the biogeochemical modelling of surficial sediments. In the Seine estuary, sediment dynamics are very high, and characterized by a succession of accumulation and erosion periods. Despite a fully dynamic diagenetic model mainly forced by the erosion and resuspension processes of surficial sediments, the profiles obtained with the model agreed relatively well with field data. In such a benthic system, this work shows that the effect of bioturbation is almost negligible, unless one focuses on a given pathway that may have a slight effect locally.

While considering only the organic inputs at the sediment–water interface, and disregarding the temporal variability in sedimentation and resuspension processes (as may be done with models commonly run in deeper systems), one may expect to obtain a higher impact of bioturbation on surficial biogeochemical processes. Such results highlight the need for diagenetic models implicitly integrating particulate dynamics at the sediment–water interface when applied in coastal systems. This is clearly the case for the Seine estuary, but one may expect that it is also of prime importance in coastal systems in which hydrodynamic forcing may seriously interfere with the steady state concept commonly adopted in diagenetic models.

ACKNOWLEDGMENTS

The authors would like to thank all the people who, by their work in the field during sampling, by their help in data processing and analyses, provided the dataset used for this modeling work. Thanks should also be addressed to Claude Amiard-Triquet who conducted the project that is the subject of this book. The model was developed in part through the Programme National d'Océanographie Côtière and the MAST3 INTRMUD project; it has been applied in the Seine Aval program (DYVA and MEDIANS projects), and was used here in the framework of the PNETOX project.

REFERENCES

Bally, G. et al. 2004. Chemical characterization of porewaters in an intertidal mudflat of the Seine estuary: relationship to erosion–deposition cycles. *Mar. Poll. Bull.* 49: 163–173.
Banta, G.T. et al. 1999. Effects of two polychaete worms, *Nereis diversicolor* and *Arenicola marina*, on aerobic and anaerobic decomposition in a sandy marine sediment. *Aquat. Microbial Ecol.* 19: 189–204.

Berg, P., S. Rysgaard, and B. Thamdrup. 2003. Dynamic modeling of early diagenesis and nutrient cycling: a case study in an arctic marine sediment. *Am. J. Sci.* 303: 905–955.

Berner, R.A. 1980. *Early Diagenesis: A Theoretical Approach*. Princeton, NJ: Princeton University Press.

Boudreau, B.P. 1996. A method-of-lines code for carbon and nutrient diagenesis in aquatic sediments: *Comput. Geosci.* 22: 479–496.

Boudreau, B.P. 1997. *Diagenetic Models and their Implementation*. Berlin: Springer.

Boudreau, B.P. et al. 1998. Comparative diagenesis at three sites on the Canadian continental margin. *J. Mar. Res.* 56: 1259–1284.

Boudreau, B.P. and B.P. Ruddick. 1991. On a reactive continuum representation of organic-matter diagenesis. *Am. J. Sci.* 291: 507–538.

Boust, D. 1999. *Fer et Manganèse: réactivités et recyclages*. Programme Scientifique Seine Aval, fascicule 9. Editions IFREMER.

Christensen, B., A. Vendel, and E. Kristensen. 2000. Carbon and nitrogen fluxes in sediment inhabited by suspension feeding (*Nereis diversicolor*) and nonsuspension feeding (*N. virens*) polychaetes. *Mar. Ecol. Prog. Ser.* 192: 203–217.

Cugier, P. and P. Le Hir. 2002. Development of a 3D hydrodynamical model for coastal ecosystem modelling: application to the plume of the Seine River. *Estuar. Coast. Shelf Sci.* 55: 673–695.

Dalsgaard, T., B. Thamdrup, and D.E. Canfield. 2005. Anaerobic ammonium oxidation (anammox) in the marine environment. *Res. Microbiol.* 156: 457–464.

Dauvin, J.C. 2008. Effects of heavy metal contamination on the macrobenthic fauna in estuaries: the case of the Seine estuary. *Mar. Poll. Bull.* 57: 160–169.

Deloffre, J. et al. 2005. Sedimentation rhythms on a fluvial estuarine mudflat: case of the macrotidal Seine estuary (France). *Estuar. Coast. Shelf Sci.* 64: 710–720.

Deloffre, J. et al. 2006. Interactions between intertidal mudflat and turbidity maximum in macrotidal estuarine context. *Mar. Geol.* 235: 151–164.

Dhakar, S.P. and D. Burdige. 1996. A coupled, non-linear, steady state model for early diagenetic processes in pelagic sediments. *Am. J. Sci.* 296: 296–330.

Durou, C., C. Mouneyrac, and C. Amiard-Triquet. 2005. Tolerance to metals and assessment of energy reserves in the polychaete *Nereis diversicolor* in clean and contaminated estuaries. *Environment. Toxicol.* 20: 65–76.

Durou, C. et al. 2007. From biomarkers to population response in *Nereis diversicolor*: assessment of stress in estuarine ecosystems. *Ecotox. Environ. Saf.* 66: 402–411.

Even, S. and B. Thouvenin. 2004. *Synthèse sur le cycle du carbone*. Rapport Seine Aval RA-2004-1-8.

Fauchald, K. and P. Jumars. 1979. The diet of worms: a study of polychaete feeding guilds. *Oceanogr. Mar. Biol. Ann. Rev.* 17: 193–284.

Fernandes, S., F.J.R. Meysman, and P. Sobral. 2006. The influence of Cu contamination on *Nereis diversicolor* bioturbation. *Mar. Chem.* 102: 148–158.

François, F. et al. 1997. A new approach for the modelling of sediment reworking induced by a macrobenthic community. *Acta Biotheor.* 45: 295–319.

François, F. et al. 2002. A functional approach to sediment reworking by gallery-forming macrobenthic organisms: modelling and application with the polychaete *Nereis diversicolor*. *Mar. Ecol. Prog. Ser.* 229: 127–136.

Froelich, P.N. et al. 1979. Early oxidation of organic matter in pelagic sediments of the eastern equatorial Atlantic: suboxic diagenesis. *Geochim. Cosmochim. Acta* 43: 1075–1090.

Garnier, J. et al. 2005. *Exploration du comportement de la matière organique dans l'estuaire aval par le modèle simplifié LIFTE: impacts des apports d'amont*. Rapport Seine Aval RA-2005-1-5.

Gilbert, F. et al. 2007. Sediment reworking by marine benthic species from the Gullmar Fjord (Western Sweden): importance of faunal biovolume. *J. Exp. Mar. Biol. Ecol.* 348: 133–144.

Gillet, P. et al. 2008. Response of *Nereis diversicolor* population (polychaeta, nereididae) to the pollution impact: Authie and Seine estuaries (France). *Estuar. Coast. Shelf Sci.* 76: 201–210.

Graf, G. 1992. Benthic-pelagic coupling: a benthic view. *Oceanogr. Mar. Biol. Ann. Rev.* 30: 149–190.

Graf, G. et al. 1982. Benthic response to sedimentation of a spring phytoplankton bloom: progress and budget. *Mar. Biol.* 67: 201–208.

Gray, J.S. 1976. The fauna of the polluted river Tees estuary. *Estuar. Coast. Mar. Sci.* 4: 653–676.

Hargrave, B.T. 1973. Coupling carbon flow through some pelagic and benthic communities. *J. Fish. Res. Board Can.* 30: 1317–1326.

Herman, P.M.J. et al. 2001. The seafloor as the ultimate sediment trap: using sediment properties to constrain benthic–pelagic exchange processes at the Goban Spur: *Deep-Sea Res. II* 48: 3245–3264.

Hinga, K.R., J.M. Sieburth, and G.R. Heath. 1979. The supply and use of organic material at the deep sea floor. *J. Mar. Res.* 37: 557–579.

Jestin, H. et al. 1998. Development of ALTUS, a high frequency acoustic submersible recording altimeter, to accurately monitor bed elevation and quantify deposition or erosion of sediments. *In Proceedings of Ocean'98 IEEC/OES Conference*, Nice, pp.189–194.

Jørgensen, B.B. 1978. A comparison of methods for quantification of bacterial sulfate reduction in coastal marine sediments. 2. Calculation from mathematical models. *Geomicrobiol. J.* 1: 29–47.

Koretsky, C.M., C. Meile, and P. Van Cappellen. 2002. Quantifying bioirrigation using ecological parameters: a stochastic approach. *Geochem. Trans.* 3: 17–30.

Lafite, R. et al. 2002. *Projet DYVA: Fonctionnement des vasières*. Rapport Seine Aval RA-2002-2-2.

Laima, M.J.C. et al. 1999. Distribution of adsorbed ammonium pools in two intertidal sedimentary structures, Marennes–Oléron Bay, France. *Mar. Ecol. Prog. Ser.* 182: 29–35.

Lecroart, P., S. Schmidt, and J.M. Jouanneau. 2007. Numerical estimation of the error of the biodiffusion coefficient in coastal sediments. *Estuar. Coast. Shelf Sci.* 72: 543–552.

Le Hir, P., P. Bassoullet, and H. Jestin. 2000. Application of the continuous modeling concept to simulate high-concentration suspended sediment in a macrotidal estuary. In *Proceedings in Marine Science 3*. Fifth International Conference on Nearshore and Estuarine Cohesive Sediment Transport (INTERCOH '98), pp. 229–247.

Le Hir, P. et al. 2001. Fine sediment transport and accumulations at the mouth of the Seine Estuary (France). *Estuaries* 24: 950–963.

Lesourd, S. et al. 2002. Seasonal variations in the characteristics of superficial sediments in a macrotidal estuary (Seine inlet, France). *Estuar. Coast. Shelf Sci.* 58: 3–16.

Luff, R. et al. 2000. Numerical modeling of benthic processes in the deep Arabian Sea. *Deep-Sea Res. II* 47: 3039–3072.

Michaud, E. et al. 2005. The functional group approach to bioturbation: effects of biodiffusers and gallery diffusers of the *Macoma balthica* community on sediment oxygen uptake. *J. Exp. Mar. Biol. Ecol.* 326: 77–88.

Middelburg, J.J. 1989. A simple rate model for organic matter decomposition in marine sediments. *Geochim. Cosmochim. Acta* 53: 1577–1581.

Ouddane, B. et al. 2008. A comparative study of mercury distribution and methylation in mudflats from two macrotidal estuaries: the Seine (France) and the Medway (United Kingdom). *Appl. Geochem.* 23: 618–631.

Osaki, S. et al. 1997. Biodiffusion of [7]Be and ^{210}Pb in intertidal estuarine sediment. *J. Envir. Radioactivity* 37: 55–71.

Pearson, T.H. and R. Rosenberg. 1987. Feast and famine: structuring factors in marine benthic communities. In *27th Symposium of the British Ecological Society, Aberystwyth (1986)*, Gee, J.H.R. and Giller, P.R., Eds. Oxford: Blackwell Scientific, pp. 373–395.

Rowe, G.T. et al. 1975. Benthic nutrient regeneration and its coupling to primary production in coastal waters. *Nature* 255: 215–217.

Rusch, A., M. Huettel, and S. Forster. 2000. Particulate organic matter in permeable marine sands: dynamics in time and depth. *Estuar. Coast. Shelf Sci.* 51: 399–414.

Scaps, P. 2001. Electrophoretic heterogeneity in *Hediste diversicolor* (Annelida: Polychaeta) within and between estuaries in northern France. *Vie et Milieu* 51: 99–111.

Scaps, P. and O. Borot. 2000. Acetylcholinesterase activity of the polychaete *Nereis diversicolor*: effects of temperature and salinity. *Comp. Biochem. Physiol.* 125: 377–383.

Schmidt, S. et al. 2007. Sedimentary processes in the Thau lagoon (France): from seasonal to century time scales. *Estuar. Coast. Shelf Sci.* 72: 534–542.

Schulz, H.D. and M. Zabel, Eds. 2006. *Marine Geochemistry*. Berlin: Springer.

Soetaert, K., P.M.J. Herman, and J.J. Middelburg. 1996. A model of early diagenetic processes from the shelf to abyssal depths. *Geochim. Cosmochim. Acta* 60: 1019–1040.

Terzaghi, K. and R.B. Peck. 1967. *Soils Mechanics in Engineering Practice*. New York: John Wiley & Sons.

Van Cappellen, P. and Y. Wang. 1996. Cycling of iron and manganese in surface sediments: a general theory for the coupled transport and reaction of carbon, oxygen, nitrogen, sulfur, iron, and manganese. *Am. J. Sci.* 296: 197–243.

Verney, R. et al. 2006. The effect of wave-induced turbulence on intertidal mudflats: impact of boat traffic and wind. *Cont. Shelf Res.* 27: 594–612.

Viollier, E. et al. 2003. Benthic biogeochemistry: state of the art technologies and guidelines for the future of *in situ* survey. *J. Exp. Mar. Biol. Ecol* 285–286: 5–31.

Westrich, J.T. and R.A. Berner. 1984. The role of sedimentary organic matter in bacterial sulphate reduction: the G-model tested. *Limnol. Oceanogr.* 29: 236–249.

Wijsman, J.W.M. et al. 2002. A model for early diagenetic processes in sediments of the continental shelf of the Black Sea. *Estuar. Coast. Shelf Sci.* 54: 403–421.

16 Conclusions

Claude Amiard-Triquet and Philip S. Rainbow

CONTENTS

This book is intended to apply the findings of a multidisciplinary comparative study of two estuarine systems to provide a future methodology for such studies. We are seeking improved risk assessment of the potential ecotoxicological effects of contamination in estuaries before a stage has been reached when remediation is no longer possible or commercially viable. Clearly this comparison of the two estuaries, the anthropogenically contaminated Seine estuary and the reference Authie estuary, has involved considerable monitoring of chemical concentrations in the key compartments of the biogeochemical cycle—water, sediments, and biota exemplified by zooplankton (*Eurytemora affinis*), a benthic macroinvertebrate (*Nereis diversicolor*), and a fish (*Platichthys flesus*) (Chapters 3, 5, 6, 10, and 11).

The importance of sediments, whether deposited or suspended in the water column, as reservoirs for contaminants has been highlighted as has their role as sources of contaminants to biota. Of course, the interaction of contaminants with the biota controls their potential ecotoxicological effects—the necessary basic consideration for any environmental risk assessment. We have analysed concentrations of the now classical anthropogenic toxic contaminants, trace metals, PAHs, and PCBs, but also considered emerging contaminants such as PBDEs (brominated flame retardants), pharmaceutical products, and synthetic endocrine disruptors such as alkylphenols. In the field, these contaminants act as mixtures, not as the single contaminants typical of many laboratory toxicity tests. Biota must be examined in real field situations. The presence of such mixtures of contaminants causes the observed ecotoxicological effects and make it so difficult to name prime suspects among the list of toxins.

It is also highly relevant that even the cleanest estuaries are characterised by physical parameters (and variations in such physical parameters) that are stressful to many aquatic organisms, not least low and fluctuating salinity levels, varying temperatures in a shallow habitat suffering cycles of immersion and emersion, and high suspended sediment loads. These "natural" factors and their variations over tidal, daily, monthly, and seasonal cycles must all be considered along with temporal variations in the availabilities of contaminants, in solution or associated with sediments, to estuarine biota. Allowance for such natural parameters must be made, although it can never be complete, when comparing estuaries, inevitably of different sizes, hydrodynamic characteristics, and other characteristics In this conclusions chapter, we highlight some of the key points that have emerged from the study, particularly those that provide lessons for future investigations.

16.1 EFFECTS OF SEDIMENTARY PROCESSES

Sediments (both suspended and deposited) constitute the main reservoirs for most chemicals introduced into aquatic environments by human activities (Gagnon and Fisher 1997; Bernes 1998; Mora 1999). Thus at both sites of interest in the present programme, particular attention has been devoted to sediment characteristics (grain size, organic matter content) in order to seek comparability between sites. The alternation of deposition and resuspension phases of the sediment has here been confirmed to be very important in a dynamic estuary in controlling the surface sediment concentrations of all contaminants (Chapter 3), and in influencing the speciation of trace metals (Chapter 5), the structure of assemblages of small-sized organisms (Chapters 12, 13, 14), and the distribution of metal-resistant bacteria on the estuarine mudflats (Chapter 7).

For many aspects of environmental monitoring, our present state of knowledge and the insufficiency of background data available mean that the use of a reference site for comparison is essential. This presents a potential problem when we try to eliminate comparative differences resulting from hydrodynamic differences between estuaries under comparison because potential reference estuaries with low perceived anthropogenic pressures are generally small, whereas the human activities responsible for the presence of many chemicals in the environment have historically developed on the banks of larger main watercourses. Thus, in parallel with chemical and biological monitoring, it is necessary, and indeed recommended, to assess the importance of sediment changes over time and distance. Depending on the specific objectives, it is possible to use specialised instruments such as the altimeter described in Chapter 2 that allow measurements of bed elevations at a high sampling frequency, with excellent resolution and accuracy. However, to monitor a large number of sampling stations, less costly procedures such as the measurement

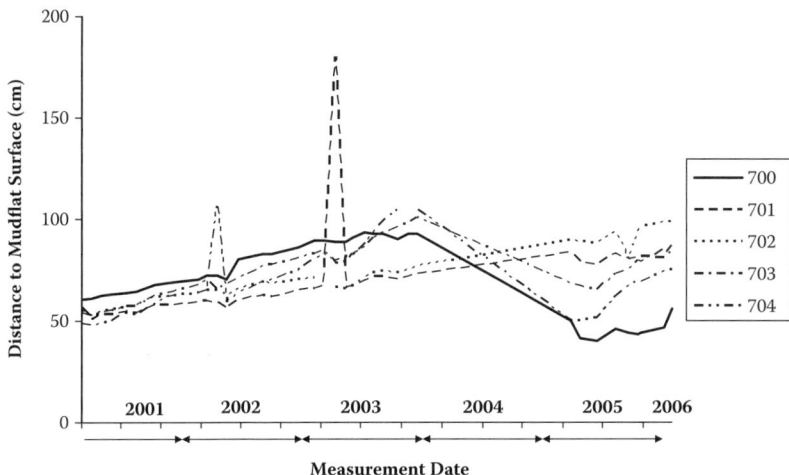

FIGURE 16.1 Changes in bed elevation along a transect near station PN in the Seine estuary (Figure 1.3). Y axis: distance between the top of the sticks and the sediment surface (increase of this distance corresponds to resuspension, decrease to deposition). Point 700: limit between schorre and slikke (C. Bessineton, personal communication).

of mudflat levels along a transect defined with a series of sticks need to be deployed (Figure 16.1).

As part of this study a successful attempt was made to model the interactions of living organisms in the sediments and the deposition and resuspension of sediments in early diagenesis, taking into account the strength of hydrodynamic forcing in the Seine estuary (Chapter 15). We found good agreement between profiles obtained from a fully dynamic diagenetic model and field data, showing that in the Seine estuary the effect of bioturbation is almost negligible. These conclusions can be extended to other estuarine and coastal areas where hydrodynamic forcing may seriously interfere with the steady state concept commonly adopted in diagenetic models.

16.2 BIOAVAILABILITY OF CHEMICALS: FROM SCIENTIFIC KNOWLEDGE TO REGULATION

With a view to sustainable development, regulatory frameworks have been designed to establish quality standards with the aim to protect human health and environmental quality. To meet these quality criteria, it is necessary to control a number of human activities responsible for the input of chemicals into the environment. Briefly, standardization is mainly based upon the relationship between the dose at which a pollutant is present in the environment and the noxious biological effect it is likely to exert.

Most standards for the control of metals in water and seafood products are expressed as total concentrations for each element, although the total dose of metal or organic compound contaminants in any compartment of the environment (water, food, sediment) may present little ecotoxicological significance. Considering both the fates and effects of metals in the environment, the available scientific data underline the major role of the physico-chemical forms of contaminants in controlling their bioavailability and hence toxicity (Tessier and Turner 1995). Thus, physico-chemistry can control the exchanges of metals between compartments of the environment, particularly their uptake into biota (bioavailability).

The toxicity of metals depends on their abilities to cross biological membranes and then to bind to intracellular ligands. Both processes are governed by the physico-chemical characteristics of both contaminants and relevant binding sites. Reciprocally, as sources of organic matter and as a consequence of pollutant biotransformation, living organisms clearly play an important role in biogeochemical cycles of contaminants, particularly in areas of high productivity such as estuarine and coastal zones.

Thirteen years after the publication of the review by Tessier and Turner (1995) and follow-up work, it is surprising that such vital physico-chemical information has not yet been integrated into environmental regulations. This is partly because the determination of total concentrations, whether in physical or biological compartments, is easy to master; assessing metal speciation is a much more difficult task. However, since Batley (1989) edited a review of analytical methods for trace element speciation, many improvements have been achieved and passive samplers such as DET and DGT have allowed measurements of particular physico-chemical characteristics of metals that govern their fate and bioavailability in both surface and interstitial waters. Thus even in sites with background levels of metals in sediments, it is possible to determine concurrently soluble and labile fractions that are of greater ecotoxicological relevance than total concentrations *per se* (Chapter 5).

More recently the technology of passive samplers has been extended to organic contaminants (Togala and Budzinski 2007; Cornelissen et al. 2008; Hunter et al. 2008). Whatever the class of contaminant sampled, passive samplers share a number of features that have been recently reviewed (ITRC 2005; Vrana et al. 2005 and 2007). A passive sampler permits the acquisition of a sample from a discrete location without active transport of the medium induced by pumping or purge techniques.

These passive technologies rely on the exposure of the sampling device to a medium in ambient equilibrium. Devices that rely on diffusion and sorption to accumulate analytes in the sampler (DET, DGT, SPMD, POSIS, SPME) provide time-integrated samples that represent physicochemical conditions at the sampling point over the entire deployment period. Spatial integration, if any, results from natural ambient flow of the sampled medium that is linked to river flow and movements of water masses during tidal cycles in estuarine areas.

As a result of the nature of their original designs, passive samplers allow the quantification of a large number of chemical contaminants and accumulation of information about their mobilities, thus providing important insights about the bioavailabilites of different contaminants. Semi-permeable membrane devices (SPMDs) sample hydrophobic molecules while polar organic compound integrative samplers (POSISs) allow investigation of hydrophilic molecules including pesticides, pharmaceutical products, and detergents. A combination of solid phase microextraction techniques (SPMEs) with gas chromatography and mass spectrometry allows the investigation in a water sample of the interactions of the compounds analysed and the dissolved organic matter (DOM) present. Thus it is possible to measure the total concentrations of pollutants in a water sample and also distinguish components that are free in solution rather than associated with organic matter (King et al. 2004; De Perre et al. 2007).

A major caveat remains when interpreting the data gained from passive samplers. Metals, for example, are taken up into biota across membranes via different routes of varying importance, although the generalisation can often be made that the availability of the dissolved free metal ion is the best model for the dissolved bioavailable form of the metal. It is tempting then to refer to the species of dissolved metal accumulated by particular passive samplers as bioavailable metal. This is unfortunate shorthand because it ignores the variety of uptake routes available to metals across the membrane, and ignores the exact nature of low molecular weight complexed metal species that may be accumulated by the sampler yet not be taken up by an organism. Nevertheless the use of passive samplers is a step forward, so long as it is remembered that the amount of dissolved metal accumulated by a sampler is a better model of what might be bioavailable in solution than the total dissolved concentration, but it is still a model and not a true measure of bioavailable dissolved metal. Furthermore, interpretation of the results provided by passive samplers requires awareness of the processes that can influence resulting measures, as detailed in Chapter 5 using DGT as an example.

According to the U.S. Interstate Technology and Regulatory Council (ITRC 2005), some of these sampling technologies are relatively mature and are accepted for appropriate applications by regulators in certain regions and states. Nonetheless, they are still considered innovative technologies and few if any specific policies governing their use have been incorporated into official regulations.

The bioavailability of contaminants from sediments depends on several factors including the biology of the organism and the physico-chemical characteristics of the sediment metals, whether in porewater or on the sediment particles, including the degree of partitioning of metals between water and sediment particles. Estuarine mudflats constitute important areas for the storage of fine sediment particles where metals are typically trapped as sulfides (Chapter 5). When sediments are resuspended in the water column (Figure 2.1, Chapter 2), reoxidation of these compounds takes place, leading to a release of metals in soluble form. Thus the presence of sulfides in sediments governs the local solubility of sediment-associated metals, and affects the dissolved bioavailability of the metals (Figure 16.2).

An infaunal animal can take up dissolved metals from any surrounding porewater in contact with an external permeable surface, and, in the case of sediment ingestors, from sedimentary particles taken into the alimentary tract. Physico-chemical factors affecting dissolved bioavailability clearly affect uptake from solution while the physico-chemical nature of the association of metals with sediment particles will affect trophic bioavailability of the ingested metals. The relative importance of the two routes of uptake depends greatly on the biology of the relevant organism such as the redox nature of the specific part of the sediment in which it lives, powers of irrigation of a burrow, selectivity of sediment for ingestion, strength of digestive processes, rates of gut passage, etc. (Figure 16.3).

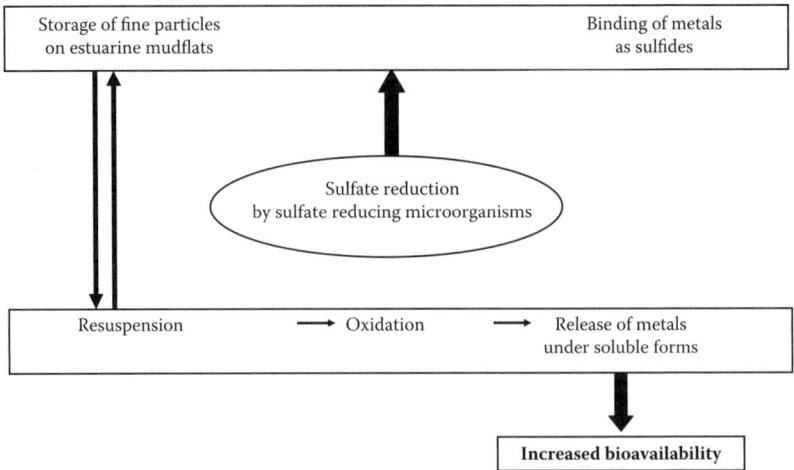

FIGURE 16. 2   Mobility and bioavailability of sediment-bound metals.

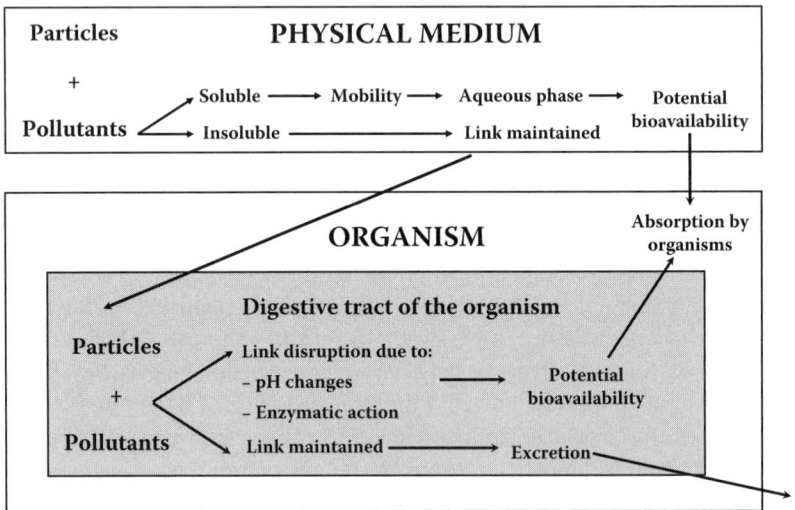

FIGURE 16.3   Fate of sediment-bound contaminants in the external medium and in the digestive tract of benthic organisms (J.C. Amiard, personal communication).

Investigation of sediment bioavailability may be approached from a mainly chemical standpoint. For example, the comparative measurement of acid volatile sulfide (AVS) and simultaneously extracted metal (SEM) in a sediment sample assesses how much sedimentary metal is bound as sulfide and considered not bioavailable and how much (remaining metal) can be considered bioavailable. Such a simple modelling approach to the assessment of bioavailable (and therefore potentially harmful) metal in sediment is attractive, but it is crucial to measure AVS and SEM in their proper vertical positions in the sediment (see Chapter 5) to reflect the true situation met by infaunal organisms.

The irrigation of burrows by most infaunal animals is a further complicating factor to this simple chemical modelling approach that must be taken into account. Macrobenthic activity and physical mixing in the field cause an increased downward flux of electron acceptors such as oxygen via irrigation activities (Langezaal et al. 2003), a pattern illustrated in Chapter 5 by the presence of an oxic microniche related to the biological activity.

Amiard et al. (2007) reported an alternative approach to investigating sediment bioavailability. Desorption from particles under the conditions prevailing in the digestive tracts of aquatic organisms has been quantified as a possible way to predict the digestive absorption from particles. Digestive conditions have been simulated using either gut juice or enzymes tested at pH levels that are known to exist in the guts of invertebrates. In many cases a clear relationship may be established between concentrations of metals that are labile at physiological pHs and bioaccumulated concentrations. Any mimicking of digestive processes establishes an upper limit on how much sediment metal may be available to sediment ingestors, since not all digestively solubilized metals may be absorbed subsequently.

Studies of the sedimentary bioavailabilities of organic contaminants have not reached the same state of development achieved for trace metals, as a result of the much shorter history of successful techniques for their analysis, prerequisite to any appropriate experimental investigation. Nevertheless studies investigating the principles underlying the bioavailability of organic contaminants in sediments are continuing (Reichenberg and Mayer 2006). In the case of PAHs, information is available about the importance of sedimentary carbon in modelling PAH bioaccumulation from sediments (Vinturella et al. 2004), the effects of quality and quantity of associated organic matter on the mobility of sediment PAHs (Oen et al. 2006), the control of PAH bioaccumulation by porewater concentrations (Lu et al. 2006), and a better correlation of accumulation of PAHs in biota with freely dissolved porewater concentrations as opposed to total sediment PAH concentrations (Cornelissen et al. 2006).

Similarly in the case of organochlorines, variations in the amounts and types of sediment particulate carbonaceous materials strongly influence the bioavailability of a PCB (McLeod et al. 2004). The point again is that total sedimentary concentrations of these contaminants do not directly represent bioavailable concentrations; bioavailability is affected by more subtle physico-chemical changes even in the absence of differences in total concentrations.

As for metals, *in vitro* extraction of sediment using the digestive fluids of deposit-feeding invertebrates as been suggested as a means to quantify the bioavailability of particle-adsorbed organic contaminants. Compared to conventional extractions employing a strong acid or exotic organic solvent, the use of natural digestive fluid solubilizes a far smaller fraction of the total particle-associated contaminant. However, as mentioned above for metals, the release of a chemical from ingested food is a prerequisite for uptake and assimilation, but only a proportion of a chemical ingested with sediment will enter the systemic circulation. Thus, the determination of lability of a sediment-bound contaminant, as measured by its extractability, can be used as an indicator of its maximum bioavailability. Such an *in vitro* extraction approach has been used successfully for PAHs in particular (Weston and Mayer 1998a and b; Voparil and Mayer 2000; Mayer et al. 2001).

Similarly to the potential effects of the physico-chemical nature of the binding of contaminants to sediment particles on their subsequent trophic bioavailabilities to sediment ingestors, the physico-chemical natures of metals accumulated by organisms preyed on by members of higher trophic levels in a food chain (along with the digestive properties of the feeder) also affect their trophic bioavailabilities. Palmqvist et al. (2006) examined the trophic transfer of the fluoranthene (Flu) PAH from two closely related polychaete species (*Capitella* sp. I and *Capitella* sp. S) differing in their PAH biotransformation abilities to the predatory polychaete *Nereis virens*. They found that *N. virens* fed on the biotransforming *Capitella* sp. I accumulated significantly more Flu equivalents compared to worms fed *Capitella* sp. S that showed very limited biotransformation ability. PCBs and PBDEs were biomagnified through trophic levels from phytoplankton through zooplankton, ultimately to predatory fish, while PAHs that are more easily biotransformable in biota showed an opposite trend. PAH concentrations in sea bass and flounder were lower than those in phytoplankton and zooplankton (Chapter 6).

For metals, the physico-chemical forms of storage also affect trophic transfer to a predator; biomineralized forms of accumulated metals are considered poorly bioavailable to predators (Nott and Nicolaidou 1990). However, depending on the exact physico-chemical nature of the relevant metal-rich granules and the variability between invertebrate digestive systems, some metals present in such insoluble form may become trophically available to some predators (Rainbow et al. 2007).

Trophic transfer of contaminants proceeds along the food chains that constitute the estuarine food web from, for example, microphytobenthos to meiofauna to macroinvertebrates or from phytoplankton to zooplankton to macroinvertebrates, with both routes potentially leading to fish. In Marennes–Oléron Bay, subject to chronic Cd pollution, Pigeot et al. (2006) estimated that most of the Cd was distributed among primary producers, mainly microphytobenthos (40%), and suspension feeders, mainly oysters (40%). Due to their rapid reproduction rate, the microphytobenthic compartment was estimated to be responsible for the largest flux of Cd (ca. 188 kg.yr^{-1}), suggesting an important role for the microphytobenthos in the biogeochemical cycle of Cd.

The difference in the nematode meiofauna assemblage structure between the Seine and Authie sites (Chapter 14) exerts potential effects on the trophic transfers of contaminants in the two estuaries. In the case of metals, Fichet et al. (1999) underlined the role of nematodes in the process of metal transfer to benthic and pelagic food webs. In different harbours of France's Atlantic coast, nematodes have higher Cd, Cu, Pb, and Zn transfer factors than benthic copepods or other benthic species. For organic contaminants, zooplanktonic copepods represent an important compartment for storage and thus transfer to higher trophic levels (Chapter 10). In the Seine estuary, total alkylphenol concentrations in the planktonic copepod *Eurytemora affinis* vary from 3423 to 6406 ng.g^{-1}. Mean concentrations of ΣPAHs and ΣPCBs were 879 and 976 ng.g^{-1} respectively, with maximum values reaching, respectively, 3866 and 1785 ng.g^{-1}. For comparison, mean concentrations in the ragworm *N. diversicolor* were 274 ng PAH.g^{-1} and 605 ng PCB.g^{-1} (Chapter 3).

The trophic transfer of a contaminant along a food chain brings with it an ecotoxicological risk, particularly in the cases of organic contaminants not transformed physiologically to a less toxic form at intermediate stages of a food chain, as is the case for trace metals along many aquatic food chains (Wang 2002). On several occasions, for example, a transfer of toxicity has been shown in parallel with the trophic transfer of organic contaminants (Rice et al. 2000; Lemière et al. 2005). Such situations may bring increased ecotoxicological risks to high trophic level predators such as birds (and even humans) at the top of a food chain based in a contaminated estuary, even without exposure of the predators to contaminated water or sediments.

Even if the chemical contamination of our environment has been efficiently reduced over recent decades, estuarine zones remain at risk because of historical contamination by very persistent compounds such as the PCBs in the Seine estuary, and also because of the presence of emerging contaminants (Chapters 3, 6, and 10). The presence of organic contaminants in estuarine food chains raises a very important question concerning food safety because the consumption of fish and shellfish (crustaceans and molluscs) from estuaries contributes importantly to human exposure to contaminants (Chapter 6).

16.3 VALIDATION OF BIOMARKERS IN AN ESTUARINE SPECIES REPRESENTATIVE OF THE SEDIMENTARY COMPARTMENT

The effects of stressors can be examined at different levels of biological organization from molecules to communities. Such recognisable biological effects or impairments have been proposed as biomarkers for revealing exposures to and/or effects of environmental contaminants. Over the past two decades, biomarker methodology has been developed mainly via biochemical and molecular means (Lagadic et al. 2000; Garrigues et al. 2001). The relevance of this approach remains controversial and Guilhermino (2007) reviewed the main criticisms. The first point is the relative importance of variability resulting from natural and contamination factors. Since Cairns (1992) underlined the importance of taking into account the ratio of signal to noise in any parameter, ecotoxicologists have carefully assessed the sources and amplitudes of natural variations. The influence of parameters such as variations in salinity, temperature, size, season, etc. may be taken into account in the experimental design of a biomarker sampling programme (or an equivalent ecotoxicological field programme), as was the case in this study. Natural confounding factors affecting biochemical and physiological biomarkers were discussed and compensated for in Chapters 4 and 8. Such factors are typically the same as those affecting accumulated chemical concentrations encountered when Mussel Watch programmes were initially designed and subsequently well implemented using appropriate sampling strategies and efficient statistical treatments (NAS 1980).

With regard to biomarkers, there is a lack of temporal series allowing the integration of seasonal fluctuations including both ecological and individual organism biological (growth, reproduction) factors. The present programme was an initial attempt to acquire such temporal data over a 3-year period through quarterly sampling (Chapter 4). Among the core biomarkers currently recommended in biomonitoring programmes (UNEP 1999), site differences in metallothionein (MT) concentrations in a chosen biological model were not applicable in the case of *Nereis diversicolor* chosen in this study. This feature is shared with other species (*Pyganodon grandis, Orchestia gamarellus*) and is probably based on the ability of this biomolecule to play its role in detoxification through increased concentration and/or increased turnover rate (Couillard et al. 1995; Mouneyrac et al. 2002). MT is envisaged as an intermediate step in the final detoxification of toxic metals in biomineralized form, in agreement with the observed important storage of metals in *N. diversicolor* as detoxified insoluble fractions in the Seine estuary and at other metal-contaminated sites (Berthet et al. 2003; Poirier et al. 2006). Increases in the activities of the glutathione S-transferase (GST) and the antioxidant enzyme catalase are also recognized as defence reactions of organisms exposed to different chemicals in their environments. In ragworms from the Seine estuary, only GST activity was significantly increased and no consistent increase of concentration of the TBARS end product of lipid peroxidation was established. All these data on biomarkers of defence are consistent with the tolerance to zinc observed in the ragworms from the Seine estuary (Chapter 7).

Despite exhibiting a number of traits indicating a potential for tolerance to metals and other compounds generating oxidative stress whose presence was recognized in the Seine estuary (Chapter 3), and given the absence of detectable genetically based selection (Chapter 7), the population of *N. diversicolor* living in the Seine estuary seems to be at some ecotoxicological risk. AChE activity, for example, was consistently lower in ragworms from the Seine estuary than in those from the reference site. AChE activity was initially considered a specific biomarker of carbamate and organophosphate pesticides but the responsiveness of AChE to contaminants other than pesticides suggests that AChE activity may be a useful biomarker of general physiological stress in aquatic organisms and in *N. diversicolor* in particular (Chapter 4).

Intersite differences related to energy metabolism were also observed, possibly as consequences of changes in energy allocation associated with the physiological cost of surviving at an impacted site (Chapter 8). In addition to the cost of tolerance, chemical stress can disrupt a number of processes involved in the energy budgets of organisms (Figure 16.4). Many reports discuss the impairment of feeding behaviours in organisms exposed to pollutants in the field or in the laboratory. Different causes may be responsible for feeding disturbances such as avoidance of contaminated foods (Wilding and Maltby 2006), reduction of the filtration rate in the presence of neurotoxicants (Reynaldi et al. 2006; Barata et al. 2007), and/or impairment of food detection (Krång and Rosenqvist 2006). Feeding rate, a key factor in energy metabolism, may be quantified relatively easily by a methodology known as post-exposure feeding depression developed by McWilliam and Baird (2002) and applied to various species by subsequent workers. In addition, metals and organic contaminants can disturb food assimilation through the impairment of digestive enzymes (Soetaert et al. 2007; Dedourge et al. 2008).

Effects on energetics are manifested in individuals by reduced growth efficiency or growth potential (scope for growth) (Widdows et al. 1990), poor condition, and finally reproductive failure. In the gastropod *Hydrobia ulvae* exposed to estuarine sediments contaminated through mining activities, both feeding and growth rates

CHEMICAL STRESS

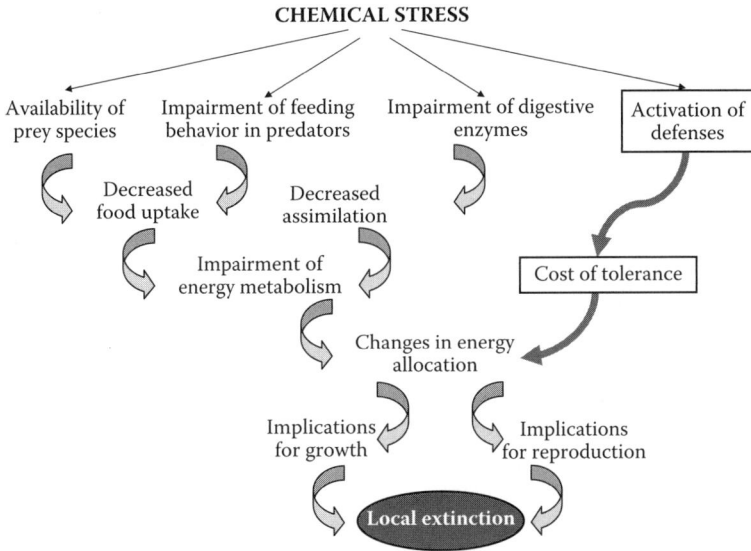

FIGURE 16.4 Cascading effects of chemical stress at different levels of biological organisation.

were decreased (Shipp and Grant 2006). In the gastropod *Helix engaddensis*, the presence of Cd or Cu in food reduced both feeding rates and growth (Swaileh and Ezzughayyar, 2000). In addition to these direct effects, indirect effects may arise from the depletion of populations of prey species (Figure 16.4).

In the present study, clear intersite differences were observed in diatom, foraminiferan, and nematode densities (at least during the spring and summer months in the last case). Whatever the causes of these differences, one consequence would be a decrease in the availability of food for primary consumers including *Nereis diversicolor*. No consistent trends, however, were depicted between the abundance of such food organisms and concentrations of energy reserves like glycogen and lipid in *N. diversicolor*. This is not really surprising since *N. diversicolor*, as an opportunistic species, is considered able to fulfil its energy needs using different diets (Meziane and Retière 2002; Olivier et al. 1995). However, the impairment of energy reserves (such as glycogen and lipids), condition, and total fecundity observed in the specimens from the Seine compared to those from the reference Authie estuary (Chapter 8) may have different origins as shown in Figure 16.4. In the ragworm, exposure to estuarine sediment originating from a contaminated site has been shown to be responsible for post-exposure feeding depression (Moreira-Santos et al. 2005; Moreira et al. 2006). In addition, activities of the digestive enzymes amylase and cellulase were lower in *N. diversicolor* from contaminated sediments including specimens from the Seine estuary (J. Kalman, personal communication).

Activation of physiological defence mechanisms against accumulated toxins by worms is well documented (Chapter 4). Such cascading effects (Figure 16.4) may lead to reduced populations as observed in the Seine (Chapter 8). For *N. diversicolor*, density was comparable in both of the estuaries of interest at the beginning of the programme ($\simeq$1000 individuals per square meter at minimum in winter); then it decreased in the Seine estuary to <300 individuals per square meter. This phenomenon was not limited to our sampling area (PN, Figure 1.3); annual sampling in the entire estuary indicated the same trend (Chapter 9). We suspect that the sustainability of a species in a given medium is linked to the size of its local population (concept of the minimum viable population). Stress is expected to exacerbate the effects of environmental fluctuations and limit the ability of populations to recover from negative oscillations (Maltby et al. 2001).

Results of biomarker studies on a fish, the flounder *Platichthys flesus* (Chapter 11), are consistent with the interpretation above for *N. diversicolor*. Fish from the Seine, with high accumulated concentrations of contaminants, showed raised activities of EROD, an enzyme involved in the biotransformation of accumulated organic compounds. These Seine fish additionally contained high concentrations of potential endocrine disruptors including nonylphenols, and indeed showed increased vitellogenin synthesis indicating endocrine disruption. Further evidence for endocrine disruption is provided by altered gonad development, gametogenesis, and fecundity. Ultimately the Seine fish showed low condition and growth indices, expected outcomes of the transfer of energy to meet the physiological costs of toxin biotransformation and/or storage detoxification.

16.4 HEALTH AND ECOLOGICAL CONSEQUENCES OF TOLERANCE

The acclimation or adaptation of organisms to chemical stress in the environment is well established, so allowing the preservation of biodiversity and the normal functioning of ecosystems. However, the evolution of tolerance to a toxic contaminant is not common. We noted in Chapter 7 that living in a contaminated environment is not alone sufficient to select for tolerance: the pollution pressure must be strong enough.

For metals, background concentrations in estuarine and coastal environments vary with precise location, season, river inputs, etc., and organisms are accustomed to maintaining intracellular essential element concentrations within optimal ranges relatively independently of external essential element concentrations (Rainbow 2007). Within the optimal exposure range of an organism to a contaminant, homeostasis as a normal life-regulating process appears not to require significant additional energy. Thus Van Tilborg and Van Assche (1998) suggested that the homeostasis under these conditions is most probably stress-free and they quoted a number of published observations (mainly in plants) to support this hypothesis. After the boundaries of homeostatic regulation capacity are exceeded, an organism experiences stress and, in response, must activate adaptation processes to survive (Van Tilborg and Van Assche 1998). The energy required for adaptation processes may interfere with allocations of energy for growth and reproduction, with expected impacts at the level of populations (Calow 1991).

The data obtained for ragworms *N. diversicolor* from the Seine estuary agree with this concept, even if other sources of energy disturbances are not excluded. The fitness cost of tolerance has been particularly well demonstrated in fish selected in the laboratory over eight generations for their resistance to Cd (Xie and Klerks 2004). In European flounders living in more or less contaminated estuaries, a relationship between stress resistance (alleles or genotypes associated with reduced DNA damage) and energetic trade-offs (reduction of fecundity, growth rate, and condition factor) has been detected (Marchand et al. 2004). This loss of fitness with a risk of extinction of local populations (Figure 16.4) is particularly important for the functioning of estuarine ecosystems in which macrofauna, at least, are characterized by a limited number of species well adapted to severe natural ecological conditions (salinity and temperature shocks, hypoxia), each species being represented by a large number of individuals (McLusky 1989). In addition, the aquatic organisms that live at the limit of their tolerance to natural changes (temperature, salinity, nutrients, food availability) are generally more sensitive to additional stresses such as exposure to toxic chemicals (Hummel et al. 1997; Heugens et al. 2001).

At the base of the food web, the microphytobenthos is very important on estuarine mudflats. In the case of photosynthetic activity, no intersite differences were apparent between the multipolluted Seine estuary and the reference Authie estuary site (Chapter 7), whereas a clear contrast was observed in diatom densities (Chapter 12). As noted above, differences in hydrodynamics were important in controlling the densities and structures of diatom assemblages. However, in another taxon, the freshwater plant *Lemna minor*, a study by Cedergreen et al. (2007) showed that for some herbicide combinations, the toxicity of the mixture as measured by pigment content did not reflect the results measured on plant population growth, emphasizing the importance of measuring growth in parallel with biomarkers.

While apparently achieving the conservation of local biodiversity, the presence of tolerant biota may also result in the presence of highly contaminated links in local food webs (see Section 16.2 above).

Based on the roles of bacteria in many aspects of ecosystem functioning, the preservation of microbial communities is, of course, important. However, crossresistance is well demonstrated and thus may contribute to the undesirable acquisition of tolerance, for example, toward antibiotics in pathogen species.

Even if tolerance is widely present in organisms chronically exposed to contaminants, it does not ensure that all populations can survive in highly polluted environments. Even between the limits allowing survival, a number of negative aspects are well documented. Thus the statement by Klerks and Weis (1987) that "It seems dangerous to relax water quality criteria on the assumption that all populations in polluted environments will evolve an increased resistance" has been reinforced by the findings of subsequent researchers.

16.5 UNDERSTANDING THE HEALTH STATUS OF ESTUARINE ECOSYSTEMS: A GLOBAL SCHEME FOR A COMPREHENSIVE APPROACH

From the integrated approach described above, we wish to propose an improved comprehensive methodology for assessing the health status of estuarine ecosystems that is relevant for many estuaries. Several authors have proposed that initially effects of pollutants may be detected by the measurement of relatively nonspecific biomarkers, usually high in the hierarchy of biological organisation, such as physiological biomarkers relating to the growth or reproduction of individual organisms or behavioral disturbances (Depledge et al. 1995; Lam and Gray 2003), or "open" genetic markers like transcriptomic variations identified by differential display (Liang and Pardee 1992) or subtractive hybridization (SSH; Diatchenko et al. 1996; Gurstaya et al. 1996).

The present study emphasizes the usefulness of biomarkers related to energy metabolism or reproduction to assess stress in estuarine environments exposed to important anthropogenic activities. Since the input of individuals within a population in terms of growth (biomass) and reproductive output (persistence) governs the maintenance of a population, energy disturbances may exert consequences at higher levels of biological organization. Thus, the final goal is to link effects of environmental impacts at subindividual (energy reserve), individual (condition indices, oxygen consumption), and population levels (Calow 1991; Maltby et al. 2001).

Within this objective, genetic tools can be useful but still need to be developed to detect the expression of genes supposed to be involved in energetic processes. Called "the missing biomarker link" by De Coen and Janssen (2003), energy is the most important field explored with a view to achieve scale changes, with the development of effect modelling based upon energy metabolism impairments (Kooijman and Bedaux 1996; Kooijman 2000).

Behavioral ecotoxicology provides another opportunity to bridge the gap between early, sensitive responses to stress at the infraorganismal level and long-term, ecologically relevant responses at the supraorganismal level because it clearly allows the linking of disturbances at biochemical level (altered neurotransmitters and thyroid hormones) to effects at the population level. These effects may be direct, such as impairment of the search for a sexual partner, care of juveniles, and avoidance of predators or pollutants. Indirect effects may include alteration of reproduction success due to impairment of feeding and thus, energy metabolism (Figure 16.4). Although they participate at a high level of biological organisation, behavioral responses are generally very sensitive and have been observed in organisms exposed to contamination in the field and in the laboratory at realistic concentrations (see review by Amiard-Triquet 2009).

Results of behavioral studies have been considered in enacting some environmental regulations, in particular in the U.S. (Little 2002), while in contrast they have been largely ignored by the European Community (EC 2000 and 2003). To convince appropriate authorities of the need to integrate biomarkers of behaviour when defining criteria for environmental standards, Little (1990, cited by Grue et al. 2002) suggested that "a few monumental field studies will be needed to confirm what we as behavioral toxicologists continually assume to be intuitively obvious: that changes in behaviour are reflected at the population and community levels." Eighteen years later, perhaps the collection of studies synthesised by Weis et al. (2001), supplemented by those reviewed by Amiard-Triquet (2009), may be considered to constitute the convincing evidence sought by Little (1990).

Endocrine disruption, genotoxicity, and immunotoxicity (see Bryan and Gibbs 1991; Vasseur et al. 2008; Fournier et al. 2005) are other impairments that most probably exert consequences at the population level, even at the community level. Lysosomal stability reflects toxicant-induced cell pathologies caused by a variety of chemicals. Correlations (Moore et al. 2006) of such lysosomal parameters with total oxyradical scavenging capacity (TOSC) and scope for growth (robust biomarkers at the physiological level of organisation) support the efficacy of the use of lysosomal biomarkers as indicative of toxic effects at higher biological levels.

Among the criticisms directed at biomarker methodologies is their lack of ecological relevance (Guilhermino 2007). On the other hand, their ability to reveal environmental disturbances well before changes appear in benthic communities is of considerable advantage. The programme developed in the Seine estuary as well as (a few) other research studies mentioned in this book indicate that impairments observed at the individual or subindividual levels may be linked to changes at the population, assemblage, and community levels. However, additional evidence is needed to convince regulatory authorities of the usefulness of core and ecologically relevant biomarkers (lysosomal parameters, energetics, behaviour, endocrine disorders, immunotoxicity, and genotoxicity).

The Water Framework Directive represents the path forward chosen by the EC to assess environmental risks in aquatic habitats. The failure of the directive to recognise a role for biomarkers in this context is regrettable. We have successfully correlated biomarkers measured at lower organisation levels (for example, biological markers such as enzyme activities, Chapter 4) in *N. diversicolor* to intermediate level (physiological) biomarkers such as levels of energy reserves (Chapter 8) and to effects at the population level (abundance and population structure) in the ragworms (Chapter 8). These biomarkers have also been coincident with observed changes at the community level for microphytobenthic diatoms (Chapter 12), meiofaunal foraminiferans (Chapter 13), and nematode communities (Chapter 14). The Water Framework Directive recommends investigation of the abundance and composition of phytoplankton, other aquatic flora, benthic invertebrate fauna, and fish (additionally investigating age structures in fish). In agreement, in this study we studied microphytobenthos (diatoms), benthic invertebrates at meiofaunal (foraminiferans, nematodes) and macrofaunal (*N. diversicolor*) levels, and the growth and fecundity of the flounder *Platichthys flesus* (Chapter 11). Additionally, we have studied benthic foraminiferans (Chapter 13), zooplanktonic copepods (Chapter 10), and microbial communities (Chapter 7). For the smaller organisms, however, we found that variations in hydrodynamic characteristics within and between the compared estuaries was a strong controlling factor given the effected physical changes in surface sediments.

The comparative study of the diatom assemblages of the two estuaries was informative (Chapter 12). A clear intersite difference in diatom densities was noted, although a strong argument can be made that the difference is a result of different hydrodynamic conditions in the estuaries rather than differential contaminant availabilities alone. In the case of the foraminiferans, differences in assemblage structure were found in a comparison of the two estuaries (Chapter 13). Although it was not possible to exclude the potential effect of differences in sedimentation processes of the estuaries, it does appear from the assemblage data for the Seine (lower density and lower total taxon number of foraminiferans, coupled with increased dominance of *Ammonia tepida,* usually associated with highly impacted habitats) that the main cause of differences in foraminiferan assemblages of the estuaries was their differential contamination.

The foraminiferan study also revealed temporal changes in both estuaries between 2003 and 2004, perhaps in reflection of a more general (climate-related?) phenomenon. Over this period, an increase in the proportion in the Authie of another species considered tolerant of adverse conditions, *Haynesina germanica,* also occurred, suggesting that even the Authie may be facing increased environmental pressure. High proportional contributions of the two "tolerant" species, *A. tepida* and *H. germanica,* in foraminiferan assemblages have considerable potential as bioindicators at a high level of biological organisation.

The differences in nematode meiofauna assemblage structures of the Seine and Authie sites (Chapter 14) also represent compelling evidence that the two systems are subject to very different ecological pressures despite the similarities in salinity, sediment granulometry, and organic loading. It is possible (arguably likely) that this reflects the combined effects of the chronic levels of multiple pollutants found in the Seine, coupled with significant and rapid periods of sediment erosion and deposition (Chapter 2).

Microphytobenthic taxa and meiofauna such as foraminiferans and nematodes are relatively abundant compared with macrofaunal taxa, so they can be investigated with small samples, relatively simple equipment, and minimised impacts on study sites. Such taxa also tend to exhibit higher diversity than macrofauna, increasing the range of potential responses to pollutants and potential sensitivity for detecting impacts. Data obtained have high information content, facilitating statistical analysis. The short generation times and low dispersal characteristics of most of these taxa are associated with rapid population responses to environmental perturbations.

Some meiofaunal taxa, particularly the nematodes, have also been found to contain species with very high tolerance to certain pollutants, such that nematodes may be the last metazoans to persist in grossly polluted systems and thus provide the only model organisms to study under such conditions. On the other hand, such microscopic taxa often require specialist taxonomic skills—expertise that is becoming less available throughout the world—unless this so-called "taxonomic impediment" can be reversed.

Similar arguments for and against also apply to the investigation of zooplankton in this context. This study has concentrated on contaminant bioaccumulation and the population dynamics of the common estuarine copepod *Eurytemora affinis* (Chapter 10). *E. affinis* is the dominant member of the zooplankton assemblage in the Seine estuary and appears well adapted to live in the oligo- and meso-haline zones of a strongly contaminated estuary, even in the presence of strong hydrodynamic processes. *E. affinis* can therefore be considered a potential bioindicator of an estuary impacted by strong anthropogenic contamination. Species of another widely distributed and ecologically important copepod genus, the harpacticoid *Tigriopus,* have been proposed as potentially viable and ecologically relevant invertebrate toxicity testing models for evaluating the impacts of marine pollution in coastal regions (Raisuddin et al. 2007). Copepods are key members of estuarine food chains, playing an important role in the transfer of both energy and aquatic contaminants, and deserve consideration in ecotoxicological programmes.

Much has already been written here about the value of the infaunal polychaete *Nereis diversicolor* in such ecotoxicological programmes, allowing the biomonitoring of contaminant exposures and the concomitant detection with biomarkers of ecotoxicological effects at levels of biological organisation from the biochemical to the population.

Studies of flounder (Chapter 11) were also successful in revealing low growth and condition indices in fish from the Seine in spite of an apparent sufficiency of food in the studied area at the mouth of the estuary. These higher order effects were correlated with increased bioaccumulation of contaminants and with observed physiological changes in processes of biotransformation of organic compounds and vitellogenin synthesis. These results echo our findings with *N. diversicolor* (Chapter 8). While the reduction of available surface area of sediment harbouring the prey species of flatfish (Chapter 9) may be a source of energy metabolism disturbance, opportunistic feeding behaviour has been observed in flatfish that are able to shift from benthic prey species to water column species such as copepods *E. affinis* and mysids *Neomysis integer*. In the Loire estuary, copepods represent the major food resources for juvenile sole in spring and are also important for juvenile flounder. Considering all size classes and all seasons, copepods were present in the gut contents of 11 to 40% of sole, depending on the collection area (oligo- and meso-haline, respectively), and mysids were present in 20 to 40% of individuals. In flounder, copepods were present in 25 to 29% of individuals, and mysids in 24 to 25% (Marchand et al. 1977).

16.6 RECOMMENDATIONS

It remains then to use the information compiled in this book and the lessons learnt therefrom to make recommendations on designing a programme of research to assess the environmental health status of an estuary suspected to have been impacted by anthropogenic contamination.

16.6.1 CHOICE OF SITES FOR COMPARISON

Based on our present state of knowledge, it is unacceptable to study the ecotoxicological status of an estuary without comparison against a reference site. Such a reference site should be chosen to be as close as possible to the first estuary in terms of salinity regime, temperature, granulometry, organic content of sediment, and other parameters to mitigate the importance of confounding factors. In practice, it is probably impossible to achieve perfect agreement between estuaries but every attempt must be made to assess the effects of all such differences where possible. Even this may prove impossible, for example for differences in hydrodynamic conditions, but a careful assessment of these differences is still necessary. The addition of more than one reference site into any comparative study, however superficially attractive, presents great resource implications.

16.6.2   CHEMICAL ASSESSMENT OF ENVIRONMENTAL QUALITY

Safety standard concentrations have usually been established by considering the toxicity of a given contaminant only in water. This soluble phase is, therefore, still the only contamination source usually recognized in regulations, despite the now established role of food, that may include sediments, as predominant sources of contaminants for many animals.

Concentrations determined in water masses are typically compared to these dissolved phase safety standards to assess environmental quality. In water, concentrations generally fluctuate, appear at low levels, and are not easily accessible to analysis. It is thus necessary to take into account chemical speciation of metals and physico-chemical characteristics of organics. In the water column and in the porewater of sedimentary compartments, new tools, the so-called passive samplers, provide opportunities to conduct relevant analyses, allowing access to characteristics that govern contaminant bioavailability, but they need careful interpretation.

Because sediments are the major sinks for most chemicals introduced by human activities into the environment, showing high concentrations and thus being relatively easy to determine, the sediment matrix has been often proposed for environmental monitoring. If chemical speciation is taken into consideration, sediments have real interest for determining presence in the case of metals. The ΣSEM/AVS ratio has been proposed as a fast tool to evaluate the potential toxicities of sediments, thus eliminating the need for many biological tests; it was cited in USEPA regulations (1995). Nevertheless this ratio must be carefully determined in the field, particularly because it is relevant only if oxic and anoxic sediments are considered separately.

It is valid to ask whether the biomonitoring of bioaccumulated concentrations of contaminants has a role in an ecotoxicological assessment of an estuary. Although such biomonitoring provides information on the temporal and spatial distribution of the availabilities of specified contaminants, it provides no information on ecotoxicological effects, and may thus be a superfluous step. The measurement of bioaccumulated concentrations of a contaminant absolutely requires nomination of that contaminant before analysis. This is easily done in the cases of toxic metals through comprehensive analytical techniques such as inductively coupled plasma spectrophotometry, and would be of considerable value if the only contaminants present are toxic metals. This is of course rarely the case in anthropogenically contaminated estuaries that inevitably are exposed to mixtures of "classical" organic contaminants such as PAHs and PCBs and also "emerging" contaminants such as PBDEs, pharmaceuticals, alkylphenols, and yet unknown organic contaminants with limited or no analytical tools available. It is arguably impossible to measure all potential organic contaminants in an organism and it may be a waste of resources to try to do so.

16.6.3   BIOMONITORING BASED ON BIOMARKERS

There is clearly value in analysing biomarkers. Biomarkers are direct manifestations of the exposure of an organism in the field to a toxin or other form of stress.

Evidence of field exposure is absolutely vital, and biomarkers are relatively effective in revealing overall toxicities of complex mixtures, particularly those at a high level of biological organization such as physiological biomarkers relating to growth or reproduction of individual organisms.

The first step may consist of investigations of impairments (lysosomal parameters, bioenergetics, behaviour, endocrine disruption, genotoxicity, immunotoxicity) that are most likely to exhibit consequences at the population or even community level simultaneously in a multipolluted estuary (such as the Seine in the present case study) and a comparatively clean estuary (such as the Authie).

When such impairments are revealed, it is necessary to identify the main classes of contaminants that may be responsible for them. At this stage, a multibiomarker approach based the on so-called "core" biomarkers is appropriate. Biomarkers of defence (SOD, HSP, catalase, MT, P-gp) and damage (AChE and TBARS) can be utilized (Van der Oost et al. 2005) to determine gene expression, induction or changes in protein concentrations and enzyme activities, or both. The data can be interpreted by using principal component analysis, as in the present study, and incorporated into integrative indices or expert systems (for details see Chapter 4).

The measurement of chemicals in physical compartments of the environment and in biota (either as total concentrations or, more interestingly, concentrations in specific fractions and metabolites) will permit the validation of hypotheses based on biomarker studies. However, in estuaries, complex mixtures of contaminants are present, including many classes of compounds (many persistent organic pollutants) that are not yet accessible to analysis or are extremely expensive to analyse. In some cases, it will remain impossible to establish a clear link between observed biological responses and specific contaminant doses.

The responses of these biomarkers (lysosomal parameters, bioenergetics, behaviour, endocrine disruption, genotoxicity, immunotoxicity) that have capacities to forecast population effects may be linked to changes at the assemblage and community levels, provided that the studied species plays a significant role in the structure and functioning of the ecosystem.

16.6.4 Choice of Biota for Sampling

Initially the choice of biota to be sampled will depend on the geomorphological nature of the estuaries under consideration, especially the nature of the substrate (rock, mud, or sand), for example. In northwest Europe the common mussel *Mytilus edulis* is a strong candidate for sampling from hard substrates, as are the worm *Nereis diversicolor* and bivalves *Scrobicularia plana* and *Macoma balthica* from soft substrates. Euryhaline zooplanktonic copepods will be present over rock or sediment when tides are in. The common European shore crab, *Carcinus maenas*, although somewhat mobile, has the advantage of living in many coastal habitats including salt marshes, sedimentary and rocky shores, and the sublittoral. Clearly the choice of biomarkers involves the utilisation of particular organisms in which the use of these biomarkers has been established, as is the case for the above-named invertebrates for one or more relevant biomarkers.

In addition to the criteria applying to biomonitoring based on chemical analysis in bioaccumulators (NAS 1980), other factors must be carefully considered: relative sensitivity of a species or a particular stage of its life cycle; role in biogeochemical cycles of both nutrients and contaminants; importance in the food web (for details see Galloway et al. 2006; Berthet 2008). By focussing on key species, it is assumed that impairments of their responses used as biomarkers will reveal a risk of cascading deleterious effects right through to the ecosystem level. Disturbances at the supraindividual levels affecting these organisms may result in ecological disorders such as changes in biogeochemical cycles and trophic relationships that govern nutrient and food availability (both as energy resource and risk of contaminant transfer) in the vital ecosystems that are estuaries.

The same concepts are useful when considering the choice of taxa to be studied at higher levels of biological organisation such as the role of bacterial communities and primary producers in biogeochemical cycles and their importance at the bases of aquatic food webs. The advantages of small-sized biota were discussed in detail above.

16.6.5 CONFOUNDING FACTORS

A large range of natural factors, both biological (size, sex, position in intertidal zone, seasonal changes in physiology) and abiotic, can influence living organisms, their health status, and even their presence in ecosystems. This is particularly true for estuarine ecosystems that are associated with significant physical fluctuations (salinity, turbidity, hypoxia) over different time scales (tidal rhythm, seasonal changes, interannual fluctuations, climate changes). A number of these factors that may interfere with the responses of biota to anthropogenic stress, particularly chemical stress, may be easily controlled as described previously in the case of biomonitoring based on chemical analyses in bioaccumulators (NAS 1980).

Factors that cannot be controlled (salinity, granulometry, hydrodynamics, food availability) must at least be measured in parallel with ecotoxicological determinations. For some parameters, it may be possible to determine their part in overall fluctuations by designing special studies that allow the measurement of variations of a biological response of interest along a gradient of salinity, temperature, etc., with the intent of establishing correction factors (Mourgaud et al. 2002; Poirier et al. 2006).

16.6.6 MODELLING

Mathematical models may be useful in complementarity with measures, field observations, and mechanism studies. In the Seine estuary, modelling was applied to many processes (Thouvenin 1999) such as multivariate models of transport concerning both dissolved and particulate constituents of the water masses; models of physical characteristics of agitation (waves); models of geochemical processes such as those linked to early diagenesis (Chapter 15) or to carbon cycle in the water column; models of biological processes including biokinetics of bioaccumulation/biomagnification (Chapter 6). Population dynamics may be also linked to energy metabolism impairments (Kooijman and Bedaux 1996; Kooijman 2000) or effects of endocrine disruptors (Rose et al. 2003; Miller et al. 2007). Models are research tools that allow us to test our understanding of a number of processes and simulate their spatial and temporal variations in an estuarine ecosystem.

As exemplified by dynamic diagenetic modelling, the improvement of such methodologies depends on the quality of field measurements. The complementarity of field measurements and modelling has a role in validating a model. From an operational view, the next step will be designing predictive models with a sufficient level of confidence for their use in environmental management.

16.6.7 SAMPLING REGIME

For this study, we chose a common (quarterly) sampling programme for all biota, clearly representing a compromise outcome of a cost–benefit analysis. Arguably a quarterly sampling regime is not optimal for macrofauna such as *N. diversicolor*. To account for temporal variability, more frequent sampling is necessary, and very often monthly sampling is utilised in ecological studies. If the intent is to assess spatial variability, the sampling design adopted will have further objectives, for example to account for variation in salinity regimes along an estuary, or for differential emersion/immersion regimes (and associated salinity and temperature changes) experienced at different vertical heights on a shore. The addition of sampling stations to allow for such spatial variability will require a reduction in temporal frequency of sampling for reasons of practicability. At a bare minimum, such sampling may be conducted annually; for benthic invertebrate macrofauna, this should be at the end of the summer when recruitment is complete and juveniles are large enough (>1 mm, the conventional lower size limit of macrofauna) to be collected.

Taxa characterized by short generation times pose a particular problem for the design of sampling programmes, specifically the frequency of sampling effort over the year, particularly against a background of natural seasonal variation. Such questions are not unique to the smaller organisms sampled but are clearly highlighted in their case. Inevitably a compromise must be made between sampling effort and likelihood of recognising real changes caused by contamination against a natural variable background. In this study, the chosen quarterly sampling regime did reveal population changes in the organisms investigated.

REFERENCES

Amiard-Triquet, C. 2009. Behavioral disturbances: the missing link between sub-organismal and supra-organismal responses to stress? Prospects based on aquatic research. *Hum. Ecol. Risk Assess* (in press).

Barata, C. et al. 2007. Combined use of biomarkers and *in situ* bioassays in *Daphnia magna* to monitor environmental hazards of pesticides in the field. *Environ. Toxicol. Chem.* 26: 370–379.

Batley, G.E., Ed. 1989. *Trace Element Speciation: Analytical Methods and Problems.* Boca Raton, FL: CRC Press.

Bernes, C. 2000. *Persistent Organic Pollutants: A Swedish View of an International Problem.* Stockholm: Swedish Environmental Protection Agency.

Berthet, B. et al. 2003. Accumulation and soluble binding of cadmium, copper, and zinc in the polychaete *Hediste diversicolor* from coastal sites with different trace metal bioavailabilities. *Arch. Environ. Contam. Toxicol.* 45: 468–478.

Berthet, B. 2008. Les espèces sentinelles. In *Les biomarqueurs dans l'évaluation de l'état écologique des milieux aquatiques*, Amiard-Triquet, C. and Amiard, J.D., Eds. Paris: Lavoisier, pp. 122–148.

Bryan, G.W. and P. E. Gibbs. 1991. Impact of low concentrations of tributyltin (TBT) on marine organisms: a review. In *Metal Ecotoxicology: Concepts and Applications*, Newman, M.C. and McIntosh, A.W., Eds. Chelsea, MI: Lewis Publishers, pp. 323–361.

Cairns, J. Jr. 1992. The threshold problem in ecotoxicology. *Ecotoxicol* 1: 3–16.

Calow, P. 1991. Physiological costs of combating chemical toxicants: ecological implications. *Comp. Biochem. Physiol.* 100C: 3–6.

EC. 2000. Directive 2000/60/EC of the European Parliament and Council of 23 October 2000 establishing a framework for policy in the field of water (JOL 327). Brussels: European Commission.

EC. 2003. Technical guidance document in support of Commission Directive 93/67/EEC on risk assessment for new notified substances, Commission Regulation 1488/94 on risk assessment for existing substances, and Commission Directive 98/8 on biocides. Brussels: European Commission.

Cedergreen, N. et al. 2007. Is mixture toxicity measured on a biomarker indicative of what happens on a population level? A study with *Lemna minor. Ecotoxicol. Environ. Saf.* 67: 323–332.

Cornelissen, G. et al. 2006. Bioaccumulation of native polycyclic aromatic hydrocarbons from sediment by a polychaete and a gastropod: freely dissolved concentrations and activated carbon amendment. *Environ. Toxicol. Chem.* 25: 2349–2355.

Cornelissen, G. et al. 2008. Field testing of equilibrium passive samplers to determine freely dissolved native polycyclic aromatic hydrocarbon concentrations. *Environ. Toxicol. Chem.* 27: 499–508.

Couillard, Y. et al. 1995. Field transplantation of a freshwater bivalve, *Pyganodon grandis*, across a metal contamination gradient. I. Temporal changes in metallothionein and metal (Cd, Cu, and Zn) concentrations in soft tissues. *Can. J. Fish Aquat. Sci.* 52: 690–702.

De Coen, W.M. and C.R. Janssen. 2003. The missing biomarker link: relationship between effects on the cellular energy allocation biomarker of toxicant-stressed *Daphnia magna* and corresponding population characteristics. *Environ. Toxicol. Chem.* 22: 1632–1641.

Dedourge, O., A. Geffard, and C. Amiard-Triquet. 2008. Origine des perturbations du métabolisme énergétique. In *Les biomarqueurs dans l'évaluation de l'état écologique des milieux aquatiques*, Amiart-Triquet, C. and Amiard, J.C., Eds. Paris: Lavoisier, pp. 241–271.

De Perre, C., H. Budzinski, and E. Parlanti. 2007. Application of solid phase microextraction (SPME) to the study of interactions between dissolved organic matter (DOM) and organic contaminants in aquatic environment. IMOG 2007, Torquay, U.K.

Depledge, M.H., A. Aagaard, and P. Györkös. 1995. Assessment of trace metal toxicity using molecular, physiological and behavioural biomarkers. *Mar. Pollut. Bull.* 31: 19–27

Diatchenko, L. et al. 1996. Suppression subtractive hybridization: a method for generating differentially regulated or tissue-specific cDNA probes and libraries. *Proc. Nat. Acad. Sci. USA* 93: 6025–6030.

Fichet, D. et al. 1999. Concentration and mobilisation of Cd, Cu, Pb and Zn by meiofauna populations living in harbour sediment: their role in the heavy metal flux from sediment to food web. *Sci. Total Environ.* 243–244: 263–272.

Fournier, M. et al. 2005. Biomarqueurs immunologiques appliqués à l'écotoxicologie. *Bull. Soc. Zool. Fr.* 130: 333–351.

Gagnon, C. and N.S. Fisher. 1997. The bioavailability of sediment-bound Cd, Co, and Ag to the mussel *Mytilus edulis*. *Can. J. Fish. Aquat. Sci.* 54: 147–156.

Galloway, T.S. et al. 2006. The ECOMAN project: a novel approach to defining sustainable ecosystem function. *Mar. Pollut. Bull.* 53: 186–194.

Garrigues, P. et al. 2001. *Biomarkers in Marine Organisms: A Practical Approach.* Amsterdam: Elsevier.

Grue, C.E., S.C. Gardner, and P.L. Gibert. 2002. On the significance of pollutant-induced alterations in the behaviour of fish and wildlife. In *Behavioural Ecotoxicology*, Dell'Omo, G., Ed. Chichester: John Wiley & Sons, pp. 1–90.

Guilhermino, L. 2007. The use of biomarkers to assess the effects of environmental contamination in coastal and estuarine ecosystems: what questions remain? An example of the Portuguese NW coast. http://www.ices.dk/products/CMdocs/CM-2007/I/I-2007.pdf.

Gurskaya, N.G. et al. 1996. Equalizing cDNA subtraction based on selective suppression of polymerase chain reaction: cloning of Jurkat cell transcripts induced by phytohemagglutinin and phorbol 12-myristate 13-acetate. *Anal. Biochem.* 240: 90–97.

Heugens, E.H.W. et al. 2001. A review of effects of multiple stressors on aquatic organisms and analysis of uncertainty factors for use in risk assessment. *Crit. Rev. Toxicol.* 31: 247–284.

Hummel, H. et al. 1997. Sensitivity to stress in the bivalve *Macoma balthica* from the most northern (Arctic) to the most southern (French) populations: low sensitivity in Arctic populations because of genetic adaptations? *Hydrobiologia* 355: 127–138.

Hunter, W. et al. 2008. Using disposable polydimethylsiloxane fibers to assess the bioavailability of permethrin in sediment. *Environ. Toxicol. Chem.* 27: 568–575.

ITRC. 2005. *Technology Overview of Passive Sampler Technologies*, DSP-4. Washington: Interstate Technology & Regulatory Council. www.itrcweb.org.

King, A.J., J.W. Readman, and J.L. Zhou. 2004. Determination of PAHs in water by solid phase microextraction gas chromatography–mass spectrometry. *Anal. Chim. Acta* 523: 259–267.

Klerks, P.L. and J.S. Weis. 1987. Genetic adaptation to heavy metals in aquatic organisms: a review. *Environ. Pollut.* 45: 173–205.

Kooijman, S.A. and J.J.M. Bedaux. 1996. *The Analysis of Aquatic Toxicity Data.* Amsterdam: VU University Press.

Kooijman, S.A. 2000. *Dynamic Energy and Mass Budgets in Biological Systems.* Cambridge: Cambridge University Press.

Krång, A.S. and G. Rosenqvist. 2006. Effects of manganese on chemically induced food search behaviour of the Norway lobster, *Nephrops norvegicus* (L.). *Aquat. Toxicol.* 78: 284–291.

Lam, K.S. and J.S. Gray. 2003. The use of biomarkers in environmental monitoring programmes. *Mar. Pollut. Bull.* 46: 182–186.

Langezaal, A.M. et al. 2003. Disturbance of intertidal sediments: the response of bacteria and foraminifera. *Estuar. Coast. Shelf Sci.* 58: 249–264

Lemière, S. et al. 2005. DNA damage measured by the single-cell gel electrophoresis (comet) assay in mammals fed with mussels contaminated by the *Erika* oil spill. *Mutat. Res.* 581: 11–21.

Liang, P. and A.B. Pardee. 1992. Differential display of eukaryotic messenger RNA by means of polymerase chain reaction. *Science* 257: 967–971.

Little, E.E. 2002. Behavioural measures of injuries to fish and aquatic organisms: regulatory considerations. In *Behavioural Ecotoxicology*, Dell'Omo, G., Ed. Chichester: John Wiley & Sons, pp. 411–431.

Lu, X., D.D. Reible, and J.W. Fleeger. 2006. Bioavailability of polycyclic aromatic hydro-carbons in field-contaminated Anacostia River (Washington, DC) sediment. *Environ. Toxicol. Chem.* 25: 2869–2874.

Maltby, L. et al. 2001. Linking individual-level responses and population-level consequences. In *Ecological Variability: Separating Natural from Anthropogenic Causes of Ecosystem Impairment*, Baird, D.J. and Burton, G.A., Jr., Eds. Pensacola, FL: SETAC, pp. 27–82.

Marchand, J., J.C. Denayer, and D. Montfort. 1977. Etude écologique de la Basse-Loire de Nantes à Saint-Nazaire (Invertébrés–Vertébrés). University of Nantes, Contract OREAM 959-76-002-04.

Marchand, J. et al. 2004. Physiological cost of tolerance to toxicants in the European flounder *Platichthys flesus* along the French Atlantic Coast. *Aquat. Toxicol.* 70: 327–343.

McLeod, P. et al. 2004. Effects of particulate carbonaceous matter on the bioavailability of benzo[a]pyrene and 2,2′,5,5′-tetrachlorobiohenyl to the clam, *Mac. Balth. Environ. Sci. Technol.* 38: 4549–4556.

McLusky, D.S. 1989. *The Estuarine Ecosystem.* Glasgow: Blackie.

McWilliam, R.A. and D.J. Baird. 2002. Post-exposure feeding depression: a new toxicity end-point for use in laboratory studies with *Daphnia magna. Environ. Toxicol. Chem.* 21: 1198–1205.

Mayer, L.M., D.P. Weston, and M.J. Bock. 2001. Benzo[a]pyrene and zinc solubilization by digestive fluids of benthic invertebrates: a crossphyletic study. *Environ. Toxicol. Chem.* 20: 1890–1900.

Meziane, T. and C. Retière. 2002. Croissance de juvéniles de *Nereis diversicolor* nourris avec des détritus d'halophytes. *Oceanol. Acta* 25: 119–124.

Miller, D.H. et al. 2007. Linkage of biochemical responses to population-level effects: a case study with vitellogenin in the fathead minnow (*Pimephales promelas*). *Environ. Toxicol. Chem.* 26: 521–527.

Moore, M.N., J.I. Allen, and A. McVeigh. 2006. Environmental prognostics: an integrated model supporting lysosomal stress responses as predictive biomarkers of animal health status. *Mar. Environ. Res.* 61: 278–304.

Mora, S.J. 1999. The oceanic environment. In *Understanding Our Environment*, Harrison, R.M., Ed. Cambridge: Royal Society of Chemistry, pp. 139–198.

Moreira, S.M. et al. 2006. Effects of estuarine sediment contamination on feeding and on key physiological functions of the polychaete *Hediste diversicolor*: laboratory and *in situ* assays. *Aquat. Toxicol.* 78: 186–201.

Moreira-Santos, M. et al. 2005. Short-term sublethal (sediment and aquatic roots of floating macrophytes) assays with a tropical chironomid based on postexposure feeding and bio-markers. *Environ. Toxicol. Chem.* 24: 2234–2242.

Mouneyrac, C. et al. 2002. Partitioning of accumulated trace metals in the talitrid amphipod crustacean *Orchestia gammarellus*: a cautionary tale on the use of metallothionein-like proteins as biomarkers. *Aquat. Toxicol.* 57: 225–242.

Mourgaud, Y. et al. 2002. Metallothionein concentration in the mussel *Mytilus galloprovincialis* as a biomarker of response to metal contamination: validation in the field. *Biomarkers* 7: 479–490.

NAS. 1980. *The International Mussel Watch. Report of a Workshop Sponsored by the Environment Studies Board.* Washington: Natural Resources Commission, National Academy of Sciences.

Nott, J.A. and A. Nicolaidou. 1990. Transfer of metal detoxification along marine food chains. *J. Mar. Biol. Assess. U.K.* 70: 905–912.

Olivier, M. et al. 1995. Behavioural responses of *Nereis diversicolor* (O.F. Müller) and *Nereis virens* (Sars) (polychaeta) to food stimuli: use of specific organic matter (algae and halophytes). *Can. J. Zool.* 73: 2307–2317.

Palmqvist, A., L.J. Rasmussen, and V.E. Forbes. 2006. Influence of biotransformation on trophic transfer of the PAH, fluoranthene. *Aquat. Toxicol.* 80: 309–319.

Pigeot, J. et al. 2006. Cadmium pathways in an exploited intertidal ecosystem with chronic cadmium inputs (Marennes–Oléron, Atlantic coast, France). *Mar. Ecol. Prog. Ser.* 307: 101–114.

Poirier, L. et al. 2006. A suitable model for the biomonitoring of trace metal bioavailabilities in estuarine sediments: the annelid polychaete *Nereis diversicolor*. *J. Mar. Biol. Assess. U.K.* 86: 71–82.

Rainbow, P.S. 2007. Trace metal bioaccumulation: models, metabolic availability and toxicity. *Environ. Int.* 33: 576–582.

Rainbow, P.S. et al. 2007. Trophic transfer of trace metals: subcellular compartmentalization in bivalve by prey, assimilation by a gastropod predator and *in vitro* digestion simulations. *Mar. Ecol. Progr. Ser.* 348: 125–138.

Raisuddin, S. et al. 2007. The copepod *Tigriopus*: a promising marine model organism for ecotoxicology and environmental genomics. *Aquat. Toxicol.* 83: 161–173.

Reichenberg, F. and P. Mayer. 2006. Two complementary sides of bioavailability: accessibility and chemical activity of organic contaminants in sediments and soils. *Environ. Toxicol. Chem.* 25: 1239–1245.

Reynaldi, S. et al. 2006. Linking feeding activity and maturation of *Daphnia magna* following short-term exposure to fenvalerate. *Environ. Toxicol. Chem.* 25: 1826–1830.

Rice, C.A. et al. 2000. From sediment bioassay to fish biomarker: connecting the dots using simple trophic relationships. *Mar. Environ. Res.* 50: 527–533.

Rose, K.A. et al. 2003. Using nested models and laboratory data for predicting population effects of contaminants on fish: a step toward a bottom-up approach for establishing causality in field studies. *Hum. Ecol. Risk Assess.* 9: 231–257.

Shipp, E. and A. Grant. 2006. *Hydrobia ulvae* feeding rates: a novel way to assess sediment toxicity. *Environ. Toxicol. Chem.* 25: 3246–3252.

Soetaert, A. et al. 2007. Molecular responses during cadmium-induced stress in *Daphnia magna*: integration of differential gene expression with higher level effects. *Aquat. Toxicol.* 83: 212–222.

Swaileh, K.M. and A. Ezzughayyar. 2000. Effects of dietary Cd and Cu on feeding and growth rates of the land snail *Helix engaddensis*. *Ecotoxicol. Environ. Saf.* 47: 253–260.

Tessier, A. and D.R. Turner, Eds. 1995. *Metal Speciation and Bioavailability in Aquatic Systems*. Chichester: John Wiley & Sons.

Togola, A. and H. Budzinski. 2007. Development of polar organic integrative samplers for analysis of pharmaceuticals in aquatic systems. *Anal. Chem.* 79: 6734–6741.

Thouvenin, B. 1999. *Les modèles: outils de connaissance et de gestion*. Programme scientifique Seine Aval. Plouzané: Editions IFREMER.

UNEP/RAMOGE. 1999. *Manual on the Biomarkers Recommended for the MED POL Biomonitoring Programme*. Athens: UNEP.

USEPA. 1995. Science Advisory Board, Ecological Processes and Effects Committee, Sediment Criteria Subcommittee. http://www.epa.gov/sab/pdf/epe95020.pdf

Van der Oost, R. et al. 2005. Biomarkers in environmental assessment. In *Ecotoxicological Testing of Marine and Freshwater Ecosystems*, den Besten, P.J. and Munawar, M., Eds. Boca Raton, FL: Taylor & Francis, pp. 87–152.

Van Tilborg, W.J.M. and F. Van Assche. 1998. Homeostatic regulation defines a stress-free concentration band for essential elements relevant for risk assessment. *SETAC Europe News* 9: 7–8.

Vasseur, P. et al. 2008. Marqueurs de génotoxicité et effets in situ, individuals et populationnnels. In *Les biomarqueurs dans l'évaluation de l'état écologique des milieux aquatiques*, Amiart-Triquet, C. and Amiard, J.E. Eds. Paris: Lavoisier, pp. 295-330.

Vinturella, A.E. et al. 2004. Importance of black carbon in distribution and bioaccumulation models of polycyclic aromatic hydrocarbons in contaminated marine sediments. *Environ. Toxicol. Chem.* 23: 2578–2586.

Voparil, I. and L.M. Mayer. 2000. Dissolution of sedimentary polycyclic hydrocarbons (PAHs) into the lugworm's (*Arenicola marina*) digestive fluids. *Environ. Sci. Technol.* 34: 1221–1228.

Vrana, B. et al. 2005. Passive sampling techniques for monitoring pollutants in water. *Trends Anal. Chem.* 24: 845–868.

Vrana, B., R. Greenwood, and G. Mills. 2007. *Comprehensive Analytical Chemistry. Passive Sampling Techniques in Environmental Monitoring*, Vol. 38. Ville: Elsevier.

Wang, W.X. 2002. Interactions of trace metals and different marine food chains. *Mar. Ecol. Progr. Ser.* 243: 295–309.

Weis, J.S. et al. 2001. Effects of contaminants on behavior: biochemical mechanisms and ecological consequences. *BioScience* 51: 209–217.

Weston, D.P. and L.M. Mayer. 1998a. *In vitro* digestive fluid extraction as a measure of the bioavailability of sediment-associated polycyclic aromatic hydrocarbons: sources of variation and implications for partitioning models. *Environ. Toxicol. Chem.* 17: 820–829.

Weston, D.P. and L.M. Mayer. 1998b. Comparison of *in vitro* digestive fluid extraction and traditional *in vivo* approaches as measures of polycyclic aromatic hydrocarbon bioavailability from sediments. *Environ. Toxicol. Chem.* 17: 830–840.

Widdows, J. et al. 1990. Measurement of physiological energetics (scope for growth) and chemical contaminants in mussel (*Arca zebra*) transplanted along a contamination gradient in Bermuda. *J. Exp. Mar. Biol. Ecol.* 138: 99–117.

Wilding, J. and L. Maltby. 2006. Relative toxicological importance of aqueous and dietary metal exposure to a freshwater crustacean: implications for risk assessment. *Environ. Toxicol. Chem.* 25: 1795–1801.

Xie, L. and P.L. Klerks. 2004. Fitness cost of resistance to cadmium in the least killifish (*Heterandria formosa*). *Environ. Toxicol. Chem.* 23: 1499–1503.

Index